D0083853

The Origins and Nature
of Sociality

The Origins and Nature of Sociality

EDITORS

Robert W. Sussman

Audrey R. Chapman

ALDINE DE GRUYTER
New York

About the Editors

Robert W. Sussman   Professor, Department of Anthropology, Washington University, St. Louis, Missouri

Audrey R. Chapman   Director, Science and Human Rights Program, American Association for the Advancement of Science, Washington, D.C.

Copyright © 2004 Walter de Gruyter, Inc., New York. All rights reserved. No part of this publication may be reproduced or transmitted in any form or by any means, electronic or mechanical, including photocopy, recording, or any information storage or retrieval system, without permission in writing from the publisher.

ALDINE DE GRUYTER
A division of Walter de Gruyter, Inc.
200 Saw Mill River Road
Hawthorne, New York 10532

This publication is printed on acid-free paper ∞

Library of Congress Cataloging-in-Publication Data
The origins and nature of sociality / editors, Robert W. Sussman, Audrey R. Chapman.
    p. cm.
"Developed from a series of workshops and symposia sponsored by the Program for Dialogue on Science, Ethics, and Religion (DoSER) of the American Association for the Advancement of Science (AAAS)"—Ack. Includes bibliographical references and index.
    ISBN 0-202-30730-1 (cloth : alk. paper) — ISBN 0-202-30731-X (pbk. : alk. paper) 1. Social interaction—Congresses. 2. Sociobiology—Congresses. I. Sussman, Robert W., 1941- II. Chapman, Audrey R.

HM1111.O75 2004 302—dc22   2003022786

Manufactured in the United States of America

10 9 8 7 6 5 4 3 2 1

Contents

III   MECHANISMS OF SOCIALITY

IV   PRIMATE SOCIALITY

V EVOLUTION OF SOCIALITY

VI PHILOSOPHICAL OVERVIEW

Acknowledgments

Like other edited volumes, this book has benefited from the input and assistance of many people. The book developed from a series of workshops and symposia sponsored by the Program for Dialogue on Science, Ethics, and Religion (DoSER) of the American Association for the Advancement of Science (AAAS). Among the goals of DoSER are to increase the engagement of scientific communities in the dialogue on science, ethics, and religion and to facilitate collaboration among scientists, ethicists, and religion scholars to address critical multidisciplinary issues related to these topics.

We would like to thank the DoSER staff, particularly Karin Sypura and Jim Miller, for their help in organizing the meetings. The participants in the workshops and symposia have made this an enjoyable and enlightening experience, and we express our appreciation to them. We would also like to acknowledge the generous support of the John Templeton Foundation, whose grant to DoSER provided the funding that made the project possible. In addition, we would like to thank Linda Sussman for copyediting the manuscript, and Lisa Kelley and Chris Shaffer for their assistance. Thanks also to Richard Koffler, our editor at Aldine de Gruyter, and his excellent production staff.

Contributors

T. K. Ahn   Department of Political Science, Florida State University, Tallahassee, FL 32306

Marc Bekoff   Department of Ecology and Evolutionary Biology, University of Colorado, Boulder, CO 80309-0334

Irwin S. Bernstein   Department of Psychology, University of Georgia, Athens, GA 30602-3013

Christopher Boehm   The Jane Goodall Research Center, Department of Anthropology, University of Southern California, Los Angeles, CA 90089

Debra R. Bolter   Department of Anthropology, University of California-Santa Cruz, Santa Cruz, CA 95064

C. Sue Carter   Brain Body Center, Department of Psychiatry, University of Illinois at Chicago, Chicago, IL 60612

Audrey R. Chapman   Director, Science and Human Rights Program, American Association for the Advancement of Science, Washington, DC 20005

James M. Cheverud   Department of Anatomy and Neurobiology, Washington University School of Medicine, St. Louis, MO 63110

Bruce S. Cushing   Brain Body Center, Department of Psychiatry, University of Illinois at Chicago, Chicago, IL 60612

Agustin Fuentes   Department of Anthropology, University of Notre Dame, Notre Dame, IN 46556-5611

Paul A. Garber   Department of Anthropology, University of Illinois, Urbana, IL 61801

Marco A. Janssen   Center for the Study of Institutions, Population, and Environmental Change, Indiana University, Bloomington, IN 47408-3799

Elinor Ostrom Workshop in Political Theory and Policy Analysis, Center for the Study of Institutions, Population, and Environmental Change, Department of Political Science, Indiana University, Bloomington, IN 47408

Stephen J. Pope Theology Department, Boston College, Chestnut Hill, MA 02467-3859

Richard Potts Human Origins Program, National Museum of Natural History, Smithsonian Institution, Washington, DC 20560-0112

Barbara B. Smuts Department of Psychology, University of Michigan, Ann Arbor, MI 48109-1109

Karen B. Strier Department of Anthropology, University of Wisconsin-Madison, Madison, WI 53706

Robert W. Sussman Department of Anthropology, Washington University, St. Louis, MO 63130

Ian Tattersall Division of Anthropology, American Museum of Natural History, New York, NY 10024

John M. Watanabe Department of Anthropology, Dartmouth College, Hanover, NH 03755-3570

Adrienne L. Zihlman Department of Anthropology, University of California-Santa Cruz, Santa Cruz, CA 95064

I
INTRODUCTION

1

The Nature and Evolution of Sociality

Introduction

Robert W. Sussman and Audrey R. Chapman

A BRIEF HISTORY OF THIS VOLUME

This book results from a symposium sponsored by the Program for Dialogue on Science, Ethics, and Religion (DoSER) of the American Association for the Advancement of Science (AAAS). The symposium, which took place in January 2001, was cochaired by the editors of this volume. Among the goals of DoSER is to increase the engagement of scientific communities in the dialogue on science, ethics, and religion, and to facilitate collaboration among scientists, ethicists, and religion scholars to address critical multidisciplinary issues related to these topics. A further goal is to increase public understanding and appreciation of science and improve the level of scientific understanding in religious communities.

Increasingly, scientific developments over the past century have been transforming our understanding of humanity's place in nature as well as the very nature of human beings. The study of humanity, carried out in a variety of disciplines, from anthropology and paleontology to genetics and the neurosciences, is shedding new light on human origins and the biological bases of human nature and culture. The findings of these sciences can have profound implications for the interpretations of human nature and the determinants of human behavior within the religious traditions of humankind. Sometimes these developments raise challenges to long-held religious beliefs.

This is certainly true in the case of sociobiology. Sociobiology, more recently referred to by many as evolutionary psychology, Darwinian anthropology, or evolutionary anthropology (Wright 1994; but see Mysterud, in press), offers a radically selfish and individualist account of human nature vaguely reminiscent of the concept of original sin (and many sociobiologists refer to this religious metaphor). But despite conceptions of human beings as inherently limited and flawed, many religious traditions also put forward a concept of the common good, an understanding of the individual as a self-in-community, and a vision of *Homo sapiens* as a basically social and interdependent species. Judaism, Islam, and Christianity generally understand persons as moral and social beings created to live in communities linked by relationships of mutual caring and responsibility. Sociobiology's portrayal of human nature strips away the potential for genuine moral and social development.

At the time we began this project, the paradigm put forward by sociobiology dominated the literature, both popular and scientific, and public conversations about evolution, animal behavior, and human nature. Proponents of sociobiology characterize human conduct as ruthlessly selfish and relentlessly driven by an intense drive to compete with others for natural resources and reproductive advantage (see, for example, Wilson 1975, 1978; Dawkins 1976; Wright 1994; Ridley 1996). Human beings, like all animals, are viewed as controlled by their genes to improve their prospects for survival and reproduction. In its more extreme versions, sociobiology proposes that genes are the main units of natural selection and reduces human persons to little more than vehicles for their genes. According to Richard Dawkins, human beings, like all animals, are basically survival machines created by our genes. Taking the license of writing about genes as if they had conscious aims, Dawkins writes that genes control the behavior of their survival machines, "not directly with their fingers on puppet strings, but indirectly like the computer programmer" (1976:56). In this model, gene selfishness gives rise to selfishness in the behavior of both animals and humans that in turn leads to continuous and necessary competition between individuals.

This reductionism and biological determinism extend to the interpretation of culture. Genes hold culture on a leash, a long leash, but a leash nevertheless (Wilson 1978:167). In Dawkins's scenario, the struggle for survival produces brains, and brains in turn give rise to a new kind of replicator termed "memes." Memes constitute units of cultural transmission or mental entities in the form of ideas, concepts, or ways of making things that are analogous to genes and follow a vaguely similar process of evolution (1976:206). And like genes, memes are selfish and try to dominate the brain's thought processes at the expense of rival memes. Dawkins treats the idea of God as a very old meme that probably originated many times by independent "mutation" and persisted because of its survival value (ibid.:207). And what about the concepts that religions and ethicists value? "Much as we might wish to believe otherwise, universal love and the welfare of the species as a whole are concepts which simply do not make evolutionary sense" (ibid.:2).

Proponents of sociobiology acknowledge that there are special circumstances in which animals and persons exhibit cooperative, friendly, even seemingly sacrificial behavior, but they seek to explain this away by reducing these social behaviors to a biological survival strategy to achieve selfish goals. To put it another way, in order to improve our ability to survive, natural selection has vested us with a capacity to be cooperative. As a strategy to maximize opportunities for reproduction, it makes sense in some circumstances for an animal to take risks to protect members of its own kind who share its genes. Thus most forms of seemingly altruistic behavior tend to be demonstrated to close relatives. Additionally, animals, and by extension humans, engage in forms of mutually advantageous reciprocal altruism—the principle "you scratch my back, I'll ride on yours" (ibid.:179). Sociobiologists generally deny that such forms of kin or reciprocal altruism are in any way comparable to traditional ethics.

Likewise, in the sociobiological account, human beings are ethical solely because morality is a strategy to promote self-interest or, more specifically, to preserve their genes. So why do we believe that we are ethical? To be effective, genes somehow have to convince us that we are behaving cooperatively because we believe it is right to do so, not because it is in our evolutionary interest, and hence the emergence of ethics (Wilson 1978, 1998; Ruse and Wilson 1985; Ruse 1987, 1991; Wright 1994; Sommer 2000). In this convoluted explanation, moral beliefs are "no more than a collective illusion fobbed off on us by our genes for reproductive ends" (Ruse 1991:508).

In its "classic" form, sociobiology claims that genes, not their carriers, are necessarily selfish. Although E. O. Wilson characterizes humankind as an aggressive species, he attributes the incidence and severity of violence not to selfish or aggressive genes or a pervasive aggressive instinct, but to factors in the environment, particularly social stress (1994:167–70). That said, however, claims that selfishness and violence are somehow biologically built into human nature are currently receiving considerable attention.

In *Demonic Males: Apes and the Origins of Human Violence*, a work that has received considerable attention, primatologist Richard Wrangham and science writer Dale Peterson resurrect the theory that hunting, killing, and extreme aggressive behaviors are inherited biological traits. Taking issue with the assumption of many scholars that human aggression is relatively unique, they reinterpret previously collected empirical data, once viewed as showing the great apes to be basically unaggressive, gentle creatures, to contend that the inclination to kill other members of one's species is a defining mark of our closest relative, chimpanzees, as well as humans (Wrangham and Peterson 1996). Since humans and chimpanzees share an inherited propensity for killing, they argue that human violence has long evolutionary roots and is likely to be fixed. Wrangham had earlier claimed, with little evidence, that chimpanzees were a conservative species likely to resemble the common ancestor from which both species evolved (Wrangham 1995). Wrangham and Peterson attribute the origins of this "demonic" tendency to the existence of unique patrilineal, male-bonded kin groups that these species form with each other to attack outsiders. They also base their belief that humans were favored by natural selection to hate and to kill their enemies on the sociobiological tenet of the selfish gene (1996:23). To explain why humans and chimpanzees are so different from bonobos, our equally close genetic cousins who are one of the most peaceful and socially oriented species, they point to the absence of male bonding, attributing differences in fundamental social patterns to variations in the nature of the food supply among the groups (ibid.:220–30). This explanation, which suggests that violence and nonviolence are equally adaptive responses to the environment, contradicts their fundamental claim that humans and chimpanzees have inherited aggressive, fixed biological traits.

Other primatologists have provided a very different interpretation of our evolutionary ancestry. As the title of Frans de Waal's book *Good Natured* (1996) implies, his characterization of chimpanzees emphasizes their sociability and incipient morality, not their inherent violence. In a spoof of Wrangham and Peterson, Sussman uses similar logic to propose a less serious, but no less feasible, theory of the evolution of human behavior—man the dancer. He points out that the evidence for a love of dancing as a motivating source of other human adaptations is certainly as good as that for hunting or killing (1999:456–57).

Sociobiology has always had its critics, among them many prominent evolutionary biologists, paleontologists, and naturalists, such as Richard Lewontin, David Hull, Stephen Jay Gould, Niles Eldredge, and Russel Gray, concerned about the lack of compelling scientific evidence offered by proponents and its simplistic reductionism and genetic determinism (Lewontin, Rose, and Kamin 1984; Hull 1988; Eldredge 1995; Gould 2000). Stephen Jay Gould, for instance, dismissed the selfish gene model as based on logically erroneous premises (2000). Niles Eldredge takes the "ultra-Darwinists" to task for their narrow scientific agenda and failure to take the existence of larger-scale entities, such as species and ecosystems, seriously or to view them as something more than simple epiphenomena borne out of competition for reproductive success (1995:223–27). Russell Gray implicates the lack of proper knowledge of molecular and developmental biology and suggests we shift our attention to the evolutionary dynamics of developmental systems (2000:184–207).

In the sociobiology literature, there is no meaningful way of looking at affiliative behavior. Yet many evolutionary biologists have recognized the importance of sociality and cooperation among members of a species or a group. In his seminal book *The Descent of Man*, Charles Darwin (1871) acknowledged that many of the most praiseworthy human qualities appear to benefit others at expense to self, such as honesty, charity, and heroism. Darwin attributed the emergence of these characteristics to the fact that natural selection sometimes acts on groups, just as it acts on individuals. As such, although an altruist may have fewer offspring than a nonaltruist, groups of altruists will have more offspring than groups of nonaltruists. In addressing the issue as to whether human evolution by natural selection allowed for moral sensibilities and social behavior, Darwin had this to say:

> It must not be forgotten that although a high standard of morality gives but a slight or no advantage to each individual and his children over the other men of the same tribe, yet that an increase in the number of well-endowed men and advancement in the standard of morality will certainly give an immense advantage to one tribe over another. (1871:166)

Anticipating by many decades the work of population geneticists and current research on the role of group selection, Darwin described the advantage to human communities with a strongly endowed moral sense, such as "patriotism, fidelity, obedience, courage and sympathy," and a willingness "to aid one another, and to sacrifice themselves for the common good," over societies with lesser acquisition of such qualities and characterized this as natural selection (ibid.).

Similarly, in his book *Mutual Aid* ([1902] 1987) written a century ago, Petr Kropotkin amassed a wealth of data showing how cooperation and mutual assistance among members of a species may act as a countervailing force against the natural competitive and aggressive instincts of individual members of an animal group or a human community. Kropotkin, a Russian nobleman, humanist, and anarchist, was a highly respected geographer, geologist, and zoologist, who was awarded a gold medal by the Russian Geographical Society for his work on the geographic structure of Asia (Montagu 1987). For Kropotkin the ability of members of a species to cooperate and share resources, whether they belonged to the "lower animals," primitive societies, or present-day human communities, conferred advantages. He viewed the capacity for mutual aid as an important and determining factor in biological and human evolution. Kropotkin attributed the evolution of moral teachings among religious, philosophical, and humanistic movements to the principle of mutual aid becoming more and more ingrained in human consciousness ([1902] 1987). Although *Mutual Aid* was widely read and influential, Montagu (1987) states that "it has also suffered some misunderstanding on the part of those who know of the book at second or third hand." Many believe that Kropotkin was anti-Darwin but this was not the case. Kropotkin was greatly influenced by Darwin's concept of natural selection but was highly critical of Huxley's evolutionary extremism and strongly social Darwinist interpretations. (See Killen and de Waal 2000 for a further discussion of Kropotkin and nineteenth-century debates about the evolution of morality.)

Despite its promise, the concept of group selection, which provided an evolutionary explanation for the emergence of altruistic and cooperative behavior, did not find much favor with the established schools of evolutionary biology. Recently, however, Elliot Sober and David Sloan Wilson have reinvigorated debate on the importance of group selection as a basis for understanding the role of altruism and cooperative behavior in biological evolution. In their landmark book, *Unto Others,* Sober and Wilson (1998) draw on four disciplines—evolutionary biology, social psychology, anthropology, and philosophy—to demonstrate that unselfish behavior is an important feature of both biological and human nature. Offering evidence from self-sacrificing parasites to insects and other animals and finally the human capacity for selflessness, they explain how altruistic behavior can evolve by natural selection. The message of their book is that natural selection is unlikely to have conferred purely egoistic motives.

In addition to the work on group selection, there is a complementary stream in the evolutionary literature with the view that there are certain emergent qualities in human evolution that cannot simply be reduced to genetic or biological factors. It recognizes that human beings and to a lesser extent primates, as well, have a capacity for affiliative behavior at least partially due to adaptive changes. Intellectual capabilities give rise to an additional overlay of cultural evolution in human communities. Researchers also have shown that some primate communities demonstrate the existence of a protoculture in the form of distinctive learned behaviors and social responses (Itani 1958; Kawai 1965; McGrew, Marchant, Scott, and Tutin 2001; Perry et al. 2003).

Theodosius Dobzhansky, the eminent population geneticist whose groundbreaking studies on the fruit fly paved the way to a comprehensive understanding of the genetic mechanisms at work in the evolutionary process, for one did not understand human evolution as a purely biological process. Instead he characterized human evolution as shaped by two components, the biological or organic and the cultural or superorganic, interacting through a series of nonlinear, feedback interactions between biological and cultural processes (Dobzhansky 1962).

Francisco Ayala, an evolutionary biologist, philosopher, and man of letters who studied with Dobzhansky, puts forward a view that acknowledges the evolutionary origins of morality, but still affirms the objective and independent basis of ethical systems. Like the sociobiologists, he accepts that the high intellectual abilities present in modern humans are an outgrowth of the process of evolution directed by natural selection. Unlike the sociobiologists, he posits that moral reasoning, that is, the proclivity to make ethical judgments by evaluating actions as either right or wrong, emerged as an outgrowth of human intellectual development and not because it conferred biological benefit (Ayala 1995, 1998). Ayala also makes the point that the justification of ethical norms on the basis of biological evolution or any other natural process is a reversion to the naturalistic fallacy, the confusion of "is" with "ought." He cautions that the confusion of evolutionary processes with morality seems to justify a morality consistent with a social Darwinism most of us would find abhorrent (Ayala 1995:126). Ayala further underscores the need to differentiate between genetic predispositions and genetic determinism. While a natural predisposition may influence our biological nature, it does not constrain or force us to behave accordingly (ibid.:128).

AAAS PROJECT

The AAAS project grew out of a desire to find alternative and more scientifically valid ways to understand the biological bases of human sociality. Believing that primatology may offer significant comparative insights into the evolutionary foundations of human nature, the coeditors decided to hold a series of research workshops linking primatology and the evolution of human behavior. Because our ancestors are extinct and the paleontological record is limited, living nonhuman primates and ethnographic studies of contemporary human foraging cultures (Boehm, chapter 13 in this volume) offer important potential models of the biological and social foundations of human nature and society. Moreover, our view of humanity has often been influenced by our interpretation of the behavior of other animals. However, the use of research on nonhuman primates to inform the human sciences on ethics and constructive or systematic religious thought about human nature raises serious methodological and interpretative issues.

To provide a setting to explore the interdisciplinary relationships between contemporary primatology, other human and biological sciences, and religious thought and ethics, the AAAS Program for Dialogue on Science, Ethics, and Religion convened a series of research workshops and symposia at the annual meetings of AAAS and the American Anthropological Association. The first was held in 1998. At the time of these meetings, A.R.C. was Director of DoSER and R.W.S. served as a consultant to the program and editor-in-chief of the *American Anthropologist*, the flagship journal of the American Anthropological Association (AAA). In this series of research symposia and conferences the contributors have begun to address four broad questions: (1) What do current studies of higher primates tell us about their nature, sociability, inclinations to violence, and rules of behavior or "protoethics"? (2) Why do researchers disagree so significantly on the interpretation of these studies? (3) What relevance, if any, do these findings have for understanding the biological foundations of human nature? (4) What is the ethical and theological relevance of these findings (e.g., do they illuminate questions of theological anthropology such as the nature of human sin)? Theological anthropology is the study or interpretation of human nature, what distinguishes us from other species, and the foundation of a moral life from a faith perspective (see, for example, Pope 1998).

Two major research workshops sponsored by DoSER were held in Washington, D.C. Each of these brought together fifteen to eighteen researchers drawn from primatology, paleoanthropology, biological and cultural anthropology, biology, psychology, genetics, philosophy of science, ethics, and theology. A core of researchers participated in both conferences.

The first was held in October 1998. The goal of this workshop was to begin a dialogue on the subject of the evolution of human nature and to examine the status of some of the various data and theories currently available on that subject, focusing especially on primatology. The participants focused on methodological issues concerning the interpretation of primatological data and their inferential applicability for understanding the biological basis of human nature. Although participants were asked to prepare presentations on specific topics, no formal papers were presented. In this initial workshop, participants refined the focus of potential future conferences and identified possible future paper topics and potential authors.

During the concluding session of the first workshop, a consensus emerged that the second meeting should focus on the evolution and nature of nonhuman and human primate sociality. It also was suggested that emphasis should be placed on careful discussion of techniques of data collection and on methodology. This book results from the second research workshop, which took place in January 2001.

THEORETICAL BACKGROUND

All diurnal primates (those active during daylight hours), from prosimians to humans, are highly social. Furthermore, with the exception of orangutans, these primates habitually form and travel in cohesive social groups. However, group-living individuals must forgo some of their individual freedoms in order to socialize within the "group." In a broad sense, the compromises that individuals make, the mechanisms they use, and the means by which they maintain these social groups are what we refer to as "sociality."

What are some of the compromises that individuals make for this sociality? Are there certain behavioral traits shared by all social primates (mammals), and are these traits homologous or analogous? Are there hormonal correlates to sociality? Are there different patterns of sociality between and among primate species? Do any of these patterns manifest themselves in human behavior? How does this relate to the evolution of human sociality? What kinds of mechanisms do various primate species and human cultures use to socialize individuals in order to maintain species typical or "normative" groups? What are the differences in these traits given different ecological and social parameters (such as group size, number of males or females in a group, patterns of migration, group structure)? These are some of the questions relevant to an inquiry into the roots of human sociality.

In this volume, available data on primates and other social animals were examined in an attempt to understand the nature of the adjustment that individuals make in order to successfully live in social groups. The amount of time primates spend in social interaction is examined along with how such interactions relate to other behaviors, such as feeding, traveling, and general maintenance behaviors. Human societies are considered with particular attention to the variable nature of human cultures as well as the behaviors that are valued cross-culturally in relation to human sociality.

These various data contribute to the formulation of questions concerning the relationships that might exist between sociality, morality, and ethics. Such questions include: What are some of the general rules that animals must follow in order to maintain sociality, and might some of these be precursors to similar rules in humans? What specific traits need to be developed by nonhuman and human primates in order to maintain sociality and are any of these similar across species? Can behavioral patterns and mechanisms used for the fostering and maintenance of nonhuman and human primate sociality be related to the evolution of morality and ethics?

In the first conference there was discussion of how many of the current theories on the evolution and biological basis of human behavior are based on presuppositions without an adequate analysis of the data. It was noted that often such theories make selective use of available data. In this volume, we pursue this methodological concern by critically examining both the data and the methods used in developing theories about sociality among humans and other primates. This we hope will lead some to reevaluate available data and to attempt to develop new theories more adequate to the data. To accomplish this will require careful attention to the terms used in developing theories related to the evolution and biological basis of human behavior. The sorts of questions needed to be asked are: How good are the data? How good are the methods employed to interpret the data? How adequate are the explanatory theories in relation to the data? Are the terms used in cross-species comparisons suitable for this purpose?

In this set of contributions, we focus on the current status of research on sociality and the evolution of social behavior, especially but not exclusively, in nonhuman and human primates. The authors examine questions related to the evolution, cultural variability, and hormonal underpinnings of human sociality and describe patterns of sociality among nonhuman primates and how they may shed light on human social behavior.

We have found that within primate groups, affiliative and cooperative behaviors are far more frequent than agonistic behaviors. The most currently popular paradigm hypothesizes that positive social interactions are a reaction to competition necessitated by group-living or that they serve as reconciliatory behaviors between competing individuals. However, if conditions favor cooperative behavior among both kin and nonkin group members, and these outweigh any negative conditions, natural selection could favor cooperative social interactions *in their own right*.

Currently there are few theories that present the case that primate and human sociality may be driven by factors other than aggression and self-interest. The basic aim of this volume is to present alternative hypotheses and to base these alternatives on what we believe to be better biological evidence and more appropriate genetic and evolutionary approaches than the sociobiology paradigm.

SYNOPSIS OF THE CONTENTS

The book is divided into six sections. In the second section, the authors describe proximate behavioral mechanisms that provide underpinnings to sociality. Zihlman and Bolter describe the importance of the mammalian system of mothering and infant care and development in the evolution of sociality. By contrasting individual life strategies and species life histories (for example, in elephant seals, elephants, and some primates), they illustrate the connection between these factors and the size and complexity of social groups, and the development of human culture. They maintain that sociality and "the integrity of the community is maintained over the long haul through the emotional and social bonds that are formed during maturation and the affiliative skills practiced throughout life, rather than through frequent aggression and fighting" (p. 37). Bekoff discusses social play behavior in mammals and its role in the evolution of cooperation, fairness, and sociality. He argues that some animals have codes of social conduct that regulate their behavior in terms of what is and what is not permissible during social interactions. Furthermore, he believes that the study of mammalian social play can help us learn more about the evolution of social morality in humans and other animals. Bernstein points out that the frequency of agonistic interactions is correlated with the time individuals spend in proximity, with aggression being more frequent among those in close proximity, including kin. Agonism is often a proximate result of sociality. He describes the normal context of many aggressive behaviors and a number of ways individuals control aggression in social groups. Bernstein, however, stresses that many of these behaviors do not have the "goal" of "peacemaking." Proximate causes and function are distinct. He emphasizes that social bonding may be less exciting than functional theories of reciprocity, bartering services for favors, punishing cheaters, and scheming like Machiavelli's prince, but it is more parsimonious than assuming that animals understand functional consequences of their behavior, plan accordingly, and consciously strive to improve their genetic fitness.

The third section of the book deals with hormonal, neurological, and genetic factors related to sociality and its evolution. Carter and Cushing describe specific hormones that influence general sociality, the capacity to form social bonds, and parental behavior. These neurochemicals and their receptors are regulated by genetics as well as epigenetic factors and may help to account for species-typical variations in social behavior. These authors summarize current understanding of these physiological mechanisms underlying mammalian sociality. Ahn, Janssen, and Ostrom focus on cooperation, mainly among humans, in situations where the temptation to defect exists. They argue that the ability of humans to use signals and symbolic systems, and the biological and mental capacities related to this, facilitate cooperation and stimulate the development of large-scale and complex social organization. It is interesting to note, however, that many of the factors described by these authors as unique to humans can be shown, at rudimentary levels, in other animals, as seen in chapters throughout this volume (see, for example, Bekoff on play signals and Watanabe and Smuts on greeting rituals). Cheverud carefully explains that Darwinian evolution occurs by the natural selection of heritable variation. He illustrates how these factors are measurable and describes some common misconceptions concerning these terms. Cheverud then develops a model of the potential for the evolution of sociality and cooperation under naturalistic selection with direct benefits for interacting parties. He stresses the need to actually measure fitness consequences for both the actor and the recipient in social interactions rather than assuming selection based on incomplete information.

The fourth section includes chapters on patterns of primate sociality. From a survey of the literature on diurnal primates, Sussman and Garber find that, even though almost all of these primates live in permanent social groups, very little time is spent in active social interaction. Furthermore, the overwhelming majority of these interactions are affiliative and only a minute proportion of them are agonistic. Given these findings, they point out the need to carefully examine patterns of sociality in the context of normal daily life and they call for a reevaluation of the conventional wisdom that primate sociality is driven by competition and aggression. Strier examines how dispersal and resident patterns directly affect the opportunity for, and patterns of, interactions among nonkin and different categories of biological kin in various nonhuman primate species. An important distinction between nonhuman and human primates has been the latter's ability to maintain both kin and nonkin relationships without coresiding in the same group by classifying and keeping track of relationships through language, although Strier takes the argument further by illustrating how demographic conditions can affect social bonds. She points out how different demographic conditions can either permit or preclude the development of certain categories of relationships, independent of whether the differences in these relationships are recognized. Fuentes examines the wide degree of variability in what is considered postconflict behavior, especially behavior considered reconciliation. He shows that it is extremely difficult to disentangle cooperative relationships and relationship histories of individuals from the conflicts in which they engage and how they behave before and after these conflicts. He concludes that rather than having an evolved set of behavioral responses to conflict, normal "patterns of cooperation and affiliative relationships may be important causal factors behind observed postconflict behavior" (p. 231).

In the fifth part, the authors focus on the evolution of human sociality. Tattersall emphasizes the importance of understanding mechanisms of macroevolution in developing theories of the evolution of human cognition and sociality, and he warns that the currently popular "adaptionist" paradigm is too simplistic. He argues that modern human behavioral characteristics are founded upon the basic higher primate (diurnal, group-living, intensely social) qualities already possessed by the common hominoid ancestor. However, their unique qualities are the product of a recent, fairly abrupt, and emergent event, resulting from a chance coincidence of innovations. Tattersall urges us to remember that adaptations do not have independent existences and that evolutionary processes work on whole organisms and taxa, not on constituent parts of individuals. Potts describes a distinctive suite of archaeologically detectable behaviors that mark the emergence of modern humans and the long period of ecological unpredictability during which modern human behavior emerged. He argues that this volatile environment impacted the social fabric of early humans and led to symbolic expression and language. These factors, he believes, impart a peculiar quality to the personal and social behavior of *Homo sapiens* and make understandable the origin of a spiritual sense.

Potts outlines "how the cultural behavior characteristic of modern humans emerged from a *paleocultural* (emphasis in original) system of earlier humans" (p. 250), and he emphasizes the need for a redefinition of the concept of culture. Boehm uses a cladistic methodology, based on a comparison of the behaviors of our closest ape relatives, to infer certain behaviors of our earliest human ancestors. Combining these inferences with ethnographic analogies based on modern human foraging cultures, he develops a preliminary model of the evolution of human sociality and the development of human morality. Watanabe and Smuts "address the relationship between continuities and transformations in the evolution of human sociality through discussion of the social cooperation and commitment" that they see as "intrinsic to both human and nonhuman primate communication" (p. 288). Further, they ask "what difference does having language and culture make in human sociality?" (p. 288). They develop the argument that language presupposes and intensifies social cooperation already present in nonhuman primates, and they use ritual greeting behavior among baboons to illustrate this point.

In the concluding essay, Pope summarizes the chapters in the volume and puts them into the context of historical, theological, and philosophical perspectives on natural law, ethics, and moral reflection. He traces how the Hobbesian perspective of the competitive nature of humans has greatly influenced the current theories of sociality from Darwin to the present. However, he points out that earlier philosophers, such as Aristotle and Aquinas, saw human sociality quite differently, as part of a classical natural law tradition. These philosophers believed, as does Pope, that human sociality is "primary and not simply derivative from instrumental purposes . . . ; it is essential to human well-being, rooted in biology as well as intelligence, and not a dispensable addition of culture" (p. 323). Pope argues that the primate studies described in this book offer an alternative to the currently dominant paradigm. He states that:

> If other primates are prone to social behavior more often than antisocial behavior, perhaps pity, empathy, and other prosocial feelings do not have to be laid on top of a substrate that is essentially antisocial. . . . An alternative position, and one that retrieves Aristotle's notion of the human being as a "political animal," can draw some help from this prosocial view of primatology in viewing society as a network of communities that make a positive contribution to human well-being. (pp. 328–329)

FUTURE PLANS

We hope that this volume will not be the only one produced from this initiative. In our original plans, we envisioned four major conferences. As described above, two of these have now been accomplished. Two future conferences are in various stages of development. In these future meetings, we will explore further many of the ideas presented in the current volume. In our third conference, we plan to focus on the origins and evolution of human cooperative and altruistic behavior and build on discussions on primate sociality. The nature of human altruism is a topic that has perplexed evolutionary biology and is one of considerable interest to both the social sciences and the religious community. For example, media coverage of the response to the September 11 tragedy documented many examples of courage, and cooperative and altruistic behavior in response to this terrible event. Yet, as discussed above, many socioecologists and sociobiologists believe that social animals, including human and nonhuman primates, are cooperative and altruistic only if they have something to gain from their actions. However, the reaction of millions of people to the September 11 event does not fit this paradigm. As the *New York Times* (Angier 2001) reported:

> Hearing of the tragedy whose dimensions cannot be charted or absorbed, tens of thousands of people across the nation storm their local hospitals and blood banks, begging for the chance to give blood, something of themselves to the hearts of the wounded.

In the literature on social primates, including humans, and other social mammals it is not difficult to find a multitude of examples of altruistic behaviors that do not fit the commonly accepted paradigm. In fact, there is a great deal of slippage in the currently popular theories of altruism. As the *New York Times* article continues: "As biologists are learning, there is more to cooperation and generosity than an investment in one's nepotistic patch of DNA." A large number of behavioral, hormonal, neurological, genetic, and computer studies and theoretical models drawn from them are beginning to offer alternative explanations for cooperative and altruistic behavior. However, scholars working on this subject from different disciplines have seldom communicated with one another, and these data have yet to be synthesized. This will be a primary goal of our next conference.

In the final research conference, we will address the topic of the origin and evolution of morality. We will consider the requirements for morality and the extent to which critical precursors of human morality are present among various groups of primates. This conference will also evaluate whether there is evidence for "protoethical" rules of behavior among primates and, if so, how these rules might be related to the development of human morality. Findings related to patterns of altruistic behavior among primates and whether the "selfish gene" hypothesis is able to explain the basis of such behavior will be yet another topic for discussion. Finally, we also hope to explore the question of whether kin-based altruism and recip-rocal altruism are a sufficient foundation for moral relationships among groups of primates and human beings, or whether there are other behav-ioral, genetic, and neurological mechanisms that can help explain human morality that is not based on purely selfish considerations.

CONCLUDING REMARKS

The authors of chapters in this volume come from a diversity of fields, including anthropology, primatology, sociology, political science, paleon-tology, biology, psychology, psychiatry, genetics, neurobiology, ethics, the-ology, philosophy, and science and religion. We believe that the book will be of interest to individuals in all of these disciplines. The volume is writ-ten so that it is accessible to both an academic and an educated popular audience. We, further, believe that the volume can be used for both under-graduate and graduate courses in a number of the above fields and pos-sibly even in high school courses. Professionals seeking alternative explanations for cooperative behavior will find the book extremely useful, and we hope that it will stimulate discussion, controversy, and an impetus for other researchers to delve into theories that are at odds with some of those currently in vogue.

As we all know from recent history, some "scientific" theories, such as social Darwinism and eugenics, can become very popular both among sci-entists and among the general public, and yet they can be very, very wrong. Thus, even though some theories may gain a great deal of scientific support and general popularity, it is important that alternative hypotheses be pre-sented in the literature. We believe that many of the current theories on sociality are often accepted as if they are truisms much like the theories of social Darwinism and eugenics were in the past. This volume presents ample evidence that there are alternative and more convincing hypotheses that may lead to better explanations and to a better understanding of pat-terns of nonhuman and human primate sociality.

REFERENCES

Angier, Natalie. 2001. "Of Altruism, Heroism and Evolution's Gifts." *New York Times*, September 18, late edition, Section F, p. 1 (also available at http://www.nytimes.com).

Ayala, Francisco, J. 1995. "The Difference of Being Human: Human Ethical Behav-ior as an Evolutionary Byproduct." Pp. 113–36, in *Biology, Ethics, and the Ori-gins of Life*, edited by Holmes Rolston, III. Boston and London: Jones and Bartlett.

Ayala, Francisco. J. 1998. "Biology Precedes, Culture Transcends: An Evolutionist's View of Human Nature." *Zygon* 33(December):507–24.

Darwin, Charles, 1871. *The Descent of Man and Selection in Relation to Sex.* London: John Murray.

Dawkins, Richard. 1976. *The Selfish Gene.* Oxford and New York: Oxford University Press.

de Waal, Frans. 1996. *Good Natured: The Origins of Right and Wrong in Human and Other Animals.* Cambridge, MA: Harvard University Press.

Dobzhansky, Theodosius. 1962. *Mankind Evolving.* New Haven, CT: Yale University Press.

Eldredge, Niles. 1995. *Reinventing Darwin: The Great Debate at the High Table of Evolutionary Theory.* New York: John Wiley & Sons.

Gould, Stephen Jay. 2000. "The Evolutionary Definition of Selective Agency, Validation of the Theory of Hierarchical Selection, and Fallacy of the Selfish Gene." Pp. 208–34 in *Thinking About Evolution: Historical, Philosophical and Political Perspectives,* edited by Rama S. Singh, Costas B. Krimbas, Diane B. Paul, and John Beatty. Cambridge: Cambridge University Press.

Gray, Russell D. 2000. "Selfish Genes or Developmental Systems?" Pp. 184–207 in *Thinking About Evolution: Historical, Philosophical and Political Perspectives,* edited by Rama S. Singh, Costas B. Krimbas, Diane B. Paul, and John Beatty. Cambridge: Cambridge University Press.

Hull, David. 1988. *Science as Process.* Chicago: University of Chicago Press.

Itani, J. 1958. "On the Acquisition and Propagation of a New Food Habit in the Troop of Japanese Monkeys of Takasakiyama." Pp. 52–65 in *Japanese Monkeys: A Collection of Translations,* edited by K. Imanishi and S. Altmann. Edmonton: University of Alberta Press.

Kawai, M. 1965. "Newly Acquired Pre-cultural Behavior of a Natural Troop of Japanese Monkeys on Koshima Island." *Primates* 6:1–30.

Killen, Melanie and Frans B. M. de Waal. 2000. "The Evolution and Development of Morality." Pp. 352–72 in *Natural Conflict Resolution,* edited by F. Aureli and F. B. M. de Waal. Berkeley: University of California Press.

Kropotkin, Petr. [1902] 1987. *Mutual Aid: A Factor in Evolution.* London: Freedom Press.

Lewontin, Richard, Steven Rose, and Leon J. Kamin. 1984. *Not in Our Genes: Biology, Ideology and Human Nature.* New York: Random House.

McGrew, W. C., L. F. Marchant, S. E. Scott, and C. E. G. Tutin. 2001. "Intergroup Differences in a Social Custom of Wild Chimpanzees: The Grooming Hand Clasp of the Mahale Mountains." *Current Anthropology* 42:148–53.

Montagu, Ashley. 1987. "Preface." In Petr Kropotkin, *Mutual Aid: A Factor in Evolution.* London: Freedom Press.

Mysterud, Iver. In press. "One Name for the Evolutionary Baby? A Preliminary Guide for Everyone Confused by the Chaos of Names." *Social Science Information.*

Perry, Susan, Mary Baker, Linda Fedigan, Julie Gros-Louis, Katherine C. Mackinnon, Joseph H. Manson, Melissa Panger, Kendra Pyle, and Linda Rose. 2003. "Social Conventions in Wild White-faced Capuchin Monkeys: Evidence for Traditions in a Neotropical Primate." *Current Anthropology* 44:241–68.

Pope, Stephen J. 1998. "The Evolutionary Roots of Morality in Theological Perspective." *Zygon* 33:545–56.

Ridley, Mark. 1996. *Evolution* (2d ed.). Cambridge, MA: Blackwell Science.

Ruse, Michael. 1987. "Darwinism and Determinism." *Zygon* 22(December):419–42.

Ruse, Michael. 1991. "The Significance of Evolution." Pp. 500–10 in *A Companion to Ethics,* edited by K. Malden. Malden, MA: Malden.

Ruse, Michael and Edward O. Wilson. 1985. "The Evolution of Ethics." *New Scientist* 108:50–52.

Sober, Elliott and David Sloane Wilson. 1998. *Unto Others: The Evolution and Psychology of Unselfish Behavior.* Cambridge, MA: Harvard University Press.

Sommer, Volker. 2000. "The Holy Wars about Infanticide. Which Side Are You On? And Why?" Pp. 9–26 in *Infanticide by Males and Its Implications,* edited by C. van Schaik and C. H. Janson. Cambridge: Cambridge University Press.

Sussman, Robert W. 1999. "The Myth of Man the Hunter, Man the Killer and the Evolution of Human Morality." *Zygon* 34(September):453–71.

Wilson, Edward O. 1975. *Sociobiology: The New Synthesis.* Cambridge, MA: Harvard University Press.

Wilson, Edward O. 1978. *On Human Nature.* Cambridge, MA: Harvard University Press.

Wilson, Edward O. 1994. "Human Decency Is Animal." Pp. 167–70 in *From Gaia to Selfish Gene* (reprinted), edited by Connie Barlow. Cambridge, MA: MIT Press.

Wilson, Edward O. 1998. *Consilience.* New York: Alfred A. Knopf.

Wrangham, Richard W. 1995. "Ape, Culture and Missing Links." *Symbols* 2:2–9, 20.

Wrangham, Richard W. and Dale Peterson. 1996. *Demonic Males: Apes and the Origins of Human Violence.* Boston: Houghton Mifflin.

Wright, Robert. 1994. *The Moral Animal: Evolutionary Psychology and Everyday Life.* New York: Vintage.

II

Origins of Sociality

2

Mammalian and Primate Roots of Human Sociality

Adrienne L. Zihlman and Debra R. Bolter

AN EVOLUTIONARY APPROACH TO SOCIALITY

Sociality, the preference for living in a community rather than in isolation, is central to human survival and reproduction. It is as much a product of evolution as is hominid bipedal posture and a large brain. It forms the foundation for maintaining traditions and for developing human language and culture. Indeed, sociality is a necessary prerequisite for culture.

The social nature of *Homo sapiens* has its roots in mammalian and primate biology and behavior. Human social life continues the mammalian system of caretaking, in which females produce milk, infants suckle, and females and young maintain contact through olfactory, tactile, and vocal modes of communication. Primate ancestry elaborates the mammalian base through extended life stages; longer infancy and juvenility, later maturity, and a long life span all increase the potential for intense and long-term social interaction. By keeping close contact with the infants they carry, females remain mobile and integrated into social groups of all age-sex classes that associate throughout life. Primate color vision and vocalizations enhance interindividual communication and group cohesion.

The human lineage further elaborates the primate base, initially through a shift to bipedal locomotion and, later, through an enlarged brain. Compared with other primates, *Homo sapiens*, originating in Africa about 150,000 years ago, added a distinct childhood stage, prolonged adolescence, and lengthened the life span. The fossil, archaeological, and molecular records provide a time dimension and a context for estimating the emergence of modern human life stages, symbolic activity, abstract material culture, and communal life ways.

In this chapter we argue that human culture cannot be disassociated from social life and therefore from humanity's mammalian and primate foundations. Our approach departs from views of human behavior and culture that bypass the individual and sociality, focus on the coevolution of genes and culture, and treat behaviors as discrete units to inherit (e.g., Boyd and Richerson 1985). In contrast, we stress individual interaction and the evolutionary heritage of sociality, its underpinnings, and development; all are integrated with survival and reproductive behaviors—locomotion, foraging, mating, and caretaking. Although competition and aggression are part of social living, we maintain that these behaviors play a less than dominant force in mammalian, primate, or human social life. Instead we emphasize the development and maintenance of social relationships throughout an individual's life and their positive emotional and social expression. In humans, we further articulate the important connection between sociality and symbolic culture, and make a distinction between tradition and culture in order to reflect the greater reach of culture through generational time and across space.

MAMMALIAN INNOVATION IN SOCIAL LIFE

Compared to reptilian ancestors, mammals express a radically new social life that emerges from novel biological structures for maternal feeding of the fetus and infant. Mammalian social life is shaped through the integration of a complex of features: (1) female mobility and higher energetic levels necessary to carry and nurture a fetus and newborn; (2) defined stages of life, including infant, juvenile, and adult; (3) infant-maternal olfactory recognition, and tactile contact through suckling, lactation, and maternal care; (4) audio-vocal communication between infant and adult; and (5) playful behavior among litter mates (Pond 1977; MacLean 1985, 1990).

Mammalian innovations in anatomy and physiology provide a basis for profound changes in energetics and, consequently, for social behavior. Mammals require ten to thirteen times as much food energy as reptiles to maintain a high, internally regulated body temperature and to sustain a high level of activity and muscular work (Radinsky 1987). The expanded mammalian brain requires continual oxygen and nourishment during fetal and infant growth and throughout life, which is achieved through body temperature regulation and steady food supply. The reproductive system and the requirements of milk production place further energetic demands on female mammals. On the other hand, mammals only grow for a limited period in life and replace teeth once, whereas most reptiles continually grow and replace teeth throughout their lives (ibid.).

The mammalian locomotor system becomes a linchpin of the mammalian reproductive system by allowing female mobility during pregnancy and nursing (Pond 1977). The underlying musculoskeletal frame houses well-anchored limbs, mobile but stable joints, distinct vertebrae and flexible spinal column, coordinated movements, and a metabolism to promote movement for effective foraging and predator avoidance.

Mammalian newborns vary in their degree of locomotor independence, but all share a dependence on suckling and milk (ibid.). Newly weaned mammals have a full complement of "temporary" (deciduous or milk) teeth and therefore can consume an adult diet. This feeding pattern offers the potential for a social life in that both immature and mature individuals can live together. In contrast, at hatching, reptiles are equipped with teeth, and prey size determines diet; as body size increases, potential prey size increases. Younger, smaller reptiles may be food for larger conspecifics; therefore young cannot safely cohabit with adults, which tends to prevent different age groups from intermingling and forming social communities (MacLean 1988).

An expanded forebrain, with the newly added mammalian structures forming the limbic system, becomes an integral part of reproduction. Anatomical and physiological features of the brain play a pivotal role in the expression of instincts, drives, and emotions, as well as the association of feelings with sensations, such as smell and sight, and the formation of memories (MacLean 1990). Through experimental studies, MacLean showed that maternal care, play, and vocalizations are impaired if the limbic cortex is damaged in hamsters. The young fail to play, and adult females do not develop appropriate nesting and caretaking behaviors. Thus, complex social behaviors associated with brain structures emerge in mammals but at a higher energy cost.

Mammalian Fossil Record

The fossil record documents the emergence of structures and a basis for inferring behavioral correlates. Dental and skeletal evidence of placental mammals dates to the early Cretaceous period about 125 million years ago, with the discovery of a complete skeleton preserved with hair halo (Qiang Ji et al. 2002). The skeleton closely resembles that of a living tree shrew and indicates a flexible locomotor system and improved foraging (Jenkins 1974). Differentiated dentition into incisors, canines, and molars with interlocking cusps and ridges between upper and lower teeth for shearing and crushing are also identifiable features in fossil mammals that suggest new patterns of feeding and food processing.

The shift of the mammalian jaw from several bones in ancestral reptiles to a single one, the dentary, not only strengthened jaw action, but also refined the sound-transmitting system. The middle ear ossicles, the incus and malleus, transformed from reptilian jaw bones, increased acuity for higher frequency sounds (Davis 1961). Compared to reptiles of similar body size, mammals have a five- to tenfold increase in brain size, as indicated by the expanded brain case of fossil mammals (Radinsky 1987).

Although there is no direct fossil evidence for mammary glands or fat stores, all living mammals, including the egg-laying monotremes (platypus and echidna) and pouched marsupials, produce some type of milk. The ubiquity of lactation suggests that it, along with fat reserves, is a fundamental mammalian feature that emerged in the earliest stages of mammalian evolution (Pond 1977).

Mammalian Stages of Life

Infancy and Emergent Sociality

Social development begins at birth and is expressed through two innovative and essential behaviors of newborns: suckling and maintaining maternal contact through olfaction, audition, and touch. An infant mammal depends on the mother for nourishment and therefore requires close physical contact with her. Maternal licking and grooming give warmth and protection, comfort and reassurance, and contribute to infant physical and emotional health and well-being.

Infant mammals concentrate suckling, swallowing, breathing, smelling, and vocalizing within a small region of the nose, mouth, and throat; a novel neural and muscular apparatus accommodates this complex set of functions (Smith 1992). The olfactory bulb and tract, part of the expanded forebrain, facilitate infant-mother contact through a keen sense of smell. Vocalizations are produced in the cingulate cortex of the limbic system and received through new middle ear bones. Using its larynx in sound production, the infant communicates its location and emotional state—such as pleasure or distress—and the vocalizations tend to be of high frequency (Gould 1983). MacLean labels these early sounds the isolation or separation call, the most primitive and basic mammalian vocalization. Almost from the moment of birth then, communication is established between young and mother and among litter mates.

Milk provides the infant's growing brain continual nourishment. In the course of evolution in some mammalian lineages, lengthened lactation and suckling created the potential for an extended immaturity, for growth of a larger brain, and for an expanded social life.

Immaturity and Social Development through Play

The duration of immaturity varies considerably among mammalian species. After weaning, young mammals follow one of a number of paths: off on their own in less social species, at one end of the continuum, and, at the other, integration into a larger social network. Social integration begins through play, a mammalian hallmark, between litter mates in the nest or when age mates come together. Whether involving physical or social objects, play expresses the higher activity and energy level of mammals. Play constitutes a forum for establishing interindividual communication that can be generalized later in life to other members of the group or population. Through play, females and males learn the social rules for mating, and males learn social rules for displays and test their physical and psychological strength against each other. These interactions have survival value in a variety of ways—for establishing and refining social skills, assessing future rivals, coordinating physical movement, and practicing behaviors in preparation for adult life (Mason 1979; Bekoff 1984, chapter 3 in this volume; Fagen 1993; Rubenstein 1993).

Adult Social Behavior and Female Parental Care

The energy requirements for female mammals are pronounced, a significant departure from reptiles or birds. Reptilian parents provide very minimal or no infant care after laying the eggs; female and male birds both provision the hatchlings. Female mammals, in contrast, provide all the nutrients to the fetus through the placenta and to newborns through milk production from mammary glands, structures not present in birds. Protection of the mammalian fetus from environmental extremes opens up the possibility for mammalian species to colonize diverse habitats.

Mammalian fat storage and mobile calcium can be put to use in both females and males. Males have nutritional storage tanks, like male elephant seals' fat stores, and calcium in cervid antlers. Mammalian females of all species maintain nutritional warehouses for reproduction (Pond 1977). Mammalian bones store calcium, which is mobilized during pregnancy and transferred through the placenta for bone and tooth formation in the fetus (Galloway 1997). After birth, calcium is transferred through milk to promote infant skeletal and dental growth.

Two Ends of the Mammalian Continuum: Elephant Seals and Elephants

The intensity of social life varies dramatically among mammals. We illustrate with two large-bodied species with different social lives to show the correlation between the length of life stages, especially the closeness between mother and young, and social complexity. Elephant seals (*Mirounga angustirostris*) are marine mammals on the central California coast that provide a stark contrast to African elephants (*Loxodonta africana*) on the eastern savannas.

Elephant seals, a "minimalist" mammal with social life confined to a short period during the annual breeding season, lie at one end of the continuum. Adults feed alone at sea but reproduce on land (LeBoeuf and Laws 1994). Each year females (500 kg) come ashore, give birth within twenty-four hours, then lactate, come into estrus, mate, and return to sea, all within a one-month period (Reiter 1997). They give birth at around age four and reproduce throughout their life until about age fourteen. Infants suckle for only three weeks and add 150 kg to their 45 kg birth weight. During this short period, infants maintain continual physical contact with the mother, and the two communicate through olfactory, vocal, and tactile cues. Weaned infants congregate on the beach for several months and have aggressive play encounters with other pups. Female pups play-fight with each other and avoid contact with male pups.

Adult males come ashore before the females and fight fiercely to establish a hierarchy during the short mating season, with only a one in ten chance of ever reproducing (ibid.). Adult males ignore the young, but may injure or even kill suckling pups with their tremendous bulk while chasing other males or moving through the pod to mate with females. Young males engage in mock fighting, foreshadowing the violent interactions among adult males. They begin to compete for territory around age eight, when they have reached adult size of 2500 kg or more (ibid.). Aggression among adult males that inflicts injury is dramatic and well studied. Such overt behavior is often overemphasized, in contrast to the more subtle interactions between mother and pup that are key to understanding the elephant seal social system.

African elephants, at the other end of the mammalian social continuum, live in closely bonded matriarchal groups connected to each other through a social network involving hundreds of conspecifics distributed over a large area (Moss 2000). Young elephants associate closely with their mothers, siblings, and other related group members from birth onward; they are weaned by age four. Play behavior shows up early, as infants find their way around the social group, while they are learning to feed and drink. Females reproduce at about age fourteen, remain in the natal group, and cooperate in taking care of their siblings and related young. Three or more generations of females reside within the family group and in the wider community. As males approach reproductive maturity, they become peripheral to the matriarchal group and join up with other young males or become solitary. When young males meet, they spar in order to assess each other's weight and strength. The older males (around thirty years of age) are the preferred mating partners. Elephant lives span fifty to sixty years.

Elephant communication is highly developed and relies on many sensory modalities (ibid.). Trunks and bodies communicate tactilely in close-up interactions; scent serves to send and receive close and distant messages. Low rumbles flow between and among closely associated individuals as contact calls, and loud trumpeting expresses excitement or distress. Noisy greetings accompany the reuniting of separated families, and infrasonic wave sounds communicate movements among social groups separated by several kilometers (Payne 1998). Elephants apparently recognize and remember hundreds of other elephants. Given four years as infants and ten years as juveniles, young elephants have ample time to learn and master the appropriate behaviors for living in their socially complex world. Their large brains, good memories, problem-solving abilities, complex ways of communicating, social knowledge, and long life span give elephants the added capability for survival and for transmitting learned traditions from individuals of the older generation to members of the younger, up-and-coming generation (Eisenberg 1973; McComb, Moss, Durant, Baker, and Sayialel 2001).

Summary

These two species highlight the connection between individual life stages and species life history and the size and complexity of their social groups. We argue that the extended period of association between adult and young, and among immature animals, provides time and experience for establishing emotional ties and for learning and practicing necessary social skills. The three-week weaning age, the short life span, and relatively solitary existence of elephant seals require only minimal sociality, communication, and interaction. In contrast, the extended immaturity of elephants, long life span, and complex communication correlate with a high degree of cohesion within groups and elaborate interaction among individuals in a wider community.

A social system includes group members that show cohesion, interdependence and permanent associations, and complex communication among individuals. Complex mammalian social systems involve significant dependence of infants on adults. Vocalizations in infant-mother interactions characteristically "resurface" in adult mammals' greeting displays and sexual behavior and become an important vehicle for facilitating social communication among all group members (Eisenberg 1973; Gould 1983; MacLean 1985). Long lives and interaction among all group members and young with adults form the foundation necessary to establish traditions, and provide the setting for diverse and complex social behavior. Social systems, such as those of elephants, other mammals, and primates, are therefore not unique to humans.

PRIMATE BEHAVIORAL FLEXIBILITY

Primates represent a diverse mammalian order with over 250 species that live in a range of habitats. For this discussion we draw on primate examples from the catarrhine phylogeny, which includes African and Asian monkeys, apes, and humans and which represents a shared ancestry with *Homo*. Primate social life builds upon mammalian biology and behavior and further expands the time and energy put into the caretaking of young. Primate locomotor and sensory systems and extended life stages underpin the diversity observed in social groups. The social structure is variable, for example, a female and her offspring as in orangutans; female/male pair bonds with dependent offspring, as in gibbons; multiple females with one or more adult males, as in colobine monkeys and gorillas; or large multifemale/multimale groups, as in baboons, macaques, and chimpanzees. All social groups build upon the primary bonds between mothers and offspring and offer something for all ages: playmates, protectors, sexual partners, friends, and possible allies for coalitions. Although they vary in size and composition within and across species, social groups usually maintain spatial integrity over time and offer the possibility for generational continuity and learned traditions.

The locomotor system provides the animal's "infrastructure" and plays a central role in helping maintain the integrity of social life through female mobility (Zihlman 1992). The social life of a female that is pregnant or carrying a suckling infant is not interrupted, and she can therefore travel and interact within a stable social network. The infant, clinging tightly to its mother's hair as she runs, jumps, and climbs, becomes a social participant very early in life with the mother's associates. Primate locomotion retains the stable and flexible mammalian skeleton and adds mobile forelimbs and hands with opposable thumbs, long digits with nails, and sensory pads (Washburn 1951). The hands not only grip branches while the primate moves through all levels of the forest, but also contribute to social life through enhanced tactile communication in infant clinging, mutual grooming, and other physical contact.

Color, stereoscopic (three-dimensional) vision, day-living, and hand-eye coordination elaborate communication using gestures, facial expressions, and body postures. The expanded forebrain, especially the sensory-motor and visual cortices, underlies the functions associated with locomotion, hand function, visual processing, memory, and communication. Primates have one of the largest brain-to-body size ratios of all mammals, and the brain grows rapidly during infancy. The larger brain-to-body size correlates with longer gestation and lactation and places additional energetic demands on female primates compared to other mammals.

Primate Fossil Record

Fossils record anatomical features in the hard tissues that serve as a basis for inferring behavior. For example, an early prosimian primate with a grasping big toe with nail is well dated at 60 million years ago (Bloch and

Boyer 2002) and fits with molecular evidence indicating that prosimian primates were present by this time (Yoder et al. 1996). Fossil sites in France and western North America yield numerous prosimian species with grasping hands and feet (Le Gros Clark 1959). By 35 million years ago fossil deposits from the Fayum, Egypt, yield evidence of early monkeys (Fleagle 1999). These monkeys have enclosed and forward-facing eye orbits, a reduced snout, expanded cranium and occipital region (with a presumed expanded visual cortex); the cranial features suggest the species emphasized stereoscopic, color vision, day living, and had a reduced sense of smell. The inference that the composition of early monkey social groups was one of mixed sexes derives from fossil deposits that contain a single species with jaws and teeth of various ages and sizes (Benefit 1994).

Primate Life Stages

Infants and an Instant Social Network

Newborn primates retain the mammalian innovations of suckling and maternal contact but add the ability to cling to the mother and to communicate visually. Although motor function is minimal, infant primates have acute senses: olfaction and audition, touch, and now vision. An infant's brain is ten times heavier relative to its body weight than that of an adult and so is equipped to process the sensory information (Grand 1977). Physically dependent for transportation and food, a newborn has hands and feet that are relatively twice the mass of adults' and is able to grip its mother's hair tightly when carried, fed, and protected by her (ibid.). The infant takes an increasingly active role in maintaining contact with its mother, keeping track of her location, and attending to her activities (Altmann 1980). Brain growth, compared to somatic and dental growth, is rapid throughout the first months and years of life, and by weaning age and first permanent molar eruption, approaches adult size (Smith 1989).

Bonding with the mother and other group members correlates socially with a long infancy and close maternal contact. As the infant moves with its mother, it acquires a ready-made social network of her older offspring, and her adult female and male associates. Visually, it monitors its physical and social surroundings and watches others. Through early social experience, young primates learn the meaning of the nuances of facial expressions, body postures, and gestures that play central roles in communicating and in promoting the development and maintenance of complex social networks.

Long-term studies over generations show that in a number of species an infant primate acquires a social rank similar to its mother, by learning the social order through observing her actions and, in turn, receiving her backup support in social encounters (e.g., Kawamura 1958; Sade 1972; Goodall 1986). Regardless of social group size or composition, the physical and emotional attachment of infants to their mothers provides the experience during early development for learning motor skills, social rules, and for establishing relations with other group members. For example, during infancy, chimpanzees learn to manipulate objects by watching their mothers and others. At five years of age, Gombe chimpanzees are finally adept at the sophisticated task of fishing for termites using modified grass stems (Goodall 1986), which demonstrates that it takes a long time to learn complex tasks.

Juveniles and Learning Social Rules

The juvenile stage, longer compared to that of other mammals, is a time of physical and social change. Juveniles have not yet reached adult size and are not reproductively mature, though they now travel and forage independently. They must compete with other group members for access to resources, making this time of life most vulnerable for a primate, with highest mortality (Pereira and Altmann 1985).

In spite of the high mortality rate in juveniles, longer immaturity must somehow benefit individuals and therefore outweigh the costs. Prolonged immaturity allows primates to counter selective pressures of the environment through behavioral flexibility (Mason 1979). Through social play young primates work out intergroup relationships and practice social exchanges (Fagen 1993). They handle other infants, groom, and practice sexual signals and skills (Walters 1987). Energetically demanding play activities decrease as juvenile primates increase in size and approach maturity (Fagen 1993).

Females and males begin to diverge during the later juvenile stage. Females often complete bone, canine tooth, and muscular growth earlier than males of the species, as they turn energy to reproduction (Bolter and Zihlman 2003). During this stage young male primates in many catarrhine species, or both sexes in some species, leave their natal social group and transfer to a nearby group (Pusey and Packer 1987; Yeager and Kool 2000). Juvenile males, although physically fertile, are more active socially in becoming integrated into a new group than in pursuing reproductive bouts (Strum 1987).

Juvenile Baboons and Vervet Monkeys in Amboseli, Kenya.   The length of the juvenile stage in primates dramatically affects adult life and survival. Comparison of vervet monkeys (*Cercopithecus aethiops*) and yellow baboons (*Papio cynocephalus*) illustrates the importance of this stage in learning and practicing skills. These closely related monkeys inhabit the same woodland areas of eastern Africa; they are also similar in dental and digestive anatomy; they differ in body size and in length of the juvenile stage. Vervet monkeys mature about two years earlier and so reach adulthood before baboons.

During the juvenile stage, these monkeys are learning to feed independently and to choose their foods. Stuart Altmann (1998) documents the adeptness of young baboons at finding and choosing quality foods; they exploit more than 250 species over their lifetimes, far more than vervets. Local vervet monkey populations have become extinct during periods of food scarcity, whereas baboons have survived (Lee and Hauser 1988). Narrow food choices and reliance on seasonally abundant foods limited the vervet foraging, especially in times of low production. Baboons, in contrast, were able to exploit less lush and less available plant foods. With a shorter juvenile stage, vervets essentially "give up" time during immaturity for learning, finish growth earlier, and begin reproducing.

Juvenile primates are not only acquiring ecological knowledge but are also learning social skills that contribute to survival and, later in life, to reproductive outcome, and that form the basis for species-typical social organization. Although the acquisition of social skills may be more difficult to measure than the quantifiable numbers of utilized plant foods, time for learning and practicing social behavior and communication is as critical for adult survival as is mastering a complex ecology (Rowell 1988a, 1988b).

Adolescence and Transition into Adult Life

This stage is usually associated with humans, but incipient adolescence, sometimes called the "subadult" phase, is present in some primates like the apes. Orangutans and chimpanzees have the most extended life stages among primates; they prolong immaturity beyond the juvenile stage into a period of adolescence when they are not yet reproducing. For adolescent orangutans, learning social skills through interaction with peers and elders is an important aspect of group contact and may be necessary for the development of normal social intercourse typical of wild adult orangutans (Sugardjito, te Boekhorst, and van Hooff 1987). During this adolescent phase in chimpanzees, subadults negotiate the social community, females establish themselves in another community, and both females and males master tool-using.

The lives of female apes depart significantly from those of female Old World monkeys in having an extended period between their first sexual swellings and their first infant. Female monkeys, like vervets, baboons, and proboscis monkeys, reach sexual maturity, complete physical growth, and have their first infant all within a period of two years or less. Female chimpanzees, in contrast, have their first estrus swellings at about age ten but do not regularly ovulate or become pregnant for one to three years, a period in which they are physiologically sterile, and do not give birth until they are thirteen to fifteen years old (Pusey 1978; Boesch and Boesch-Achermann 2000; Nishida et al. 2003).

Although catarrhine males typically migrate to a new group, in some species, females as well as males also leave their birth group to seek out a new social community as adults. In a departure from these catarrhine migratory patterns, only the females in chimpanzee societies leave their community of birth; during this adolescent period of subfecundity and the transition to adulthood, females travel to another community where they will join another group, become pregnant, give birth, and likely spend the rest of their lives (Goodall 1986; Kano 1992; Boesch and Boesch-Achermann 2000).

Male chimpanzees, like male monkeys, reach adult body and canine tooth development later than females; during this time male chimpanzees establish themselves socially within the natal community and participate fully in the social arena as adults. They begin to produce viable sperm at about age eight, but are not fully grown or integrated into the male net-work; males do not achieve alpha status in the community until about age fifteen or older (Goodall 1986; Nishida 1990; Watts and Pusey 1993; Boesch and Boesch-Achermann 2000). This preadult period provides time to begin to build alliances, experiment with risk-taking, and establish a niche in the social fabric.

Chimpanzees live in social communities with fifty to eighty individuals. While foraging and traveling on the ground around the home range, they congregate temporarily in subgroups of different sizes and age-sex compositions and then go their separate ways in what has been described as "fission-fusion." This description is a partial misnomer because the "fusion" is not consistent; that is, small groups can consist of different individual members depending on time, space, or resources. This socially flexible group structure contrasts with the relatively cohesive groups of baboons, vervets, other Asian and African monkeys, and mountain gorillas, for example. Elaborate chimpanzee greeting behaviors include vocalizations, gestures, touching, facial expressions, and body postures that express recognition and reconnection to each other, reinforce friendships and social bonds, and dampen tension among individuals who do not see each other on a daily or regular basis (Goodall 1986). During the extended immaturity of adolescence, chimpanzees—as we noted also for orangutans—have time to build and practice the repertoire of social and communicative skills, develop flexible responses to social interactions, and network into a new community.

Adolescence also gives chimpanzees time to learn ecological skills, such as nut-cracking, a quantifiable ability. Chimpanzees of the Tai Forest, Ivory Coast, employ hammers and anvils to break through hard shells to gain access to nutritious nutmeats (Boesch and Boesch-Achermann 2000). Not until ten years of age do they reach the first level of skill in cracking open hard-shelled nuts; up to this time their social access to the best tools is limited. In contrast, Gombe chimpanzees in Tanzania at age five have mastered the use of grass stems for drawing termites from the mound (Teleki 1974; Goodall 1986).

Adults and Expanding Networks and Skills

Primate adult lives center around social life and reproduction (Fedigan 1998). Reproduction is a defining feature for adult females, and they accommodate biologically and behaviorally (Zihlman 1997). Compared to other mammals, female primates invest considerable energy in nurturing and carrying a relatively large-brained fetus and infant. During lactation females drop to their lowest body weight and are most vulnerable (Bercovitch 1987). A primate female counters this reproductive cost by storing fat that promotes her ability to conceive, carry the fetus to term, and support the neonate and infant; increased fat reserves provide an energy warehouse to fuel the demands of lactation (McFarland 1997). Daily travel and activity level may decrease during lactation, giving a female more time to feed and rest (Altmann 1980). A female's social life picks up when her infant is weaned; she then spends more time grooming and interacting with other group members to maintain her social network (Altmann 1980; Dunbar and Dunbar 1988).

Adult male mammals rarely live together in the same social group in the presence of sexually active females, whereas male catarrhines in many species do. Adult primate males may change groups several times during life and each time must establish themselves in their new home. They become integrated through making friends with females and through minimizing rather than maximizing aggression (Smuts 1985; Strum 1987). In the relatively cohesive baboon and macaque groups, for example, elaborate visual and vocal communication greases the social wheels to minimize disruption and promote affiliation. Male guenon monkeys living in less cohesive groups with one central male monitor each other closely rather than interact (Rowell 1988a). Social and communication skills are essential for male survival, and a long life is a key variable associated with male lifetime reproductive success (Fedigan and Zohar 1997).

Adult Chimpanzee Hunting and Nut-Cracking of the Tai Forest. Two examples in chimpanzees (*Pan troglodytes*) illustrate that some skills, hunting and nut-cracking, are mastered during the adult stage. Although female chimpanzees also prey upon small animals, they have not been observed hunting in coordinated group efforts as males have. Similarly, males use tools adeptly, but do not achieve the high level of skill in nut-cracking observed in females, nor do they act as teachers. Success in each of these activities relies upon an individual's long apprenticeship, keen observation powers, and ability to send and receive subtle communications.

Hunting.   The pursuit of red colobus monkeys is a collaborative effort among adult chimpanzee males of the Tai Forest. The activity occurs most frequently during the rainy season when branches are slippery and the prey more likely to stumble, rather than during a time of food scarcity. The "pursuit-hunting" style of chimpanzees includes a "driver" and "blockers" who herd their prey into traps, and the "ambushers" who must anticipate the directions of fleeing prey and be able to predict the responses of the other male chimpanzees, even when they are not visible. Males also cooperatively "split" the carcass based on the role of each during the hunt (Boesch and Boesch-Achermann 2000).

Among Tai chimpanzees, learning to hunt begins at about nine to ten years of age, after a male leaves his mother's company and enters male social life, and after he seems to have overcome his fear of being wounded by prey. The skills and strategies of hunting are acquired over the next twenty years. Only the oldest males over age thirty are able to fully anticipate the long-distance movements of their prey correctly, coordinate movements with hunting partners, and successfully station themselves in trees for the future ambush. One young five-year-old male orphan was adopted by the most skilled hunter of the group and became his apprentice. Over the next years of his life, the young male never left his surrogate parent's side, learned sophisticated hunting skills through observation and practice, and became an adept hunter earlier than his age-mates (Boesch and Boesch-Achermann 2000).

Some Tai males never successfully accomplish this cooperative hunting with roles and strategies because individual ability, memory, and experience all contribute. It is likely that the most important aspect of hunting functions to bond males together through cooperation and less as a means to obtain a source of food or sex with females (Mitani and Watts 2001).

Nut-Cracking.   Both females and males of the Tai Forest reach the first level of skill in nut-cracking, that is, in the number of hits required to crack open one nut. The second level of skill is a measure of efficiency, in the number of nuts processed per minute. Only adult females achieve this level (Boesch and Boesch 1983a, 1983b). In nut-cracking as well as hunting, adults maximize these skills.

The spread of tool traditions proceeds through generational time, most frequently from mothers to young daughters and sons. Females may even teach their young by demonstrating how to hold the hammer or by placing the nuts properly on the anvil (Boesch 1991). Female chimpanzees also transmit tool-using skills across space. When females change groups they take their learned ecological skills into a new community. In field experiments in Bossou, Guinea, Matsuzawa (1994) documented that a female about age thirty immediately recognized the coula nuts that he had provided. The Bossou population cracks oil-palm nuts but not coula nuts. When she began cracking the nuts, other group members ignored her, except for two six-year-old juveniles who also began cracking the unfamiliar nuts. Matsuzawa hypothesized that the female had immigrated from the Mount Nimba community some 15 miles away where she had learned the tradition and brought it with her to Bossou.

Summary

By combining a long period of mother-infant association and long life spans, group living promotes nongenetic learning of social traditions between generations. Infant primates bond with the mother through physical contact and multimodal communication, with siblings and peers through play, and with other group members through early contact. Juvenile primates hone their social skills and become a bridge to the next generation, in establishing social ties and lifetime networks that ensure group cohesion over time. Adolescents ("subadults") acquire competence in the subtle aspects of social communication and develop population-level traditions that are mastered during adulthood. Relative to other mammals, primate groups integrate males, and females invest more energy in each offspring. As part of multigenerational groups, adults have a wide social network, a characteristic of large-brained, long-lived, highly social species that includes elephants as well as primates (Eisenberg 1973).

Social life of adult females as well as males requires energy and skill in communicating and living together effectively. The integrity of the community is maintained over the long haul through the emotional and social bonds that are formed during maturation and the affiliative skills practiced throughout life, rather than through frequent aggression and fighting. Variation in social structure across species results from different patterns of socialization within a species.

Traditions or Culture?

Many social mammals including primates have traditions. The sweet potato washing of Japanese macaque monkeys represents a well-known example from catarrhine primates. Apes, particularly chimpanzees, have more elaborated traditions that vary across populations, for example in communicative gestures, learning-teaching mechanisms, types of objects used as tools, and levels of skill (Boesch and Boesch 1990; McGrew 1992; Whiten et al. 1999). Although this variation has been labeled as "culture," we maintain that a distinction between culture and tradition is essential in order to characterize, understand, and explain human adaptation.

Many aspects of chimpanzee communication, sociality, foraging, and tool-using skills foreshadow human culture. Chimpanzees share with humans several basic mechanisms for intergenerational transmission and innovation of tools, for example, observational learning and imitation, independent practice, playful experimentation with objects, and collaborative learning. These commonalities between chimpanzees and humans imply that these abilities may have been present in the common ape-human ancestor of five million years ago (Greenfield, Maynard, Boehm, and Schmidtling 2000).

The addition of arbitrary symbols commonly used by humans to teach a technological skill is not observed in chimpanzees, and this abstract dimension probably emerged in the human lineage after the ape-human split (ibid.). Traditions can be transmitted through direct observation and interaction without the use of symbols; whether hunting or using tools, chimpanzees learn in the presence of conspecifics. In contrast, cultural learning and transmission of knowledge require symbols—an item, a gesture, or a word with an abstract referent—that chimpanzees and other mammals habitually lack. By incorporating symbols, cultural learning can take place in the physical absence of a living model. Human sociality and culture derive from a primate base of extended life stages and time for learning and practice. The transmission of traditions over time and across space, we maintain, is a necessary but not sufficient base for the emergence of human culture.

HOMINIDS: COMMUNICATORS THROUGH TIME AND SPACE

Our species, *Homo sapiens,* further expands the repertoire of mammalian and primate biology and behavior. Habitual bipedalism distinguishes hominids from other primates, and a new life stage, childhood, extends the time of immaturity after weaning. Hominid life is founded upon locomotor endurance, the ability to walk long distances, to forage over a large area, to collect and carry food and implements, and to hold babies (Zihlman 1997). Implements fashioned from a variety of materials increased the range of foods gathered and prepared. Childhood allows more time for the young to learn, for the mother to reproduce again, and for the wider social network to accommodate child care. Modes of communication—tactile, visual, vocal—now add a symbolic and abstract dimension, transforming traditions into cultural practices. Cultural symbols make possible new ways to communicate, so that social interactions can go beyond direct face-to-face exchanges and kin relationships.

At present there is no way to reconstruct exactly when or in what sequence these changes took place during the past four to five million years, about the time hominids may have separated from the chimpanzee lineage (Ruvolo 1997). The combination of morphological, genetic, paleontological, and ethnographic information suggests to us that childhood, expanded adolescence, language origins, and cultural complexity all probably arose in Africa between 100,000 and 150,000 years ago. The prehistoric record, along with a growing body of molecular evidence (mtDNA, Y chromosome, nuclear DNA), points to an African origin of *Homo sapiens* around 150,000 years ago (Cann et al. 1987; Vigilant, Stoneking, Harpending, Hawkes, and Wilson 1991). After 100,000 years ago, *Homo sapiens* left Africa and eventually replaced other hominid species around the world.

Hominid Fossil Record

The course of human evolution spans about five million years. Numerous hominid species lived during this time. The earliest unequivocal fossil evidence for two-legged hominids dates to 3.5–4 million years ago and consists of fossilized footprints from Laetoli, Tanzania, and lower limb bones from the Turkana Basin, Kenya (Leakey 1984; Leakey and Walker 1997). These early members of the human family (australopithecines) had brains only slightly larger than those of chimpanzees (400–500 cc; Falk et al. 2000). Studies of fossil teeth of immature individuals indicate the timing of eruption of the first permanent molar resembles that of apes rather than that of *Homo sapiens* (Bromage 1987; Beynon and Dean 1988).

Between 2 and 2.5 million years ago, several new species belonging to the genus *Homo* appear, along with stone tools, now recognizable in the archaeological record (Ambrose 2001). Between 1.5 and 2 million years ago, hominids left Africa in several waves. Their remains are found in western Asia, Southeast Asia, and southern Spain (cf. Zihlman and Lowenstein 1999). These species of *Homo* have smaller teeth and larger brains (700–1000 cc). About 600,000 years ago, according to molecular data, the genus *Homo* (perhaps a species labeled *Homo heidelbergensis*) split into two lineages, one leading to *Homo sapiens*, evolving in Africa, the other to *Homo neanderthalensis,* evolving in Europe (Krings et al. 1997; Ward and Stringer 1997; Rightmire 1998; Stringer 2002). The Neanderthals became widespread in Europe and extended their reach west into the Ukraine, but became extinct by around 30,000 years ago.

Although Neanderthals have brains as large as those of modern humans, they are distinct from modern humans in cranial, mandibular, and dental features, limb bones, and pelvis (Stringer and Gamble 1993; Rak et al. 2002). The Neanderthal brain approaches adult size at first permanent molar eruption, about age four years (Stringer, Dean, and Martin 1990; Ponce de León and Zollikofer 2001). This timing is similar to that of wild chimpanzees (4 years) and *Homo erectus* (4.2–4.5 years) and distinct from that of *Homo sapiens* (6 to 7 years) (Dean et al. 2001; Zihlman and Bolter 2003). In contrast to Neanderthals, modern humans require a longer time to grow a somewhat smaller brain; we see this growth pattern as a major developmental shift in *Homo sapiens* (Zihlman and Bolter 2003).

Fossil and archaeological sites in Africa provide glimpses of the transition to *Homo sapiens.* The stratified site of Klasies River Mouth on the Indian Ocean in South Africa dates to 120,000 years ago and reveals some of the oldest physical remains of modern humans (Deacon and Deacon 1999). They are found along with circular hearths and food remains of plants, shellfish, and animal bones that document a way of life much like that of modern hunter-gatherers. Red ochre, a natural pigment in a crayonlike material, is present at many sites of this age and suggests its use in some type of body decoration or ritual behavior. At Blombos Cave, South Africa, an ochre stone with etchings further supports the conclusion that the capacity for creative and symbolic thinking was present in this population (Henshilwood, d'Errico, Marean, Milo, and Yates 2001). Formal bone working and finely made bifacial points represent advances in motor and cognitive skills (Potts, chapter 12 in this volume; Tattersall, chapter 11 in this volume).

Recent studies of a gene associated with word articulation indicate that language as we know it may have appeared with the origin of *Homo sapiens* (Enard et al. 2002). The *FOXP2* gene, contributing to improved communication through refined vocal (mouth) and facial movements, may have become fixed in *Homo sapiens* between 200,000 and 120,000 years ago. The combination of carved objects, items of personal adornment, and refined lithic technology implies language development and new dimensions of social complexity and tradition, abilities taken with *Homo sapiens* as they dispersed out of Africa and later elaborated in well-known cave art in western Europe about 30,000 years ago.

It appears then that the origin and later worldwide success of *Homo sapiens* are marked by new anatomical, behavioral, and developmental patterns that originated in Africa prior to dispersal elsewhere.

Hominid Stages of Life

Infants and Intensified Helplessness

Bipedal locomotion and an enlarged brain profoundly affected the hominid mother-infant interaction. At birth, the infants' special senses (tactile, auditory, taste, olfactory, visual) are well developed, although their brains are only 25 percent of adult size (in contrast to chimpanzees at 45 percent). Motor function is very underdeveloped. Human infants lack the strength in their hands to cling tightly; they must be fully supported when carried.

Like other primates, human infants are also born into a social world. Infant communication builds on the mammalian auditory, tactile, and visual modes of interaction and refines and intensifies primate face-to-face communication. With limited locomotor skills, infants must solicit and engage mothers during suckling with facial expressions and vocalizations. The development of smiling, reaching and pointing, cooing, and crying helped to ensure contact with mothers and other caretakers (Bowlby 1969). Perhaps there was natural selection at this time for elaboration of visual and vocal modes of infant communication (Borchert and Zihlman 1990). The repertory of gestures could expand now that the hands were no longer used for clinging to a caretaker.

By the end of infancy and weaning at about three years of age, human infants are still small in size, dependent, and vulnerable.

Childhood, a New Stage of Life

In all primates except humans the transition from infant to juvenile is marked by independence in travel, foraging, and feeding. In contrast, human infants at weaning—unlike other primates—do not have any permanent dentition or locomotor efficiency. Although they are able to move bipedally unassisted, they cannot walk far and must be carried on occasion. Motor development is reached at seven to eight years, and first permanent molars erupt at around age six to seven. This extended time of growth between four and seven years is what Bogin (1999) refers to as childhood. Children can feed themselves, but are unable to get food on their own and are dependent on others to provide it. This lifestage shift compared to all other primates may lie at the root of the human-unique adaptation as it opens up socialization beyond the mother-infant dyad, as found in nonhuman primates. This wider social learning and communication during development may account for the more intense variation in human social structure and organization cross-culturally than seen among the primates.

Social groups become critical for childhood survival because of the dependence on others for food and protection during this stage (Bogin and

Smith 1996). Caretaking pressures on the female parent are reduced, while energy devoted to child care may be assumed by others. For example, as in many human societies, during foraging bouts contemporary African hunter-gatherer women often leave children at camp where others keep watch (Draper 1976). Caretakers other than the biological mother also interact and bond with human infants—bonds that may persist into the childhood stage and provide additional support for childhood survival (Peacock 1991). Sometimes caretakers for children are not available, as in a Nepalese farming society during the intense planting season requiring everyone's labor. Mothers carry infants to the fields, but must leave three- to six-year-olds home alone, making children the most vulnerable age group in terms of nutrition (Panter-Brick 1997, 1998).

The childhood stage forestalls the time when an immature must compete with adults for direct access to resources, while still providing a long time frame for learning the social rules. Childhood allows for cognitive development through apprenticeship with other group members beyond the mother, much like the case with the orphaned chimp who excelled at hunting, and enables individuals to learn to communicate and interact with others beyond the immediate family (Rogoff 1990, 2003). Childhood also provides added developmental time in a group setting for the social maturation of the human mind (Wertsch 1985). In horticultural societies, for example, it is during childhood that gender roles are acquired (Morelli 1997). Cross-culturally, as they transition from childhood to the juvenile stage, humans become functioning members of their community (Rogoff et al. 1975).

During this life stage, humans become proficient in language as their brain matures (Vygotsky [1934] 1962). Language allows individuals to classify their surroundings, communicate abstract thoughts (in time, space, and meaning), and engage in more complex social relationships. Human language occurs in social settings and takes meaning in the space between participants (Vološinov [1929] 1998). Through spoken language, children learn overt and subtle rules of multigenerational living, skills through play, and interaction with individuals of all ages. Progressive mastery of key components of life—language, social roles and rules, cultural markers, motor function—characterizes this new stage of life.

Juveniles: A Time of Learning and Helping

After childhood, the juvenile period becomes a time of refining the lessons for life, through play and practice. Still sexually immature, juveniles continue learning complex cognitive schemes. Juvenile individuals, although self-sufficient in traveling and feeding themselves, have yet to reach adult size and the level of skill needed to be fully independent from their immediate family group. In the context of the social group, juveniles can seek out their own interactions rather than have them defined by the mother or caregivers and develop relationships with playmates and other adults. Juveniles in modern human populations help care for infants and children and participate in daily work (Nag et al. 1978; Monroe et al. 1984; Panter-Brick 1997). A growing facility with language and rules integrates juveniles into wider networks.

Adolescence: Preparation for Adult Life

Physical changes in young women and men between thirteen and eighteen years of age transform them into adults. During adolescence, males and females diverge significantly in physical and social maturation. In the biological transition of boys to men, viable sperm precede genital development, followed by significant skeletal growth in height, added muscularity, and weight gain (Short 1976; Dixson 1998). Physical development of male chimpanzees is similar, although the skeletal growth spurt that defines human adolescence according to Bogin (1999) does not seem to be present.

Human females extend the time of sexual maturation even longer than do chimpanzee females, who, like humans and unlike other primates, do not make an immediate transition from first sexual swellings to pregnancy. In the transition of girls to women, breast development occurs first, then skeletal growth in height and pelvic dimensions, weight gain (mostly body fat), and finally regular ovulation and fertility (Short 1976; Dixson 1998). Changes in body composition at this stage fill out the distinctly female shape; body fat is added to the trunk and hips, around the center of gravity, so that locomotor function is not compromised (Zihlman 1997). This pattern of development ensures that women are prepared for the demands of reproduction—giving birth to, feeding, and carrying large-brained infants for several successive years (Lancaster 1984).

The biological and physical transitions are often celebrated cross-culturally. Through cultural rituals, ceremonies, and rites of passage, adolescents prepare for full participation and integration into the adult social community. The age at which social adulthood is recognized varies across human societies and embodies cultural meaning and, therefore, may not mirror adult biological markers, such as body size and fertility.

Adult: Social Responsibilities and Physical Demands

Three elements of human adulthood are emphasized in our discussion: the reproductive demands on female adults, the caregiving role of adult group members, and the increased adult life span. Each is considered in light of prolonged human immaturity and how social adaptations as adults may have coevolved with new developmental changes in human young.

Human ancestors were nomadic, living as hunter-gatherers where women collect food while pregnant, lactating, and carrying infants (Lee 1979). Therefore, a reproductive woman must maintain locomotor mobility, produce milk to feed a large-brained infant, and sustain her own energy and nutritional needs. Intense infant suckling and women's travel and work effort help suppress human ovulation (Ellison 1990) and so contribute to maintaining a four-year spacing (Howell 1979; Lee 1979). Without the insertion of a childhood stage where survival of children can be assured through additional care from others, a female's birth spacing would be closer to six or seven years (Bogin 1999), an interval that would significantly limit the number of children any female could produce. We maintain, like Bogin and Smith (1996) and Parker (2000), that childhood could only have emerged within a social system friendly to children. In this system, foraging women maintained a birth interval of three to five years, similar to that of chimpanzees. Childhood extends the period of immaturity to about seven years and reduces the period of intense nurturing for the mother. Other members of the social group must contribute to the children, thereby ameliorating the impact of women giving birth to and nurturing large-brained babies who remain dependent for several years.

We depart from Bogin and Smith and others who emphasize a nuclear family, maternal associates, and other biologically related individuals who provision and protect the young. Instead we emphasize the importance of a wide social network that goes beyond biology. A long period of immaturity benefits the whole community by giving young individuals more time to learn and master skills that will be applied later in life. On the downside, this learning period demands considerable investment of time and effort, and raising the next generation demands additional help from others. The basis for recruitment and participation of others as caregivers and models, we argue, is rooted in a network of symbolic relationships specified by human language and kinship terminology.

Kinship is a cross-cultural universal and may be a fundamental component of human society that specifies roles and obligations among individuals. Although kin terms derive from biological relationships (e.g., mother, brother, uncle, grandmother), in practice, they classify individuals into social roles regardless of actual shared genes and reflect cross-cultural variation in kinship rules and expressions. Kinship systems are often founded on postmarriage residence rules rather than on true biological affinities (Sahlins 1976), and they emphasize the importance of older generations in family groups. In this kind of environment, individuals rely upon each other to maintain the community. The development of prescribed roles, through the use of symbols, may have evolved to accommodate hominid immaturity.

An extended adult life span and the presence of three or more generations promote the transmission of cultural information through language and oral traditions, stories, and folklore (Diamond 2001). Older members of the group can serve as caretakers and as repositories for information and ideas. In the absence of direct experience of "seeing" and "doing," language becomes the means to hand down information over generations. This expansion of tradition goes well beyond multigenerational learning of primates and other mammals. Culture, predicated on language, opens up more possibilities for sociality through assigning abstract and symbolic relationships whose meanings are understood by members of the society.

Summary

Sociality as we know it in *Homo sapiens* has its roots in mammalian biology and behavior and grew from the primate branch. With the evolution of childhood, older children as well as juveniles no longer depended exclusively on their mothers but claimed membership in the wider social group. The community, along with an elaborate communication system encapsulated in symbolic language, plays a vital role in being human.

Culture emerges from long-established social living that encompasses mutual recognition, interaction with, and knowledge of others. Human learning and experience transcend the immediacy of sensory impressions through abstract expressions of art, speech, and writing. These dramatic transformations of earlier primate behavior were established in *Homo sapiens* before they left Africa more than 100,000 years ago.

Shaped as all life is by millions of years of natural selection, modern humans, with refinements in the past 150,000 years, dispersed throughout the world with an armamentarium of unique adaptations for inhabiting all climates and conditions, including those of outer space. Fundamental to the success of the species is an elaborate network of social connections, an ability to communicate by complex language, and thought processes that are free from the tyranny of the here-and-now and make it possible to contemplate and learn from the past and imagine and plan for the future.

ACKNOWLEDGMENTS

We thank Robert Sussman for the opportunity to contribute to the volume and for his comments on the manuscript. We also appreciate the comments from our colleagues Don Brenneis, Catherine Handschuh, Jerold Lowenstein, Robin Mc-

Farland, Shannon McFarlin, Barbara Rogoff, Carolyn Martin Shaw, and Joanne Tanner. Research support from the Leakey Foundation and from the Social Sciences Division, University of California, Santa Cruz, is gratefully acknowledged.

REFERENCES

Altmann, J. 1980. *Baboon Mothers and Infants.* Cambridge, MA: Harvard University Press.

Altmann, S. A. 1998. *Foraging for Survival: Yearling Baboons in Africa.* Chicago: University of Chicago Press.

Ambrose, S. H. 2001. "Paleolithic Technology and Human Evolution." *Science* 291:1748–53.

Bekoff, M. 1984. "Social Play Behavior." *BioScience* 34:228–33.

Benefit, B. 1994. "Phylogenetic, Paleodemographic, and Taphonomic Implications of *Victoriapithecus* Deciduous Teeth from Moboko, Kenya." *American Journal of Physical Anthropology* 95:277–331.

Bercovitch, F. B. 1987. "Female Weight and Reproductive Condition in a Population of Olive Baboons (*Papio anubis*)." *American Journal of Primatology* 12:189–95.

Beynon, D. and M. Dean. 1988. "Distinct Dental Development Patterns in Early Fossil Hominids." *Nature* 335:509–14.

Bloch, J. and D. Boyer. 2002. "Grasping Primate Origins." *Science* 298:1606–10.

Boesch, C. 1991. "Teaching in Wild Chimpanzees." *Animal Behaviour* 41:530–33.

Boesch, C. and H. Boesch. 1983a. "Possible Causes of Sex Differences in the Use of Natural Hammers by Wild Chimpanzees." *Journal of Human Evolution* 13:414–40.

Boesch, C. and H. Boesch. 1983b. "Optimization of Nut-Cracking with Natural Hammers by Wild Chimpanzees." *Behaviour* 83:265–86.

Boesch, C. and H. Boesch. 1990. "Tool Use and Tool Making in Wild Chimpanzees." *Folia Primatologica* 54:86–99.

Boesch, C. and H. Boesch-Achermann. 2000. *The Chimpanzees of the Tai Forest: Behavioural Ecology and Evolution.* Oxford: Oxford University Press.

Bogin, B. 1999. *Patterns of Human Growth* (2d ed.). Cambridge: Cambridge University Press.

Bogin, B. and B. H. Smith. 1996. "Evolution of the Human Life Cycle." *American Journal of Physical Anthropology* 8:703–16.

Bolter, D. and A. L. Zihlman. 2003. "Morphometric Analysis in Wild-Collected Vervet Monkeys (*Cercopithecus aethiops*), with Implications for Growth Patterns in Old World Monkeys, Apes and Humans." *Journal of Zoology, London* 260:99–110.

Borchert, C. and A. L. Zihlman. 1990. "The Ontogeny and Phylogeny of Symbolizing." Pp. 15–44 in *The Life of Symbols,* edited by M. LeC. Foster and L. J. Botscharow. Boulder, CO: Westview.

Bowlby, J. 1969. *Attachment.* New York: Basic Books.

Boyd, R. and P. J. Richerson. 1985. "Culture and the Evolutionary Process." Chicago: University of Chicago Press.

Bromage, T. 1987. "The Biological and Chronological Maturation of Early Hominids." *Journal of Human Evolution* 16:257–72.

Cann, R., M. Stoneking, and A. C. Wilson. 1987. "Mitochondrial DNA and Human Evolution." *Nature* 325:32–36.

Davis, D. 1961. "On the Origin of the Mammalian Feeding Mechanism." *American Zoologist* 1:229–34.

Deacon, H. and J. Deacon. 1999. *Human Beginnings in South Africa: Uncovering the Secrets of the Stone Age.* Cape Town: David Philip.

Dean, M. C., M. G. Leakey, D. Reid, F. Schrenk, G. T. Schwartz, C. Stringer, and A. Walker. 2001. "Growth Processes in Teeth Distinguish Modern Humans from *Homo erectus* and Earlier Hominids." *Nature* 104:627–31.

Diamond, J. 2001. "Unwritten Knowledge: Preliterate Societies Depend on the Wise Words of Older Generations." *Nature* 410:521.

Dixson, A. 1998. *Primate Sexuality.* Oxford: Oxford University Press.

Draper, P. 1976. "Social and Economic Constraints on Child Life among the !Kung." Pp. 199–217 in *Kalahari Hunter-Gatherers. Studies of the !Kung San and Their Neighbors,* edited by R. B. Lee and I. DeVore. Cambridge, MA: Harvard University Press.

Dunbar, R. I. M. and P. Dunbar. 1988. "Maternal Time Budgets of Gelada Baboons." *Animal Behaviour* 36:970–80.

Eisenberg, J. F. 1973. "Mammalian Social Systems: Are Primate Social Systems Unique?" Pp. 232–49 in *Precultural Primate Behavior, Symposium of the Fourth International Congress on Primatology,* Vol. 1, edited by E. Menzel. Basel: Karger.

Ellison, P. T. 1990. "Human Ovarian Function and Reproductive Ecology: New Hypotheses." *American Anthropologist* 92:933–52.

Enard, W., M. Przeworski, S. E. Fisher, C. S. L. Lai, V. Wiebe, T. Kitano, A. P. Monaco, and S. Pääbo. 2002. "Molecular Evolution of FOXP2, a Gene Involved in Speech and Language." *Nature* 418:869–72.

Fagen, R. 1993. "Primate Juveniles and Primate Play." Pp. 183–96 in *Juvenile Primates,* edited by M. Pereira and L. Fairbanks. New York: Oxford University Press.

Falk, D., G. C. Conroy, W. Recheis, G. W. Weber, and H. Seidler. 2000. "Early Hominid Brain Evolution: New Look at Old Endocasts." *Journal of Human Evolution* 38:695–717.

Fedigan, L. M. 1998. *Primate Paradigms* (2d ed.). Chicago: University of Chicago Press.

Fedigan, L. M. and S. Zohar. 1997. "Sex Differences in Mortality of Japanese Macaques: 21 Years of Data from the Arashiyama West Population." *American Journal of Physical Anthropology* 102:161–75.

Fleagle, J. G. 1999. *Primate Adaptation and Evolution* (2d ed). New York: Academic Press.

Galloway, A. 1997. "The Cost of Reproduction and the Evolution of Postmenopausal Osteoporosis." Pp. 132–46 in *The Evolving Female,* edited by M. E. Morbeck, A. Galloway, and A. L. Zihlman. Princeton, NJ: Princeton University Press.

Goodall, J. 1986. *The Chimpanzees of Gombe: Patterns of Behavior.* Cambridge, MA: Harvard University Press.

Gould, E. 1983. "Mechanisms of Mammalian Auditory Communication." Pp. 265–342 in *Advances in the Study of Animal Behavior*, Special Publications No. 7, edited by J. F. Eisenberg and D. G. Kleiman. Shippensberg, PA: American Society of Mammalogists.

Grand, T. I. 1977. "Body Weight: Its Relation to Tissue Composition, Segment Distribution, and Motor Function II. Development of *Macaca mulatta*." *American Journal of Physical Anthropology* 47:241–47.

Greenfield, P. M., A. E. Maynard, C. Boehm, and E. Y. Schmidtling. 2000. "Cultural Apprenticeship and Cultural Change. Tool Learning and Imitation in Chimpanzees and Humans." Pp. 237–77 in *Biology, Brains, and Behavior: The Evolution of Human Development*, edited by S. T. Parker, J. Langer, and M. L. McKinney. Santa Fe, NM: School of American Research Press.

Henshilwood, C. S., F. d'Errico, C. W. Marean, R. G. Milo, and R. Yates. 2001. "An Early Bone Tool Industry from the Middle Stone Age at Blombos Cave, South Africa: Implications for the Origins of Modern Human Behaviour, Symbolism and Language." *Journal of Human Evolution* 41(6):631–78.

Howell, N. 1979. *Demography of the Dobe !Kung.* New York: Academic Press.

Jenkins, F. A. 1974. "Tree Shrew Locomotion and Primate Arborealism." Pp. 85–115 in *Primate Locomotion*, edited by F. A. Jenkins. New York: Academic Press.

Kano, T. 1992. *The Last Ape. Pygmy Chimpanzee Ecology and Behavior.* Stanford, CA: Stanford University Press.

Kawamura, S. 1958. "Matriarchal Social Order in the Minoo-B Group: A Study on the Rank System of Japanese Macaques." *Primates* 1:149–56. [Reprinted in S. A. Altmann (Ed.), *Japanese Monkeys. A Collection of Translations* (selected by K. Imanishi). University of Alberta: Editor.]

Krings, M., A. Stone, R. W. Schmitz, H. Krainitzki, M. Stoneking, and S. Pääbo. 1997. "Neanderthal DNA Sequences and the Origin of Modern Humans." *Cell* 90:19–30.

Lancaster, J. B. 1984. "Evolutionary Perspectives on Sex Differences in the Higher Primates." Pp. 3–27 in *Gender and the Life Course*, edited by A. S. Rossi. Hawthorne, NY: Aldine de Gruyter.

Le Gros Clark, W. E. 1959. *The Antecedents of Man.* Edinburgh: Edinburgh University Press.

Leakey, M. D. 1984. *Disclosing the Past.* New York: Doubleday.

Leakey, M. and A. Walker. 1997. "Early Hominid Fossils from Africa." *Scientific American* 276:74–79.

LeBoeuf, B. J. and R. M. Laws. 1994. *Elephant Seals, Population, Behavior, and Physiology.* Berkeley: University of California Press.

Lee, P. C., and M. D. Hauser. 1988. "Long-Term Consequences of Changes in Territory Quality on Feeding and Reproductive Strategies of Vervet Monkeys." *Journal of Animal Ecology* 67:347–58.

Lee, R. B. 1979. *The !Kung San: Men, Women, and Work in a Foraging Society.* Cambridge: Cambridge University Press.

MacLean, P. D. 1985. "Brain Evolution Relating to Family, Play, and the Separation Call." *Archives of General Psychiatry* 42:405–17.

MacLean, P. D. 1988. "Evolution of Audiovocal Communication as Reflected by the Therapsid-Mammalian Transition and the Limbic Thalamocingulate Division." Pp. 185–201 in *The Physiological Control of Mammalian Vocalization*, edited by J. D. Newman. New York: Plenum.

MacLean, P. D. 1990. *The Triune Brain in Evolution*. New York: Plenum.

Mason, W. 1979. "Ontogeny of Social Behavior." Pp. 1–28 in *Handbook of Behavioral Neurobiology*, edited by P. Marler and J. G. Vandenbergh. New York: Plenum.

Matsuzawa, T. 1994. "Field Experiments on Use of Stone Tools by Chimpanzees in the Wild." Pp. 351–70 in *Chimpanzee Cultures*, edited by R. W. Wrangham, W. C. McGrew, F. B. M. de Waal, and P. G. Heltne. Cambridge, MA: Harvard University Press.

McComb, K., C. Moss, S. M. Durant, L. Baker, and S. Sayialel. 2001. "Matriarchs as Repositories of Social Knowledge in African Elephants." *Science* 292:491–94.

McFarland, R. 1997. "Female Primates: Fit or Fat?" Pp. 163–75 in *The Evolving Female*, edited by M. E. Morbeck, A. Galloway, and A. L. Zihlman. Princeton, NJ: Princeton University Press.

McGrew, W. C. 1992. *Chimpanzee Material Culture. Implications for Human Evolution*. Cambridge: Cambridge University Press.

Mitani, J. C. and D. P. Watts. 2001. "Why Do Chimpanzees Hunt and Share Meat?" *Animal Behaviour* 61:915–24.

Monroe, R. H., R. L. Munroe, and H. S. Shimmin. 1984. "Children's Work in Four Cultures: Determinants and Consequences." *American Anthropologist* 86:369–79.

Morelli, G. 1997. "Growing Up Female in a Farmer Community and a Forager Community." Pp. 209–19 in *The Evolving Female*, edited by M. E. Morbeck, A. Galloway, and A. L. Zihlman. Princeton, NJ: Princeton University Press.

Moss, C. 2000. *Elephant Memories. Thirteen Years in the Life of an Elephant Family*. Chicago: University of Chicago Press.

Nag, M., B. N. F. White, and R. C. Peet. 1978. "An Anthropological Approach to the Study of the Economic Value of Children in Java and Nepal." *Current Anthropology* 19:293–306.

Nishida, T. 1990. *The Chimpanzees of the Mahale Mountains: Sexual and Life History Strategies*. Tokyo: University of Tokyo Press.

Nishida, T., N. Corp, M. Hamai, T. Hasegawa, M. Hiraiwa-Hasegawa, K. Hosaka, K. D. Hunt, N. Itoh, K. Kawanaka, A. Matsumota-Oda, J. C. Mitani, M. Nakamura, K. Norikoshi, T. Sakamaki, L. Turner, S. Uehara, and K. Zumma. 2003. "Demography, Female Life History, and Reproductive Profiles among the Chimpanzees of Mahale." *American Journal of Primatology* 59:99–121.

Panter-Brick, C. 1997. "Women's Work and Energetics: A Case Study from Nepal." Pp. 220–32 in *The Evolving Female*, edited by M. E. Morbeck, A. Galloway, and A. L. Zihlman. Princeton, NJ: Princeton University Press.

Panter-Brick, C. 1998. *Biosocial Perspectives on Children*. Cambridge: Cambridge University Press.

Parker, S. T. 2000. "*Homo erectus* Infancy and Childhood: The Turning Point in the Evolution of Behavioral Development in Hominids." Pp. 279–318 in *Biology, Brains and Behavior: The Evolution of Human Development*, edited by S. T. Parker, J. Langer, and M. L. McKinney. Santa Fe, NM: School of American Research Press.

Payne, K. 1998. *Silent Thunder. In the Presence of Elephants*. New York: Simon and Schuster.

Peacock, N. 1991. "Rethinking the Sexual Division of Labor: Reproduction and Women's Work among the Efe." Pp. 339–60 in *Gender at the Crossroads of Knowledge*, edited by M. di Leonardo. Berkeley: University of California Press.

Pereira, M. E. and J. Altmann. 1985. "Development of Social Behavior of Free-Living Nonhuman Primates." Pp. 217–309 in *Nonhuman Primate Models for Human Growth and Development*, edited by E. S. Watts. New York: Alan R. Liss.

Ponce de León, M. S. and C. P. E. Zollikofer. 2001. "Neanderthal Cranial Ontogeny and Its Implications for Late Hominid Diversity." *Nature* 412:534–38.

Pond, C. 1977. "The Significance of Lactation in the Evolution of Mammals." *Evolution* 31:177–99.

Pusey, A. 1978. *The Physical and Social Development of Wild Adolescent Chimpanzees (Pan troglodytes schweinfurthii)*. Ph.D. thesis, Stanford University, Palo Alto, CA.

Pusey, A. E. and C. Packer. 1987. "Dispersal and Philopatry." Pp. 250–66 in *Primate Societies*, edited by B. Smuts, D. L. Cheney, R. M. Seyfarth, R. W. Wrangham, and T. T. Struhsaker. Chicago: University of Chicago Press.

Qiang Ji, Zhe-xi Luo, Chong-xi Yuan, J. R. Wible, Juan-Ping Zhang, and J. A. Georgi. 2002. "The Earliest Known Eutherian Mammal." *Nature* 416:816–22.

Radinsky, L. B. 1987. *The Evolution of Vertebrate Design*. Chicago: University of Chicago Press.

Rak, Y., A. Ginzberg, and E. Geffen. 2002. "Does *Homo neanderthalensis* Play a Role in Modern Human Ancestry? The Mandibular Evidence." *American Journal of Physical Anthropology* 119:199–204.

Reiter, J. 1997. "Life History and Reproductive Success of Female Northern Elephant Seals." Pp. 46–52 in *The Evolving Female*, edited by M. E. Morbeck, A. Galloway, and A. L. Zihlman. Princeton, NJ: Princeton University Press.

Rightmire, G. P. 1998. "Human Evolution in the Middle Pleistocene: The Role of *Homo heidelbergensis*." *Evolutionary Anthropology* 6(6):218–27.

Rogoff, B. 1990. *Apprenticeship in Thinking*. New York: Oxford University Press.

Rogoff, B. 2003. *The Cultural Nature of Human Development*. Oxford: Oxford University Press.

Rogoff, B., M. J. Seller, S. Pirotta, N. Fox, and S. H. White. 1975. "Age of Assignment of Roles and Responsibilities to Children: A Cross-Cultural Survey." *Human Development* 18:353–69.

Rowell, T. 1988a. "The Social System of Guenons, Compared with Baboons, Macaques, and Mangabeys." Pp. 439–51 in *A Primate Radiation: Evolutionary Biology of the African Guenons*, edited by A. Gautier-Hion, F. Bourlière, J.-P. Gautier, and J. Kingdon. Cambridge: Cambridge University Press.

Rowell, T. 1988b. "Beyond the One Male Group." *Behaviour* 104(parts 3–4):189–201.

Rubenstein, D. 1993. "On the Evolution of Juvenile Life-Styles in Mammals." Pp. 38–56 in *Juvenile Primates*, edited by M. Pereira and L. Fairbanks. New York: Oxford University Press.

Ruvolo, M. 1997. "Molecular Phylogeny of the Hominoids: Inferences from Multiple Independent DNA Data Sets." *Molecular Biology and Evolution* 14:248–65.

Sade, D. 1972. "A Longitudinal Study of Social Relations of Rhesus Monkeys." Pp. 378–98 in *Functional and Evolutionary Biology of Primates*, edited by R. Tuttle. Chicago: Aldine-Atherton.

Sahlins, M. 1976. *The Use and Abuse of Sociobiology.* Ann Arbor: University of Michigan Press.

Short, R. V. 1976. "The Evolution of Human Reproduction." *Proceedings of the Royal Society of London* 195:3–24.

Smith, B. 1989. "Dental Development as a Measure of Life History in Primates." *Evolution* 43:683–88.

Smith, K. K. 1992. "The Evolution of the Mammalian Pharynx." *Zoological Journal of the Linnaean Society* 104:313–49.

Smuts, B. 1985. *Sex and Friendship in Baboons.* Hawthorne, NY: Aldine de Gruyter.

Stringer, C. B. 2002. "Modern Human Origins: Progress and Prospects." *Philosophical Transactions of the Royal Society of London B* 357:563–79.

Stringer, C. B., M. C. Dean, and R. D. Martin. 1990. "A Comparative Study of Cranial and Dental Development within a Recent British Sample and among Neanderthals." Pp. 115–52 in *Primate Life History and Evolution,* edited by E. Watts. New York: Wiley-Liss.

Stringer, C. and C. Gamble. 1993. *In Search of the Neanderthals.* New York: Thames and Hudson.

Strum, S. C. 1987. *Almost Human.* New York: Random House.

Sugardjito, J., I. J. A. te Boekhorst, and J. A. R. A. M. van Hooff. 1987. "Ecological Constraints on the Grouping of Wild Orang-utans (*Pongo pygmaeus*) in the Gunung Leuser National Park, Sumatra, Indonesia." *International Journal of Primatology* 8:17–41.

Teleki, G. 1974. "Chimpanzee Subsistence Technology: Materials and Skills." *Journal of Human Evolution* 3:575–94.

Vigilant, L., M. Stoneking, H. Harpending, K. Hawkes, and A. C. Wilson. 1991. "African Populations and the Evolution of Human Mitochondrial DNA." *Science* 253:1503–07.

Vološinov, V. N. [1929] 1998. *Marxism and the Philosophy of Language.* Cambridge, MA: Harvard University Press.

Vygotsky, L. S. [1934] 1962. *Thought and Language.* Cambridge, MA: MIT Press and New York: Wiley & Sons.

Walters, J. R. 1987. "Transition to Adulthood." Pp. 358–69 in *Primate Societies,* edited by B. Smuts, D. Cheney, R. Seyfarth, R. Wrangham, and T. T. Struhsaker. Chicago: University of Chicago Press.

Ward, R. and C. Stringer. 1997. "A Molecular Handle on the Neanderthals." *Nature* 388:225–26.

Washburn, S. L. 1951. "The Analysis of Primate Evolution with Particular Reference to the Origin of Man." Pp. 67–78 in *Cold Spring Harbor Symposia on Quantitative Biology,* Vol. 15: *Origin and Evolution of Man,* edited by S. L. Washburn and T. Dobzhansky. Cold Spring, NY: Cold Spring Harbor Lab. Press.

Watts, D. P., and A. E. Pusey. 1993. "Behavior of Juvenile and Adolescent Great Apes." Pp. 148–71 in *Juvenile Primates,* edited by M. Pereira and L. Fairbanks. New York: Oxford University Press.

Wertsch, J. V. 1985. *Vygotsky and the Social Formation of Mind.* Cambridge, MA: Harvard University Press.

Whiten, A., J. Goodall, W. C. McGrew, T. Nishida, V. Reynolds, Y. Sugiyama, C. E. G. Turin, R. W. Wrangham, and C. Boesen. 1999. "Cultures in Chimpanzees." *Nature* 399:683–85.

Yeager, C. and K. Kool. 2000. "The Behavioral Ecology of Asian Colobines." Pp. 496–521 in *Old World Monkeys,* edited by C. Jolly and P. Whitehead. Cambridge: Cambridge University Press.

Yoder, A. D., M. Cartmill, M. Ruvolo, K. K. Smith, and R. Vilgalys. 1996. "Ancient Single Origin for Malagasy Primates." *Proceedings of the National Academy of Science* 93:5122–26.

Zihlman, A. L. 1992. "Locomotion as a Life History Character: The Contribution of Anatomy." *Journal of Human Evolution* 22:315–25.

Zihlman, A. L. 1997. "Women's Bodies, Women's Lives: An Evolutionary Perspective." Pp. 185–97 in *The Evolving Female,* edited by M. E. Morbeck, A. Galloway, and A. L. Zihlman. Princeton, NJ: Princeton University Press.

Zihlman, A. L. and D. Bolter. 2003. "Growth and Development in Wild Chimpanzees, Fossil Hominids and *Homo sapiens.*" Paper presented at the International Anthropological Conference of Aleš Hrdlička, Prague, Czech Republic, May.

Zihlman, A. L. and J. Lowenstein. 1999. "From Eternity to Here." *California Wild* 52(3):16–21.

3

Wild Justice, Cooperation, and Fair Play

Minding Manners, Being Nice, and Feeling Good

Marc Bekoff

SOCIAL MORALITY, MANNERS, AND COOPERATION IN ANIMALS: DOING WHAT COMES NATURALLY

> Those communities which included the greatest number of the most sympathetic members would flourish best and rear the greatest number of offspring.
> —Charles Darwin [1871]1936:163

> I believe that at the most fundamental level our nature is compassionate, and that cooperation, not conflict, lies at the heart of the basic principles that govern our human existence. . . . By living a way of life that expresses our basic goodness, we fulfill our humanity and give our actions dignity, worth, and meaning.
> —His Holiness the Dalai Lama 2002:68

> My thesis is that justice is first of all a natural sentiment, an inborn sense of our connectedness with others and our shared interests and concerns.
> —Solomon 1995:153

There are many areas in which researchers and nonresearchers can pursue interesting, important, challenging, and interdisciplinary questions that center on the interface of science, religion, ethics, and the place of humans in the world. One such area concerns the evolution of social

53

morality, manners if you will, and the negotiation and enforcement of cooperation, fairness, social norms, and etiquette (for wide-ranging discussion see Solomon 1995; de Waal 1996; Ridley 1996, 2001; Mitchell 1998; Field 2001; Hinde 2002; Wilson 2002; de Waal and Tyack 2003). Recently there has been a resurgence of interest in the notion of fairness in humans (Douglas 2001; Riolo et al. 2001; Sigmund and Nowak 2001; Bowles and Gintis 2002; Fehr and Gächter 2002; Sigmund et al. 2002; Bewley 2003; Fehr and Rockenbach 2003). Researchers are interested in learning about how individuals from different cultures share resources, and if they share them equitably even if they are not compelled to do so. Much research shows that human beings are more generous and more fair than game-theory and other models predict. There seems to be a set of core values that are learned through social interactions with others, and these values influence moral decisions. There also is evidence that people will punish free-riders in the absence of personal gain, and that cooperation is sustained by such "altruistic punishment" (Bowles and Gintis 2002; Fehr and Gächter 2002). Taken together, cross-cultural data suggest that there may well be an innate drive to be fair.

But what about nonhuman animal beings (hereafter animals)? Many animals live in fairly stable social groups that resemble those of ancestral humans. There are divisions of labor, food sharing, communal care of young, and inter- and intrasexual dominance hierarchies. Many animals, especially mammals, also share with humans neuroanatomical structures in the amygdala and hypothalamus and neurochemicals (dopamine, serotonin, oxytocin) that form the neural bases for the expression and experience of emotions and empathy (Panksepp 1998; Preston and de Waal 2002). A wide variety of social behavior patterns in animals have also been influenced by living in small groups. If one is a good Darwinian, it seems premature to claim that *only* humans can be empathic and moral beings. As animals share their emotions with us it becomes increasingly difficult to deny their existence.

In this chapter I consider various aspects of the evolution of cooperation and fairness using social play behavior in animals, especially mammals, as my exemplar of an activity in which one would expect to see ongoing negotiations of cooperation and agreements to behave fairly because the social dynamics of play require that players agree to play and not to fight or to mate with one another. I am specifically concerned with the notion of "behaving fairly." By "behaving fairly" I use as a working guide the notion that animals often have social expectations when they engage in various sorts of social encounters the violation of which constitutes being treated unfairly because of a lapse in social etiquette. I conclude that social play might be a "foundation of fairness." I also argue that it is through social cooperation that groups (communities) are built from individuals agreeing to work in harmony with other individuals. Whether or not individuals lose various "freedoms" when balanced against the benefits that accrue when they work for the "good of a group" is unknown and needs to be studied more carefully in various species. Further, based on recent research on the neurobiology of human cooperation, I argue that "being fair" may feel good for animals as well. Lastly, I stress that in our efforts to learn more about the evolution of social morality we need to broaden our comparative research to include animals other than nonhuman primates.

Researchers from many disciplines have debated the evolutionary origins of social morality, asking if some animals have codes of social conduct that regulate their behavior in terms of what is permissible and what is not permissible during social encounters. They want to know just what the moral capacities of animals are; are they moral agents with a moral sense who are able to live in moral communities? In a recent issue of the *Journal of Consciousness Studies* [7(1/2), 2000], researchers from many disciplines debated the evolutionary origins of morality. These scholars were interested in discussing animal roots on which human morality might be built, even if it is not identical to animal morality. Charles Darwin's (1859, [1872] 1998) ideas about evolutionary continuity, that behavioral, cognitive, emotional, and moral variations among different species are differences in *degree* rather than differences in *kind,* are often invoked in such exercises. This view argues that there are shades of gray among different animals and between nonhumans and humans, that the differences are not black and white with no transition stages or inexplicable jumps (Gruen 2002; Güzeldere and Nahmias 2002; see also many other essays in Bekoff, Allen, and Burghardt 2002). Current work in evolutionary biology and anthropology suggests that linear scales of evolution in which there are large gaps between humans and at least some animals are simplistic views of the evolutionary process. Further, as I will discuss below, models and explanations that exclude group-selection in deference to individual selection also need to be revisited.

THE EVOLUTION OF SOCIAL MORALITY: CONTINUITY, PROTOMORALITY, AND QUESTIONS OF HUMAN UNIQUENESS

> Different as they are from language-using human beings, they are able to form relationships not only with members of their own species, but also with human beings, while giving expression to their own intentions and purposes. So that the relationships are far more clearly analogous to human relationships than some of the philosophical theorizing that I have discussed would allow. Some human beings indeed

> and some nonhuman animals pursue their respective
> goods in company with and in cooperation with each
> other. And what we mean by 'goods' in saying this is
> precisely the same, whether we are speaking of human
> or dolphin or gorilla.
>
> —Macintyre 1999:61

Evolutionary reconstructions of social behavior often depend on educated guesses (some better than others) about the past social (and other)
environments in which ancestral beings lived. In the same sense that
other's minds are private, so is evolution. Often it is difficult to know with
a great deal of certainty very much about these variables and how they may
have figured into evolutionary scenarios. It is an understatement to note
that it is extremely difficult to study the evolution of morality in any animal species, and the very notion of animal morality itself often makes for
heated discussions. Bernstein's concern that "morality in animals might lie
outside of the realm of measurement techniques available to science"
(2000:34) needs to be taken seriously. *Nonetheless, it seems clear that detailed
comparative analyses of social behavior in animals can indeed provide insights into
the evolution of social morality.* To be sure, these sorts of studies are extremely
challenging, but the knowledge that is gained is essential in our efforts to
learn more about the evolution of sociality and social morality and to learn
more about human nature and perhaps human uniqueness.

Many discussions of the evolution of morality center on the development of various sorts of models (e.g., Axelrod 1984; Ridley 1996, 2001;
Skyrms 1996; Dugatkin 1997; Sober and Wilson 1998, 2000; *Journal of
Consciousness Studies* 2000). While these models are very useful for stimulating discussion and further research, they do not substitute for available
data (however few) that may bear on animal morality [see, for example,
some essays in Aureli and de Waal (2000) for additional comparative
information].

The study of the evolution of morality, specifically cooperation and fairness, is closely linked to science, religion, theology, spirituality, and perhaps even different notions of God, in that ideas about continuity and
discontinuity (the possible uniqueness of humans and other species),
individuality, and freedom need to be considered in detail. Furthermore, it
is important to discuss relationships among science, religion, and God
because spirituality and the notion of one form of God or another had
strong influences on the evolution of our ancestors, their cognitive, emotional, and moral lives.

Recently, Peterson (2000; see also Peterson 1999) has pondered the evolutionary roots of morality (stages that he refers to as "quasi-morality" and
"proto-morality" in animals) and religion in relation to the roles played by
cognition and culture. He also stresses the importance of recognizing con-

tinuities and discontinuities with other animals, arguing ultimately (and speciesistically) that while some animals might possess protomorality [they are able "to rationally deliberate actions and their consequences" (2000:475)], none other than humans is "genuinely moral," because to be able to be genuinely moral requires higher emergent levels of cognition as well as culture and the world view that culture provides, namely, religion. Peterson claims that

> quasi-moral and proto-moral systems do not require a global framework that guides decision making. They are always proximate and pragmatic. In these systems, there is no long-term goal or ideal state to be achieved. Yet, genuine morality is virtually inconceivable without such conceptions. (ibid.:478)

Peterson also claims that any sociobiological account (based on selfishness or combativeness) of human morality is incomplete. I agree and also argue that this is so for some nonhuman animals as well. To be sure, Peterson's views are very stimulating. I cannot go into detail here but suffice it to say, and I hope that it becomes clear later on, when animals are studied in their own worlds they may indeed have their own form of genuine morality, there might indeed be long-term goals and ideal states to be achieved. Our anthropocentric view of other animals, in which humans are so taken with themselves, is far too narrow. The worlds and lives of other animals are not identical to those of humans and may vary from species to species and even within species. The same problems arise in the study of emotions (Bekoff 2002a), if we believe that emotions in animals are going to be identical to or even recognizably similar among different species. There is also variability among humans in what some might view as long-term goals and ideal states, and it would of course be premature to conclude that there is one set of long-term goals and ideal states that characterize, or are essential to, the capacity to be genuinely moral. We really are not experts about ourselves. To cash out stages of moral evolution as does Peterson, it looks like quasi-morality and protomorality are less than genuine morality. This view could lead to linear hierarchical views of evolution, whether or not it is Peterson's intent to go this route.

COOPERATION AND FAIRNESS ARE NOT BY-PRODUCTS OF AGGRESSION AND SELFISHNESS

In my view, cooperation is not merely always a by-product of tempering aggressive and selfish tendencies [combating Richard Dawkins' (1976) selfish genes] and attempts at reconciliation. Rather, cooperation and fairness can evolve on their own because they are important in the formation and maintenance of social relationships. This view, in which nature is sanitized, contrasts with those who see aggression, cheating, selfishness, and perhaps amorality as driving the evolution of sociality. The combative Hobbesian world in which individuals are constantly at one another's throats is not the natural state of affairs, nature is not always red in tooth and claw, and altruism is not always simply selfishness disguised. Dawkins (2001), himself, has been quoted as saying "A pretty good definition of the kind of society in which I don't want to live is a society founded on the principles of Darwinism."

DOES IT FEEL GOOD TO BE FAIR?

Are some animals capable of the emotions and empathy that underlie morality? Watching animals in action has convinced many researchers, including myself, that they possess various emotions upon which a moral sense is built. We know that in humans these feelings are located in the brain's amygdala and hypothalamus and mediated by neurotransmitters such as dopamine, serotonin, and oxytocin. We also know that many animals, especially mammals, share with humans the same neurological structures and chemicals (Panksepp 1998; Bekoff 2002a). Of course, this does not necessarily mean that animals share our feelings, but careful observation of individuals during social encounters suggests that at least some of them do. And their feelings are not necessarily identical to ours, but this is of little or no concern because it is unlikely that they should be the same as ours. Indeed, it is unlikely that any two humans share precisely the same feelings when a given emotion is expressed.

In a recent review Preston and de Waal (2002) reported that empathy is more widespread among animals than has previously been recognized (see also Kuczaj et al. 2001). In one classic study, Wechlin, Masserman, and Terris (1964) showed that a hungry rhesus monkey would not take food if doing so subjected another monkey to an electric shock. In similar situations rats will also hold back when they know their actions would cause pain to another individual. In another study, Diana monkeys were trained to insert a token into a slot to obtain food (Markowitz 1982). A male was observed helping the oldest female who had failed to learn the task. On three occasions he picked up the tokens she had dropped, put them into the machine, and allowed her to have the food. His behavior seemed to have no benefits for him at all; there did not seem to be any hidden agenda.

Elephants also may show concern for others. Joyce Poole (1998), who has studied African elephants for decades, was told a story about a teenage female who was suffering from a withered leg on which she could put no weight. When a young male from another group began attacking the injured female, a large adult female chased the attacking male, returned to the young female, and touched her crippled leg with her trunk. Poole argues that the adult female was showing empathy and sympathy.

While good stories are not enough to make a compelling argument, there are so many such anecdotes that can be used to provide a solid basis for further detailed empirical research. Ignoring them is to ignore a rich database. I have argued elsewhere that the plural of anecdote is data (Bekoff 2002a).

We will probably never know whether these rats, monkeys, and elephants were feeling empathy as we do. But there are ways in which we can start comparing what is going on in animal brains to what happens in our own. Neuroimaging techniques are shedding new light on human emotions, and it likely will not be long before we start doing similar studies with other animals.

It is important to consider the possibility that it feels good to be fair to others, to cooperate with them and to treat them fairly, to forgive them for their mistakes and shortcomings. Recent neural imaging research on humans by Rilling and his colleagues (2002) has shown that the brain's pleasure centers are strongly activated when people cooperate with one another, that we might be wired to be fair or nice to one another. (I do not want to argue here that "being fair" always means "being nice.") This is extremely significant research for it posits that there is a strong neural basis for human cooperation and that it feels good to cooperate, that being nice is rewarding in social interactions and might be a stimulus for fostering cooperation and fairness. This sort of noninvasive research is precisely what is needed on other animals. Studies of the evolution of social morality need to consider seriously the rich cognitive ("intellectual") and deep emotional lives of other animals (Bekoff 2000, 2002a, 2002b) and how these capacities and a sense of self figure into a moral sensibility and the ability to make moral judgments. Truth be told, we really do not know much about these capacities even in our primate relatives despite claims that we do (Bekoff 2002c, 2003).

ANIMAL PLAY: LESSONS IN COOPERATION, FAIRNESS, AND TRUST

"Happiness is never better exhibited than by young animals, such as puppies, kittens, lambs, &c., when playing together, like our own children." So wrote Charles Darwin in his book *The Descent of Man and Selection in Relation to Sex*.

Animal play is obvious, but animal social morality is not (for definitions of social play see Bekoff and Byers 1981, 1998; Fagen 1981; Power 2000; Burghardt 2002). Social play in animals is an exhilarating activity in which to engage and to observe. The rhythm, dance, and spirit of animals at play is incredibly contagious. Not only do their animal friends want to join in or find others with whom to romp, but I also want to play when I see animals chasing one another, playing hide-and-seek, and wrestling with reckless abandon. My body once tingled with delight as I watched a young elk in Rocky Mountain National Park, Colorado, running across a snow field, jumping in the air and twisting his body while in flight, stopping to catch his breath, and then jumping and twisting over and over and again. There was plenty of grassy terrain around but he chose the snow field. Buffaloes will also follow one another and playfully run onto and slide across ice, excitedly bellowing "Gwaaa" as they do so. And, of course, we all know that dogs and cats love to play, as do many other mammals. Birds also playfully soar across the sky chasing, diving here and there, and frolicking with one another.

I think of play as being characterized by what I call the "five s's of play," its spirit, symmetry, synchrony, sacredness, and soulfulness. The spirit of play is laid bare for all to see as animals prodigally run about, wrestle, and knock one another over. The symmetry and synchrony of play are reflected in the harmony of the mutual agreements to trust one another— individuals share intentions to cooperate with one another to prevent play from spilling over into fighting. This trust is sacred. Finally, there is a deepness to animal play in that the players are so immersed in play that they are the play. Play is thus a soulful activity, perhaps the essence of the individuals' being at the moment as they play from deep in their hearts. As Aquinas noted, play is about being; there are no why's in play.

There is also a feeling of incredible freedom and creativity in the flow of play. So it is important also to keep in mind the six f's of play, its flexibility, freedom, friendship, frolic, fun, and flow. As they run about, jump on one another, somersault, and bite one another animals create mind-boggling scenarios. Behavior patterns that are observed in mating are intermixed in flexible kaleidoscopic sequences with actions that are used during fighting, looking for prey, and avoiding being eaten.

The unmistakable emotions associated with play—joy and happiness— drive animals into becoming one with the activity. One way to get animals (including humans) to do something is to make it fun, and there is no doubt that animals enjoy playing. Studies of the chemistry of play support the claim that play is fun. Dopamine (and perhaps serotonin and norepinephrine) are important in the regulation of play. Rats show an increase in dopamine activity when anticipating the opportunity to play (Siviy 1998) and enjoy being playfully tickled (Panksepp 1998, 2000). There is also a close association between opiates and play (Panksepp 1998).

Neurobiological data are essential for learning more about whether play truly is a subjectively pleasurable activity for animals as it seems to be for humans. Siviy's and Panksepp's findings suggest that it is. In light of these neurobiological ("hard") data concerning possible neurochemical bases for various moods, in this case joy and pleasure, skeptics who claim that animals do not feel emotions might be more likely to accept the idea that enjoyment could well be a motivator for play behavior.

IT BEGINS WITH AN HONEST "BOW":
"I WANT TO PLAY WITH YOU"

"Would you care to play?" asks one wolf of another. "Yes, I would." After each individual agrees to play and not to fight, prey on, or mate with the other, there are ongoing rapid and subtle exchanges of information so that their cooperative agreement can be fine-tuned and negotiated on the run, so that the activity remains playful. Incorporated into explanations of social play are such notions as trusting, behaving fairly, forgiving, apologizing, and perhaps justice, behavioral attributes that underlie social morality and moral agency. Recent research by Okamoto and Matsumara (2000) suggests that punishment and apology play a role in maintaining cooperation between individual nonhuman primates.

When individuals play they typically use action patterns that are also used in other contexts, such as predatory behavior, antipredatory behavior, and mating. These actions may not vary much across different contexts, or they may be hard to discriminate even for the participants. How do animals know that they are playing? How do they communicate their desires or intentions to play or to continue to play? How is the play mood maintained?

Because there is a chance that various behavior patterns that are performed during ongoing social play can be misinterpreted, individuals need to tell others "I want to play," "This is still play no matter what I am going to do to you," or "This is still play regardless of what I just did to you." An agreement to play, rather than to fight, mate, or engage in predatory activities, can be negotiated in various ways. Individuals may use various behavior patterns—play markers—to initiate play or to maintain a play mood (Bekoff 1975, 1977a, 1995; Bekoff and Allen 1992, 1998; Allen and Bekoff 1997) by punctuating play sequences with these actions when it is likely that a particular behavior may have been, or will be, misinterpreted (it is also possible that there are auditory, olfactory, and tactile play markers; Bekoff and Byers 1981; Fagen 1981). I found that play signals in infant canids (domestic dogs, wolves, and coyotes) were used nonrandomly, especially when biting was accompanied by rapid side-to-side shaking of the head (Bekoff 1995). Biting accompanied by rapid side-to-side shaking of the head is performed during serious aggressive and predatory encounters and can easily be misinterpreted if its meaning is not modified by a play signal. There also is little evidence that play signals are used to deceive others in canids or other species.

Play signals are an example of what ethologists call "honest signals." There is little evidence that social play is a manipulative or "Machiavellian" activity. Play signals are rarely used to deceive others in canids or other species. There are no studies of which I am aware that actually look at the relative frequencies of occurrence of honest and deceptive play signaling, but my own long-term observations indicate that deceptive signaling is so rare that I cannot remember more than a few occurrences in thousands of play sequences. Cheaters are unlikely to be chosen as play partners because others can simply refuse to play with them and choose others. Limited data on infant coyotes show that cheaters have difficulty getting other young coyotes to play (personal observations). It is not known if individuals select play partners based on what they have observed during play by others.

In domestic dogs there is little tolerance for noncooperative cheaters. Cheaters may be avoided or chased from play groups. There seems to be a sense of what is right, wrong, and fair. While studying dog play on a beach in San Diego, California, Alexandra Horowitz (2002) observed a dog she called Up-ears enter into a play group and interrupt the play of two other dogs, Blackie and Roxy. Up-ears was chased out of the group, and when she returned Blackie and Roxy stopped playing and looked off toward a distant sound. Roxy began moving in the direction of the sound and Up-ears ran off following their line of sight. Roxy and Blackie immediately began playing once again. Even in rats fairness and trust are important in the dynamics of playful interactions. Sergio Pellis (2002), a psychologist at the University of Lethbridge in Canada, discovered that sequences of rat play consist of individuals assessing and monitoring one another and then fine-tuning and changing their own behavior to maintain the play mood. When the rules of play are violated, when fairness breaks down, so does play.

Individuals might also know that they are playing because the actions that are performed differ when they are performed during play when compared to other contexts (Hill and Bekoff 1977), or the order in which motor patterns are performed differs from, and might be more variable than, the order in which they are performed during the performance of, for example, serious aggressive, predatory, or reproductive activities (Bekoff and Byers 1981).

Individuals also engage in role-reversing and self-handicapping (Bekoff and Allen 1998; Bauer and Smuts 2002; Horowitz 2002) to maintain social play. Each can serve to reduce asymmetries between the interacting animals and foster the reciprocity that is needed for play to occur. Self-handicapping happens when an individual performs a behavior pattern that might compromise herself. For example, a coyote might not bite her play partner as hard as she can, or she might not play as vigorously as she can. Watson and Croft (1996) found that red-neck wallabies adjusted their play to the age of their partner. When a partner was younger, the older animal adopted a defensive, flat-footed posture, and pawing rather than sparring occurred. In addition, the older player was more tolerant of its partner's tactics and took the initiative in prolonging interactions.

Role-reversing occurs when a dominant animal performs an action during play that would not normally occur during real aggression. For example, a dominant animal might voluntarily not roll over on his back during fighting, but would do so while playing. In some instances role-reversing and self-handicapping might occur together. For example, a dominant individual might roll over while playing with a subordinate animal and inhibit the intensity of a bite.

From a functional perspective, self-handicapping and role-reversing, similar to using specific play invitation signals or altering behavioral sequences, might serve to signal an individual's intention to continue to play. In this way there can be mutual benefits to each individual player because of their agreeing to play and not fight or mate. This might differentiate cooperative play from the situation described above in which a male Diana monkey helped a female get food when she could not learn the task that would bring her food. There seemed to be no benefit to the male to do so. (I thank Jan Nystrom for marking this distinction.)

CAN ANIMALS FORGIVE?

Even for the behavior of forgiving, which is often attributed solely to humans, the renowned evolutionary biologist David Sloan Wilson shows that forgiveness is a complex biological adaptation. In his book *Darwin's Cathedral: Evolution, Religion, and the Nature of Society*, Wilson concludes that "forgiveness has a biological foundation that extends throughout the animal kingdom" (2002:195). And further, "Forgiveness has many faces— *and needs to*—in order to function adaptively in so many different contexts" (ibid.:212). While Wilson concentrates mainly on human societies, his views can easily be extended—and responsibly so—to nonhuman animals. Indeed, Wilson points out that adaptive traits such as forgiveness might not require as much brain power as once thought. This is not to say that animals aren't smart, but rather that forgiveness might be a trait that is basic to many animals even if they don't have especially big and active brains. Perhaps if we try to learn more about forgiveness in animals and how it functions in play, we will also learn to live more compassionately and cooperatively with one another.

FINE-TUNING PLAY: WHY COOPERATE AND PLAY FAIRLY?

For years I tried to figure out why play evolved as it did. Why do animals carefully use play signals to tell others that they really want to play and not try to dominate them, why do they engage in self-handicapping and role-reversing? One morning, while hiking with my companion dog, Jethro, I had one of those infamous "aha" experiences and the puzzle was solved. It dawned on me that during social play, while individuals are having fun in a relatively safe environment, they learn ground rules that are acceptable to others—how hard they can bite, how roughly they can interact—and how to resolve conflicts. There is a premium on playing fairly and trusting others to do so as well. There are codes of social conduct that regulate actions that are and are not permissible, and the existence of these codes likely speaks to the evolution of social morality. What could be a better atmosphere in which to learn social skills than during social play, where there are few penalties for transgressions? Individuals might also generalize codes of conduct learned in playing with specific individuals to other group members and to other situations such as sharing food, defending resources, grooming, and giving care. (Social morality does not mean other animals are behaving unfairly when they kill for food, for example, for they have evolved to do this.)

Playtime generally is safe time—transgressions and mistakes are forgiven and apologies are accepted by others, especially when one player is a youngster who is not yet a competitor for social status, food, or mates. There is a certain innocence or ingenuousness in play. Individuals must cooperate with one another when they play—they must negotiate agreements to play (Bekoff 1995). Fagen noted that "levels of cooperation in play of juvenile primates may exceed those predicted by simple evolutionary arguments" (1993:192). The highly cooperative nature of play has evolved in many other species (Fagen 1981; Bekoff 1995; Bekoff and Allen 1998; Power 2000; Burghardt 2002). Detailed studies of play in various species indicate that individuals trust others to maintain the rules of the game (Bekoff and Byers 1998). While there have been numerous discussions of cooperative behavior in animals (e.g., Axelrod 1984; Ridley 1996; de Waal 1996; Dugatkin 1997; Hauser 2000; *Journal of Consciousness Studies* 2000 and references therein), none has considered the details of social play—the requirement for cooperation and reciprocity—and its possible role in the evolution of social morality, namely behaving fairly.

Individuals of different species seem to fine-tune ongoing play sequences to maintain a play mood and to prevent play from escalating into real aggression. Detailed analyses of film show that in canids there are subtle and fleeting movements and rapid exchanges of eye contact that suggest that players are exchanging information on the run, from moment to moment, to make certain everything is all right—that this is still play. Aldis (1975) suggested that in play, there is a 50:50 rule so that each player "wins" about 50 percent of its play bouts by adjusting its behavior to accomplish this (for further discussion and details on rodent play, see Pellis 2002).

Why might animals fine-tune play? While play in most species does not take up much time and energy (Bekoff and Byers 1998; Power 2000), and in some species only minimal amounts of social play during short windows of time early in development are necessary to produce socialized individuals [two 20-minute play sessions with another dog, twice a week, are sufficient for domestic dogs from three to seven weeks of age (Scott and Fuller 1965)], researchers agree that play is very important in social, cognitive, and/or physical development, and may also be important for training youngsters for unexpected circumstances (Spinka, Newberry, and Bekoff 2001). While there are few data concerning the actual benefits of social play in terms of survival and reproductive success, it generally is assumed that short- and long-term functions (benefits) vary from species to species and among different age groups and between the sexes within a species. No matter what the functions of play may be, there seems to be little doubt that play has *some* benefits and that the absence of play can have devastating effects on social development (Power 2000; Burghardt 2002).

In canids and many other mammals (and some birds) there is a small time window during early development when individuals can play without being responsible for their own well-being. This time period is generally referred to as the "socialization period," for this is when species-typical social skills are learned most rapidly. It is important for individuals to engage in at least *some* play. All individuals need to play and there is a premium for playing fairly if one is to be able to play at all. If individuals do not play fairly they may not be able to find willing play partners. In coyotes, for example, youngsters are hesitant to play with an individual who does not play fairly or with an individual whom they fear (Bekoff 1977b). In many species individuals also show play partner preferences, and it is possible that these preferences are based on the trust that individuals place in one another.

FAIRNESS AND FITNESS

I believe that a sense of fairness is common to many animals, because without it there could be no social play, and without social play individual animals and entire groups would be at a disadvantage. If I am correct, morality evolved because it is adaptive in its own right, and not because it is merely an antidote to competition or aggression. Behaving fairly helps many animals, including humans, to survive and flourish in their particular social environment. I fully realize that this may sound like a radical idea, particularly if one views morality as uniquely human and a sort of mystical quality that sets us apart from other animals. But if you accept my argument that play and fairness are inextricably linked then it is just a short move to showing that individual animals might well benefit from these behaviors.

My own fieldwork on coyotes has revealed one direct cost paid by animals that fail to engage fully in play. I have found that coyote pups who do not play much are less tightly bonded to other members of their group and are more likely to strike out on their own (Bekoff 1977b). Life outside the group is much more risky than within it. In a seven-year study of coyotes living in the Grand Teton National Park outside Moose, Wyoming, we found that more than 55 percent of yearlings who drifted away from their social group died, whereas fewer than 20 percent of their stay-at-home peers did (Bekoff and Wells 1986).

THE EVOLUTION OF FAIRNESS: A GAME-THEORETICAL MODEL

Much research on the evolution of cooperation has been modeled using game-theoretic approaches. Lee Dugatkin and I (Dugatkin and Bekoff 2003) used a similar technique to analyze four possible strategies that an individual could adopt over time (for species in which fairness can be expressed during two different developmental stages), namely, being fair (F) and at a later date being fair (F/F), being fair and then not fair (F/NF), being not fair and then fair (NF/F), and being not fair and then not fair (NF/NF). Of these, only F/F was an evolutionarily stable strategy (ESS) that could evolve under the conditions of the model. None of the other three strategies were ESSs, and when no strategy was an ESS all four could coexist. There are two clear predictions from our results. First, always acting fairly should be more common than never acting fairly in species in which fairness can be expressed during two different developmental stages. Second, there should be many more cases in which none of the strategies we modeled would be an ESS, but all four could coexist at significant frequencies. That F/NF is not an ESS is of interest because this strategy could be conceived as a form of deceit. This finding fits in well with what is known about play signals, for as I mentioned above, there is little evidence that play signals are used to deceive others at any stage of development (Bekoff 1977a; Bekoff and Allen 1998). Our ideas are certainly testable in principle by following identified individuals and recording how they distribute fairness across different activities as they mature.

NEUROBIOLOGICAL BASES OF SHARING INTENTIONS AND MIND-READING: SOME POSSIBLE CONNECTIONS AMONG ACTING, SEEING, FEELING, AND FEELING/KNOWING

How might a play bow (or other action) serve to provide information to its recipient about the sender's intentions? Is there a relationship among acting, feeling, seeing, and feeling/knowing? Perhaps one's own experiences with play can promote learning about the intentions of others. Perhaps the recipient shares the intentions (beliefs, desires) of the sender based on the recipient's own prior experiences of situations in which she performed play bows. Recent research suggests a neurobiological basis for sharing intentions. "Mirror neurons," found in macaques, fire when a monkey executes an action and also when the monkey observes the same action performed by another monkey (Gallese 1998; Gallese and Goldman 1998; Motluk 2001).

Research on mirror neurons is truly exciting, and the results of these efforts will be very helpful for answering questions about which species of animals may have "theories of mind" or "cognitive empathy" about the mental and emotional states of others. Gallese and the philosopher Alvin Goldman suggest that mirror neurons might "enable an organism to detect certain mental states of observed conspecifics . . . as part of, or a precursor to, a more general mind-reading ability" (1998:493). Laurie Carr and her colleagues at the University of California at Los Angeles discovered, by using neuroimaging in humans, similar patterns of neural activation both when individuals observed a facial expression depicting an emotion and when they imitated the facial expression. This research suggests a neurobiological underpinning of empathy (Laurie Carr, personal communication). Frith and Frith (1999) report the results of neural imaging studies in humans that suggest a neural basis for one form of "social intelligence," understanding others' mental states (mental state attribution).

More comparative data are needed to determine if mirror neurons (or functional equivalents) are found in other taxa and if they might actually play a role in the sharing of intentions or feelings—perhaps keys to empathy—between individuals engaged in an ongoing social interaction such as play. Neuroimaging studies will also be especially useful.

LEVELS OF SELECTION

I am sure that close scrutiny of social animals will reveal more evidence that having a sense of fairness benefits individuals. More controversially, I also believe that a moral sense benefits groups as a whole because during social play group members learn rules of engagement that influence their decisions about what is acceptable behavior when dealing with each other. Such an understanding is essential if individuals are to work in harmony to create a successful group able to outcompete other groups. Following the lines of Sober and Wilson's (1998:135ff.) discussion concerning the choice of social partners, it may be that behaving fairly is a group adaptation, but once a social norm evolves it becomes individually advantageous to behave fairly for there are costs to not doing so (Elliott Sober, personal communication). We still need somehow to figure out how to test rigorously extant ideas about levels of selection—group selection "versus" individual selection—and studies of the evolution of social morality are good places to focus for expanding our views (e.g., Boehm 1999; Leigh 1999; see also Aviles 1999; Bradley 1999; Gould and Lloyd 1999; Kitchen and Packer 1999; Mayr 2000).

SPECIESISM AND THE TAXONOMIC DISTRIBUTION OF MORAL CAPACITY: THE IMPORTANCE OF STUDYING SOCIAL CARNIVORES

We simply do not have enough data to make hard and fast claims about the taxonomic distribution among different species of the cognitive skills and emotional capacities necessary for being able to empathize with others, to behave fairly, or to be moral agents. Recently, Marler concluded his review of social cognition in nonhuman primates and birds as follows: "I am driven to conclude, at least provisionally, that there are more similarities than differences between birds and primates. Each taxon has significant advantages that the other lacks" (1996:22). Tomasello and Call summarized their comprehensive review of primate cognition by noting that "the experimental foundation for claims that apes are 'more intelligent' than monkeys is not a solid one, and there are few if any naturalistic observations that would substantiate such broad-based, species-general claims" (1997:399–400). While Flack and de Waal's (2000) and others' focus is on nonhuman primates as the most likely animals to show precursors to human morality, others have argued that we might learn as much or more about the evolution of human social behavior by studying social carnivores (Schaller and Lowther 1969; Tinbergen 1972; Thompson 1975; Drea and Frank 2003), species whose social behavior and organization resemble that of early hominids in a number of ways (divisions of labor, food sharing, care of young, and inter- and intrasexual dominance hierarchies).

What we really need are long-term field studies of social animals for which it would be reasonable to hypothesize that emotions and morality have played a role in the evolution of sociality, that emotions and morality are important in the development and maintenance of social bonds that allow individuals to work together for the benefit of all group members (see also Gruen 2002).

IMPORTANCE OF PREDICTION: A LITMUS TEST FOR KNOWING?

The ability to make accurate predictions about what an individual is likely to do in a given social situation may be closely linked with one's having extensive experience with specific individuals. Of course, extensive formal ("scientific") experience watching animals is not necessary for being able to make accurate predictions. Also, while I cannot know with absolute certainty that any of the animals about whom I have written (or others) have beliefs, desires, or intentions, I also cannot know with absolute certainty if they have a sense of "right" or "wrong" or if they are merely acting "as if" they are moral beings. They perform what can be called "moral behavior" but it might have no bearing on what they are thinking or feeling. However, the inescapable uncertainty associated with these claims does not mean that I do not know quite a lot about what is happening in their minds. It seems fair to ask skeptics to do more than say "'as if' is not enough" and to assume some responsibility for studying these questions in more rigorous ways.

In *Species of Mind*, Colin Allen and I (1997) argued that there are a number of reasons that cognitive explanations that entail beliefs, desires, or intentions may be the best explanations to which to appeal because they help us come to terms with questions centering on the comparative and evolutionary study of animal minds. First, the explanatory power of our theorizing is increased. Second, it is obvious that a cognitive approach can generate new ideas that can be tested empirically, help in evaluations of extant explanations, lead to the development of new predictive models, and, perhaps, lead to the reconsideration of old data, some of which might have resisted explanation without a cognitive perspective. Third, cognitive explanations account for observed flexibility in behavior better than do less flexible stimulus-response accounts, which stipulate do "this" in "this" situation or "that" in "that" situation (Bekoff 1996). Fourth, cognitive explanations might help scientists come to terms with larger sets of available data that are difficult to understand. Fifth, cognitive explanations may also be more parsimonious and less cumbersome than explanations that require numerous and diverse stimulus-response contingencies (Bekoff 1996; Allen and Bekoff 1997; Bekoff and Allen 1997; see also de Waal 1991).

The ability to predict what an individual is likely to do next in a social encounter might be a useful litmus test for what is happening in that individual's brain. This is not to say that the ability to predict ongoing behavior will ever be as accurate as, say, astronomical predictions concerning the positions of stars in the sky. Nonetheless, researchers and others who have spent much time watching individual animals are rather good at predicting their behavior, and many of these predictions are tied in with attributions of beliefs, desires, or intentions. This is the case for my own extensive experience of watching canids signal their intentions to engage in and to maintain social play. Intentional or representational explanations are important to my making accurate predictions about future behavior.

All I want to put out on the table here is the idea that the ability to predict behavior with a high degree of accuracy might also be a good reason to favor cognitive explanations in certain situations. Accurate prediction might be used as one measure of what a human observer "knows" about the behavior of the animals he or she is studying. So, before skeptics adamantly claim that prediction is not a viable candidate for accepting, in some cases, that intentional or representational explanations can be reliable accounts of what animals might believe, desire, or intend, they should pay attention to the veracity (and parsimony) of the predictions that are offered.

WILD JUSTICE, SOCIAL PLAY, AND SOCIAL MORALITY: WHERE TO FROM HERE?

> Justice presumes a personal concern for others. It is first of all a sense, not a rational or social construction, and I want to argue that this sense is, in an important sense, natural.
>
> —Solomon 1995:102

> It is not difficult to imagine the emergence of justice and honor out of the practices of cooperation.
>
> —Damasio 2003:162

To stimulate further comparative research (and the development of models) on a wider array of species than has previously been studied, I offer the hypothesis that social morality, in this case behaving fairly, is an adaptation that is shared by many mammals, not only by nonhuman and human primates. Behaving fairly evolved because it helped young animals acquire social (and other) skills needed as they matured into adults. A focus on social cooperation is needed to balance the plethora of research that is devoted to social competition and selfishness (for further discussion see Boehm 1999; Singer 1999; Wilson 2002). I often wonder if our view of the world would have been different had Charles Darwin been a female, if some or many of the instances in which competition is invoked were viewed as cooperation. Women tend to "see" more cooperation in nature than do men. Adams and Burnett (1991) discovered that female ethologists working in East Africa use a substantially different descriptive vocabulary than do male ethologists. Of the nine variables they studied, those concerning cooperation and female gender were the most important in discriminating women's and men's word use. They concluded that "The variable COOPERATION demonstrates the appropriateness of feminist claims to connection and cooperation as women's models for behaviour, as divergent from the traditional competitive model" (ibid.:558). Why women and men approach the same subject from a different perspective remains largely unanswered.

Group-living animals may provide many insights into animal morality. In many social groups individuals develop and maintain tight social bonds that help to regulate social behavior. Individuals coordinate their behavior—some mate, some hunt, some defend resources, some accept subordinate status—to achieve common goals and to maintain social stability. Consider, briefly, pack-living wolves. For a long time researchers thought pack size was regulated by available food resources. Wolves typically feed on such prey as elk and moose, each of which is larger than an individual wolf. Hunting such large ungulates successfully takes more than one wolf, so it made sense to postulate that wolf packs evolved because of the size of wolves' prey. Defending food might also be associated with pack-living. However, long-term research by Mech (1970) showed that pack size in wolves was regulated by *social* and not food-related factors. Mech discovered that the number of wolves who could live together in a coordinated pack was governed by the number of wolves with whom individuals could closely bond ("social attraction factor") balanced against the number of individuals from whom an individual could tolerate competition ("social competition factor"). Codes of conduct and packs broke down when there were too many wolves. Whether or not the dissolution of packs was due to individuals behaving unfairly is unknown, but this would be a valuable topic for future research in wolves and other social animals. Solomon (1995:143) contends that "A wolf who is generous can expect generosity in return. A wolf who violates another's ownership zone can expect to be punished, perhaps ferociously, by others." These claims can easily be studied empirically. [For interesting studies of the "social complexity hypothesis," which claims "that animals living in large social groups should display enhanced cognitive abilities" when compared to those who do not, see Bond, Kamil, and Balda (2003:479) and Drea and Frank (2003).]

In social groups, individuals often learn what they can and cannot do, and the group's integrity depends upon individuals agreeing that certain rules regulate their behavior. At any given moment individuals know their place or role and that of other group members. As a result of lessons in social cognition and empathy that are offered in social play, individuals learn what is "right" or "wrong"—what is acceptable to others—the result of which is the development and maintenance of a social group that operates efficiently. The absence of social structure and boundaries can produce gaps in morality that lead to the dissolution of a group (Bruce Gottlieb, personal communication).

In summary, I argue that mammalian social play is a useful behavioral phenotype on which to concentrate in order to learn more about the evolution of fairness and social morality. (While birds and individuals of other species engage in social play, there are too few data from which to draw detailed conclusions about the nature of their play.) There is strong selection for playing fairly because most if not all individuals benefit from adopting this behavioral strategy (and group stability may be also be fostered). Numerous mechanisms (play invitation signals, variations in the sequencing of actions performed during play when compared to other contexts, self-handicapping, role-reversing) have evolved to facilitate the initiation and maintenance of social play in numerous mammals—to keep others engaged—so that agreeing to play fairly and the resulting benefits of doing so can be readily achieved.

Ridley (1996) points out that humans seem to be inordinately upset about unfairness, but we do not know much about other animals' reaction to unfairness. He suggests that perhaps behaving fairly pays off in the long run. Dugatkin's and my model of the development and evolution of cooperation and fairness (Dugatkin and Bekoff 2003) suggests it might. Hauser (2000) concluded that there is no evidence that animals can evaluate whether an act of reciprocation is fair. However, he did not consider social play in his discussion of animal morality and moral agency. De Waal (1996) remains skeptical about the widespread taxonomic distribution of cognitive empathy after briefly considering social play, but he remains open to the possibility that cognitive empathy might be found in animals other than the great apes (see Preston and de Waal 2002). It is premature to dismiss the possibility that social play plays some role in the evolution of fairness and social morality or that animals other than primates are unable intentionally to choose to behave fairly because they lack the necessary cognitive skills or emotional capacities. We really have very little information that bears on these questions.

Let me stress that I am not arguing that there is a gene for fair or moral behavior. As with any behavioral trait, the underlying genetics is bound to be complex, and environmental influences may be large. No matter. Provided there is variation in levels of morality among individuals, and provided virtue is rewarded by a greater number of offspring, then any genes associated with good behavior are likely to accumulate in subsequent generations. And the observation that play is rarely unfair or uncooperative is surely an indication that natural selection acts to weed out those who do not play by the rules.

Future comparative research that considers the nature and details of the social exchanges that are needed for animals to engage in play—reciprocity and cooperation—will undoubtedly produce data that bear on the questions that I raise in this brief essay and also help to "operationalize" the notion of behaving fairly by informing us about what sorts of evidence confirm that animals are behaving with some sense of fairness. In the absence of this information it is premature to dismiss the possibility that social play plays some role in the evolution of fairness and social morality or that animals other than primates are unable intentionally to choose to behave fairly because they lack the necessary cognitive skills or emotional capacities. These are empirical questions for which the comparative database is scant.

Gruen (2002) also rightfully points out that we still need to come to terms with what it means to be moral. She also suggests that we need to find out what cognitive and emotional capacities operate when humans perform various moral actions, and to study animals to determine if they share these capacities or some variation of them. Even if it were the case that available data suggested that nonhuman primates do not seem to behave in a specific way, for example, playing fairly, in the absence of comparative data this does not justify the claim that individuals of other taxa cannot play fairly. [At a meeting in Chicago in August 2000 dealing with social organization and social complexity, it was hinted to me that while my ideas about social morality are interesting, there really is no way that social carnivores could be said to be so decent—to behave (play) fairly—because it was unlikely that even nonhuman primates were this virtuous.]

Learning about the taxonomic distribution of animal morality involves answering numerous and often difficult questions. Perhaps it will turn out that the best explanation for existing data in some taxa is that some individuals do indeed on some occasions modify their behavior to play fairly.

Play may be a unique category of behavior in that asymmetries are tolerated more so than in other social contexts. Play cannot occur if the individuals choose not to engage in the activity and the equality (or symmetry) needed for play to continue makes it different from other forms of seemingly cooperative behavior (e.g., hunting, caregiving). This sort of egalitarianism is thought to be a precondition for the evolution of social morality in humans. Whence did it arise? Truth be told, we really do not know much about the origins of egalitarianism. Armchair discussions, while important, will do little in comparison to our having direct experiences with other animals. In my view, studies of the evolution of social morality are among the most exciting and challenging projects that behavioral scientists (ethologists, geneticists, evolutionary biologists, neu-robiologists, psychologists, anthropologists), theologians, and religious scholars face. We need to rise to the *extremely* challenging (and frustrating) task before us rather than dismiss summarily and unfairly, in a speciesis-tic manner, the moral lives of other animals. *Fair is fair*.

There is no doubt that studying and learning about animal play can teach us to live more compassionately with heart and love. Keep in mind the spirit, symmetry, synchrony, sacredness, and soul of play. Learning about the evolution of cooperation, fairness, trust, and social morality goes well beyond traditional science and can be linked to religion, theology, and perhaps even different notions of God because ideas about continuity and discontinuity (the possible uniqueness of humans and other species) and individuality have to be taken into account. I cannot think of a more exciting field of inquiry, and I feel lucky that I am able to pursue the daunt-ing questions that are concerned with the evolution of social morality.

MORALITY AND HUMAN NATURE: THE PRECAUTIONARY PRINCIPLE

Just what role does human morality play in defining "human nature"? We do not really know despite strong claims to the contrary. Using animal models to rationalize cruelty, divisiveness, warfare, territoriality, and self-ishness is a disingenuous use of much available information on animal social behavior. While animals surely can be nasty, this does not explain much of the behavior that is expressed to other individuals. Animals *do* make choices to be nice and to be fair.

Ecologists and environmentalists have developed what they call the "precautionary principle," which is used for making decisions about envi-ronmental problems. This principle states that a lack of full scientific cer-tainty should not be used as an excuse to delay taking action on some issue. The precautionary principle can be easily applied in studies of the evolution of social morality. To wit, I claim that we know enough to war-rant further comparative studies of the evolution of social morality in ani-mals other than nonhuman primates, and that until these data are available we should keep an open mind about what individuals of other taxa can and cannot do.

It is important for us to learn more about the evolution of social moral-ity and how this information can be used to give us hope for the future rather than accepting a doomsday view of where we are all heading "because it's in our nature." Accepting that competition, selfishness, and cheating are what drives human and animal behavior leaves out a lot of the puzzle of how we came to be who we are. Cooperation and fairness can also be driving forces in the evolution of sociality.

The importance of interdisciplinary collaboration and cooperation in studies of animal cognition, cooperation, and moral behavior cannot be emphasized too strongly. There really *is* a paradigm shift in our studies of the evolution of morality. It is clear that morality and virtue didn't suddenly appear in the evolutionary epic beginning with humans. While fair play in animals may be a rudimentary form of social morality, it still could be a forerunner of more complex and more sophisticated human moral systems. It is self-serving, anthropocentric speciesism to claim that we are the *only* moral beings in the animal kingdom. It is also a simplistic and misleading view to assume that humans are merely naked apes.

The origins of virtue, egalitarianism, and morality are more ancient than our own species. Humans also aren't necessarily morally superior to other animals. Indeed, it might just be that animal morality is purer than human morality because animals likely don't have as sophisticated notions of right and wrong. Wouldn't that be something? But, we will never learn about animal morality if we close the door on the possibility that it exists. It is still far too early to draw the uncompromising conclusion that human morality is different in *kind* from animal morality and walk away smugly in victory, fooling ourselves along the way.

ACKNOWLEDGMENTS

Much of this discussion appeared in two of my previous papers (Bekoff 2001a, 2001b) and has been updated where possible. I thank Bob Sussman for comments on an earlier draft of this paper and all of the people at the meeting at which this paper was discussed for their valuable and interdisciplinary insights.

REFERENCES

Adams, E. R. and G. W. Burnett. 1991. "Scientific Vocabulary Divergence among Female Primatologists Working in East Africa." *Social Studies of Science* 21:547–60.

Aldis, O. 1975. *Play-Fighting.* New York: Academic Press.

Allen, C. and M. Bekoff. 1997. *Species of Mind: The Philosophy and Biology of Cognitive Ethology.* Cambridge, MA: MIT Press.

Aureli, F. and F. B. M. de Waal (Eds.). 2000. *Natural Conflict Resolution.* Berkeley: University of California Press.

Aviles, L. 1999. "Cooperation and Non-Linear Dynamics: An Ecological Perspective on the Evolution of Sociality." *Evolutionary Ecology Research* 1:459–77.

Axelrod, R. 1984. *The Evolution of Cooperation.* New York: Basic Books.

Bauer, E. B. and B. B. Smuts. 2002. "Role Reversal and Self-Handicapping during Playfighting in Domestic Dogs, *Canis familiaris.*" Paper presented at the meetings of the Animal Behavior Society, University of Indiana.

Bekoff, M. 1975. "The Communication of Play Intention: Are Play Signals Functional?" *Semiotica* 15:231–39.

Bekoff, M. 1977a. "Social Communication in Canids: Evidence for the Evolution of a Stereotyped Mammalian Display." *Science* 197:1097–99.

Bekoff, M. 1977b. "Mammalian Dispersal and the Ontogeny of Individual Behavioral Phenotypes." *American Naturalist* 111:715–32.

Bekoff, M. 1995. "Play Signals as Punctuation: The Structure of Social Play in Canids." *Behaviour* 132:419–29.

Bekoff, M. 1996. "Cognitive Ethology, Vigilance, Information Gathering, and Representation: Who Might Know What and Why?" *Behavioural Processes* 35:225–37.

Bekoff, M. (Ed.). 2000. *The Smile of a Dolphin: Remarkable Accounts of Animal Emotions.* New York: Discovery Books/Random House.

Bekoff, M. 2001a. "Social Play Behaviour, Cooperation, Fairness, Trust and the Evolution of Morality." *Journal of Consciousness Studies* 8(2):81–90.

Bekoff, M. 2001b. "The Evolution of Animal Play, Emotions, and Social Morality: On Science, Theology, Spirituality, Personhood, and Love." *Zygon* 36:615–55.

Bekoff, M. 2002a. *Minding Animals: Awareness, Emotions, and Heart.* New York: Oxford University Press.

Bekoff, M. 2002b. "Virtuous Nature." *New Scientist* (13 July):34–37.

Bekoff, M. 2002c. "Self-Awareness." *Nature* 419:255.

Bekoff, M. 2003. "Consciousness and Self in Animals: Some Reflections." *Zygon* 38:229–45.

Bekoff, M. and C. Allen. 1992. "Intentional Icons: Towards an Evolutionary Cognitive Ethology." *Ethology* 91:1–16.

Bekoff, M. and C. Allen. 1997. "Cognitive Ethology: Slayers, Skeptics, and Proponents." Pp. 313–34 in *Anthropomorphism, Anecdote, and Animals: The Emperor's New Clothes?* edited by R. W. Mitchell, N. Thompson, and L. Miles. Albany, NY: SUNY Press.

Bekoff, M. and C. Allen. 1998. "Intentional Communication and Social Play: How and Why Animals Negotiate and Agree to Play." Pp. 97–114 in *Animal Play: Evolutionary, Comparative, and Ecological Perspectives,* edited by M. Bekoff and J. A. Byers. Cambridge and New York: Cambridge University Press.

Bekoff, M., C. Allen, and G. M. Burghardt (Eds.). 2002. *The Cognitive Animal.* Cambridge, MA: MIT Press.

Bekoff, M. and J. A. Byers. 1981. "A Critical Reanalysis of the Ontogeny of Mammalian Social and Locomotor Play: An Ethological Hornet's Nest." Pp. 296–337 in *Behavioral Development: The Bielefeld Interdisciplinary Project,* edited by K. Immelmann, G. W. Barlow, L. Petrinovich, and M. Main. New York: Cambridge University Press.

Bekoff, M. and J. A. Byers (Eds.). 1998. *Animal Play: Evolutionary, Comparative, and Ecological Approaches.* New York: Cambridge University Press.

Bekoff, M. and M. C. Wells. 1986. "Social Behavior and Ecology of Coyotes." *Advances in the Study of Behavior* 16:251–338.

Bernstein, I. S. 2000. "The Law of Parsimony Prevails: Missing Premises Allow Any Conclusion." *Journal of Consciousness Studies* 7:31–34.

Bewley, T. 2003. "Fair's Fair." *Nature* 422:125–26.

Boehm, C. 1999. *Hierarchy in the Forest: The Evolution of Egalitarian Behavior.* Cambridge, MA: Harvard University Press.

Bond, A., A. C. Kamil, and R. P. Balda. 2003. "Social Complexity and Transitive Inference in Corvids." *Animal Behaviour* 65:479–87.

Bowles, S. and H. Gintis. 2002. *"Home reciprocans." Nature* 415:125–28.

Bradley, B. J. 1999. "Levels of Selection, Altruism, and Primate Behavior." *Quarterly Review of Biology* 74:171–94.

Burghardt, G. M. 2002. *The Genesis of Play.* Cambridge, MA: MIT Press.

Damasio, A. 2003. *Looking for Spinoza: Joy, Sorrow, and the Feeling Brain.* New York: Harcourt.

Darwin C. 1859. *On the Origin of Species by Means of Natural Selection.* London: Murray.

Darwin, C. [1871] 1936. *The Descent of Man and Selection in Relation to Sex.* New York: Random House.

Darwin, C. [1872] 1998. *The Expression of the Emotions in Man and Animals,* 3rd edition (with an Introduction, Afterword, and Commentaries by Paul Ekman). New York: Oxford University Press.

Dawkins, R. 1976. *The Selfish Gene.* New York: Oxford University Press.

Dawkins, R. 2001. "Sustainability Doesn't Come Naturally: A Darwinian Perspective on Values." <www.environmentfoundation.net/richard-dawkins.htm>.

de Waal, F. 1991. "Complementary Methods and Convergent Evidence in the Study of Primate Social Cognition." *Behaviour* 18:297–320.

de Waal, F. 1996. *Good-Natured: The Origins of Right and Wrong in Humans and Other Animals.* Cambridge, MA: Harvard University Press.

de Waal, F. and P. L. Tyack (Eds.). 2003. *Animal Social Complexity: Intelligence, Culture, and Individualized Societies.* Cambridge, MA: Harvard University Press.

Douglas, K. 2001. "Playing Fair." *New Scientist* (2281, 10 March):38–42.

Drea, C. M. and L. G. Frank. 2003. "The Social Complexity of Spotted Hyenas." Pp. 121–48 in *Animal Social Complexity: Intelligence, Culture, and Individualized Societies,* edited by F. de Waal and P. L. Tyack. Cambridge, MA: Harvard University Press.

Dugatkin, L. A. 1997. *Cooperation among Animals: An Evolutionary Perspective.* New York: Oxford University Press.

Dugatkin, L. A. and M. Bekoff. 2003. "Play and the Evolution of Fairness: A Game Theory Model." *Behavioural Processes* 60:209–14.

Fagen, R. 1981. *Animal Play Behavior.* New York: Oxford University Press.

Fagen, R. 1993. "Primate Juveniles and Primate Play." Pp. 183–96 in *Juvenile Primates: Life History, Development, and Behavior,* edited by M. E. Pereira and L. A. Fairbanks. New York: Oxford University Press.

Fehr, E. and S. Gächter. 2002. "Altruistic Punishment in Humans." *Nature* 415: 137–40.

Fehr, E. and B. Rockenbach. 2003. "Detrimental Effect of Sanctions on Human Altruism." *Nature* 422:137–40.

Field, A. 2001. *Altruistically Inclined? The Behavioral Sciences, Evolutionary Theory, and the Origins of Reciprocity.* Ann Arbor: University of Michigan Press.

Flack, J. C. and F. de Waal. 2000. "Any Animal Whatever: Darwinian Building Blocks of Morality in Monkeys and Apes." *Journal of Consciousness Studies* 7:1–29.

Frith, C. D. and U. Frith. 1999. "Interacting Minds: A Biological Basis." *Science* 286:1692–95.

Gallese, V. 1998. "Mirror Neurons: From Grasping to Language." *Consciousness Bulletin* (Fall):3–4.

Gallese, V. and A. Goldman. 1998. "Mirror Neurons and the Simulation Theory of Mind-Reading." *Trends in Cognitive Science* 2:493–501.

Gould, S. J. and E. A. Lloyd. 1999. "Individuality and Adaptation across Levels of Selection: How Shall We Name and Generalize the Unit of Darwinism." *Proceedings of the National Academy of Sciences* 96:11904–09.

Gruen, L. 2002. "The Morals of Animal Minds." Pp. 437–42 in *The Cognitive Animal*, edited by M. Bekoff, C. Allen, and G. M. Burghardt. Cambridge, MA: MIT Press.

Güzeldere, G. and E. Nahmias. 2002. "Darwin's Continuum and the Building Blocks of Deception." Pp. 353–62 in *The Cognitive Animal*, edited by M. Bekoff, C. Allen, and G. M. Burghardt. Cambridge, MA: MIT Press.

Hauser, M. 2000. *Wild Minds*. New York: Henry Holt.

Hill, H. L. and M. Bekoff. 1977. "The Variability of Some Motor Components of Social Play and Agonistic Behaviour in Eastern Coyotes, *Canis latrans* var." *Animal Behaviour* 25:907–9.

Hinde, R. A. 2002. *Why Good Is Good: The Sources of Morality*. New York: Routledge.

His Holiness the Dalai Lama. 2002. "Understanding Our Fundamental Nature." Pp. 66–80 in *Visions of Compassion: Western Scientists and Tibetan Buddhists Examine Human Nature*, edited by R. J. Davidson and A. Harrington. New York: Oxford University Press.

Horowitz, A. C. 2002. "The Behaviors of Theories of Mind, and a Case Study of Dogs at Play." Ph.D. dissertation, University of California, San Diego.

Journal of Consciousness Studies. 2000. Volume 7(1/2).

Kitchen, D. M. and C. Packer. 1999. "Complexity in Vertebrate Societies." Pp. 176–96 in *Levels of Selection in Evolution*, edited by L. Keller. Princeton, NJ: Princeton University Press.

Kuczaj, S., K. Tranel, M. Trone, and H. Hill. 2001. "Are Animals Capable of Deception or Empathy? Implications for Animal Consciousness and Animal Welfare." *Animal Welfare* 10:S161–73.

Leigh, E. G., Jr. 1999. "Levels of Selection, Potential Conflicts, and Their Resolution: Role of the 'Common good.'" Pp. 15–30 in *Levels of Selection in Evolution*, edited by L. Keller. Princeton, NJ: Princeton University Press.

Macintyre, A. 1999. *Dependent Rational Animals: Why Human Beings Need the Virtues*. Chicago: Open Court.

Markowitz, H. 1982. *Behavioral Enrichment in the Zoo*. New York: Van Reinhold.

Marler, P. 1996. "Social Cognition: Are Primates Smarter Than Birds?" Pp. 1–32 in *Current Ornithology*, Volume 13, edited by V. Nolan, Jr., and E. D. Ketterson. New York: Plenum.

Mayr, E. 2000. "Darwin's Influence on Modern Thought." *Scientific American* 283:67–71.

Mech, L. D. 1970. *The Wolf*. Garden City, NY: Doubleday.

Mitchell, L. E. 1998. *Stacked Deck: A Story of Selfishness in America*. Philadelphia: Temple University Press.

Motluk, A. 2001. "Read My Mind." *New Scientist* 169:22–26.

Okamoto, K. and S. Matsumara. 2000. "The Evolution of Punishment and Apology: An Iterated Prisoner's Dilemma Model." *Evolutionary Ecology* 14:703–20.

Panksepp, J. 1998. *Affective Neuroscience*. New York: Oxford University Press.

Panksepp, J. 2000. "The Rat Will Play." Pp. 146–47 in *The Smile of a Dolphin: Remarkable Accounts of Animal Emotions*, edited by M. Bekoff. New York: Random House/Discovery.

Pellis, S. 2002. "Keeping in Touch: Play Fighting and Social Knowledge." Pp. 421–27 in *The Cognitive Animal*, edited by M. Bekoff, C. Allen, and G. M. Burghardt. Cambridge, MA: MIT Press.

Peterson, G. R. 1999. "The Evolution of Consciousness and the Theology of Nature." *Zygon* 34:283–306.

Peterson, G. R. 2000. "God, Genes, and Cognizing Agents." *Zygon* 35:469–80.

Poole, J. 1998. "An Exploration of a Commonality between Ourselves and Elephants." *Etica & Animali* 9:85–110.

Power, T. G. 2000. *Play and Exploration in Children and Animals*. Hillsdale, NJ: Lawrence Erlbaum Associates.

Preston, S. D. and F. B. M. de Waal. 2002. "Empathy: Its Ultimate and Proximate Bases." *Behavioral and Brain Sciences* 25:1–72.

Ridley, M. 1996. *The Origins of Virtue: Human Instincts and the Evolution of Cooperation*. New York: Viking.

Ridley, M. 2001. *The Cooperative Gene*. New York: Free Press.

Rilling, J. K., D. A. Gutman, T. R. Zeh, G. Pagnoni, G. S. Berns, and C. D. Kitts. 2002. "A Neural Basis for Cooperation." *Neuron* 36:395–405.

Riolo, R. L., M. D. Cohen, and R. Axelrod. 2001. "Evolution of Cooperation without Reciprocity." *Nature* 414:441–43.

Schaller, G. B. and G. R. Lowther. 1969. "The Relevance of Carnivore Behavior to the Study of Early Hominids." *Southwestern Journal of Anthropology* 25: 307–41.

Scott, J. P. and J. L. Fuller. 1965. *Genetics and the Social Behavior of the Dog*. Chicago: University of Chicago Press.

Sigmund, K. and M. A. Nowak. 2001. "Evolution: Tides of Tolerance." *Nature* 414:403–5.

Sigmund, K., E. Fehr, and M. A. Nowak. 2002. "The Economics of Fair Play." *Scientific American* 286(1):83–87.

Singer, P. 1999. *A Darwinian Left: Politics, Evolution, and Cooperation*. New Haven, CT: Yale University Press.

Siviy, S. 1998. "Neurobiological Substrates of Play Behavior: Glimpses into the Structure and Function of Mammalian Playfulness." Pp. 221–42 in *Animal Play: Evolutionary, Comparative, and Ecological Perspectives*, edited by M. Bekoff and J. A. Byers. New York: Cambridge University Press.

Skyrms, B. 1996. *Evolution of the Social Contract*. New York: Cambridge University Press.

Sober, E. and D. S. Wilson. 1998. *Unto Others: The Evolution and Psychology of Unselfish Behavior*. Cambridge, MA: Harvard University Press.

Sober, E. and D. S. Wilson. 2000. Summary of *Unto Others: The Evolution and Psychology of Unselfish Behavior*. *Journal of Consciousness Studies* 7:185–206.

Solomon, R. 1995. *A Passion for Justice: Emotions and the Origins of the Social Contract*. Lanham, MD: Rowman & Littlefield.

Spinka, M., R. C. Newberry, and M. Bekoff. 2001. "Mammalian Play: Training for the Unexpected." *Quarterly Review of Biology* 76:141–68.

Thompson, P. R. 1975. "A Cross-Species Analysis of Carnivore, Primate, and Hominid Behavior." *Journal of Human Evolution* 4:113–24.

Tinbergen, N. 1972. *Introduction to Hans Kruuk, The Spotted Hyena.* Chicago: University of Chicago Press.

Tomasello, M. and J. Call. 1997. *Primate Cognition.* New York: Oxford University Press.

Watson, D. M. and D. B. Croft. 1996. "Age-Related Differences in Playfighting Strategies of Captive Male Red-Necked Wallabies (*Macropus rufogriseus banksianus*)." *Ethology* 102:333–46.

Wechlin, S., J. H. Masserman, and W. Terris, Jr. 1964. "Shock to a Conspecific as an Aversive Stimulus." *Psychonomic Science* 1:17–18.

Wilson, D. S. 2002. *Darwin's Cathedral: Evolution, Religion, and the Nature of Society.* Chicago: University of Chicago Press.

4

Management of Aggression as a Component of Sociality

Irwin S. Bernstein

Social living may be beneficial under certain conditions, but it always imposes costs. Competition will be fostered when similar individuals with similar needs simultaneously encounter limited resources. Conflicts will arise when individuals that want to stay together also want to do different things, or go to different places, at the same time. Social living demands a degree of accommodation to the needs and desires of others. If competition and conflict stimulate aggression to impose a solution on others, then there will be definite costs to social living. Aggressive behavior must be modified and controlled to permit social animals to benefit from sociality and to make possible joint action on the environment. The management of aggression, stimulated by social living, is essential to every society. The mechanisms that control aggression are an integral part of sociality.

Adolph Schultz (1969) said that there are three attributes that explain the success of the order Primates: (a) they are smart, (b) they are social, and (c) they have long periods of biological dependency. These features are not exclusive to the primates but are nearly universal and reach extreme expression in the order. Being smart means that they can modify their behavior as a function of experience. Being social means that learning is likely to take place in a social setting and that joint action on the environment is possible. Long periods of biological dependency mean that there will be ample opportunity for the transmission of information from one generation to the next, by nongenetic means. Although the order has few anatomical specializations, social learning and joint action on the environment account for the success of this order of generalized opportunists. Our own species, *Homo sapiens,* may be regarded as showing the most extreme expressions of these features. The possibility of transmitting learned information, from generation to generation, allows for the development of traditions and the establishment of culture.

Social organizations are characterized by (1) a measure of spatial proximity, (2) a degree of temporal stability and/or continuity, (3) recognition of group members (either as individuals or as members of the group), (4) communication leading to coordination of activities, and (5) some division of labor, such that every individual in a group does not show all of the behavior seen in the group. Spatial proximity over time not only allows for joint activities in response to external challenges, but also means that individuals with similar needs will come upon resources at the same time, setting the stage for competition. Recognition of group members means that group members and non–group members will be treated differently. Coordination of activities requires individuals to engage in similar activities at the same time, such that each individual cannot go for food, water, or shelter independently. Simultaneous different initiatives may be incompatible with the maintenance of proximity, thereby leading to conflicts. Division of labor means that different roles will be played by different individuals. Behavior will never be perfectly synchronized. A measure of division of labor suggested by Wilson (1975) is the principle of minimum specification: the number of individuals that must be specified in order to see all of the activity types that occur in a group.

Most diurnal primates live all of their lives as members of groups that meet the criteria for social organizations. When we ask why primates are social, however, the answer can be provided at any of the levels specified in Tinbergen's (1951) "Why?" questions. We can try to list the adaptive functions of sociality, mindful of the fact that there must be some negatives since so many animals are not social. We can argue, in a somewhat circular fashion, that since the social primates are social, evolution must have favored sociality. We can try to identify the current consequences of social living, and try to explain why the benefits of social living outweigh the costs with regards to feeding, finding mates, escaping from predators, and similar selective pressures. We can deal with the proximal cause by saying that primates are social because they like one another. We do not need to assume that they understand the benefits of sociality or that they are motivated to be social in expectation of the functional benefits. Those that did not like others left their groups to live solitary lives. Since so few are solitary, we assume that, in species that are social, individuals that elect this option suffer a loss of genetic fitness. Ontogenetically, we can ask how individuals are integrated into societies, and what processes bond individuals to particular social units at various times.

Given that so many primates do live in social groups, and that natural selection has favored diurnal primates that live in social groups, we seek to identify mechanisms that exist to maintain groups in the face of the various forces that disrupt group living. Sources of disturbance external to the group, including other groups of conspecifics, can be dealt with by avoidance, by aggression to repel or negate the threat, or by some other means of accommodation. Aggressive responses are more effective when multiple group members act in concert against a common opponent; group cohesion and coordination thus can serve to protect a group against such challenges. Conflicts within the group, such as some wanting to travel one way to water while others want to travel another way to food, can be resolved by compromises, by temporary fissioning, by active herding, or other means. All will require some form of communication. Aggressive resolution of such conflicts may occur, but the costs will include the time and energy involved, and the risk of potential injury to the actor, as well as to the group members targeted. Aggressive solutions to conflicts among group members may reduce the ability of the group to deal with extra-group challenges. Injured group members, and individuals uneasy with each other following a recent fight, may be expected to be less effective allies in confronting extragroup challenges, and may also impair group coordination and speed in travel and other activities. Competition can be resolved using nonaggressive techniques, such as scramble competition, but when the value of the incentive is greater than the costs of contest competition, aggression may be used as an instrumental act to achieve an end.

In social groups, conflict and competition are inevitable. Kurland (1977) noted that in primate groups the frequency of agonistic interactions was directly correlated with time spent in proximity. Even when group members are close kin, animals that associate with each other more engage in more agonistic interactions. For example, infants are far more likely to receive aggression (including bites) from their mothers than from other kin, and from kin rather than from unrelated group members (Bernstein and Ehardt 1986). The overall frequency of aggressive expression is, nonetheless, low and biased toward the least consequential expressions (Bernstein, Williams, and Ramsey 1983). The more closely associated two individuals are, however, the more likely are they to experience conflicts and competition. Since aggressive contest can be costly to all involved, it is to the advantage of both individuals to minimize the frequency and severity of agonistic interactions with one another, and thereby reduce costs, whenever possible. Ritualized threats and warnings, and procedures to assess the motivation and abilities of opponents, will benefit both parties. If opponents can assess current and anticipated costs of the contest, and each can weigh the costs against the value that each places on the incentive, then decisions can be made that will reduce the severity and costs of aggressive contests. If the information available at the start of an encounter is insufficient to make a decision whether to continue or terminate the contest, then additional costs must be incurred to obtain that information. Costs include the time and energy of signals, as well as the injuries that may be received when escalation includes forms of contact aggression. Even the winner of a fight is likely to receive some damage during the combat. Avoiding that damage, while still attaining the incentive, is most desirable. Avoiding damage when not attaining the incentive is also desirable. The most dangerous situations occur when opponents are nearly evenly matched and it is extremely difficult for either to see a clear advantage favoring the other. The costs of assessment in such cases may include the costs of a long and damaging fight. If, however, during the assessment process, one opponent withdraws, there is no profit in continuing aggression toward the opponent. The incentive is now no longer being disputed and the opponent has ceased threatening. Even when group members join to confront a predator, if the predator withdraws, there is little point in continuing the attack. A predator unwilling to risk injury to obtain a meal from a group of blustering primates may still be quite capable of killing all of them, if attacked by them.

If animals are smart, then the assessment process can involve memories of past assessments—a learned component. If an individual acts in the present based on what happened in the past, then no assessment costs or escalation need take place to resolve the present contest. A learned component can also involve the power of generalization. An individual can classify a new opponent as a member of a class of former opponents and decide if the new opponent, as a member of that class, is too formidable to contest. This allows for the formation of dominance relationships, territoriality, and similar learned relationships with a minimum of assessment costs. Contests between familiar individuals are resolved based on a past history of such resolutions. Contests with unfamiliar individuals can be resolved by identifying an unfamiliar individual as a member of a class with which a learned relationship has been established. When generalizations are applied to new individuals, it can appear as if no learning is involved and that an innate relationship is being elicited. The ability to make stimulus generalizations may have a genetic basis but the generalizations are based on learning and experience.

Dominance will influence contest competitions. If aggressive contest always determines competitions then the individual that always loses will suffer the least consequence if it submits at the start of any competition (described as a *despotic* dominance style; Preushoft and van Schaik 2000). If, in contrast, competitions are only sometimes resolved by aggressive contest, then the individual that submits at the start of an agonistic contest, the subordinate, should wait for a signal that aggressive contest will take place before submitting. The dominant will have to signal when it is going to use aggression to obtain the incentive. (This is described as a *tolerant* dominance style.) A decision to use aggressive contest to obtain an incentive may depend on the value of the incentive and the expected cost of the contest. If the value of the incentive is small (to the individual concerned rather than absolutely) then even the individual that can gain the incentive using aggression (which always involves some cost or risk) may not contest it. Scramble competition may be prevalent. Note that since the value is specific to the individual, the winner of a competition may not be the individual that would be expected to win an agonistic contest, the dominant individual. If a satiated alpha male fails to respond to a hungry juvenile racing for a piece of food, the alpha male has not lost rank! Likewise, an alpha male mildly annoyed by an infant may cease threatening the infant if its mother indicates her vigorous determination to defend her infant. She would most likely do so using a combination of aggressive and submissive signals. Including submissive signals reduces the likelihood of the male interpreting her aggressive signals as a sign of imminent attack such that the male would have to defend himself. The female's aggressive signals simultaneously signal the high incentive value that she places on the infant. The male may choose not to compete and withdraw from the infant, in this case. The female has not become dominant to the male.

In describing tolerant, despotic, and egalitarian dominance styles, a suite of behavioral characteristics is hypothesized to covary (de Waal 1989; Thierry 1997, 2000). Despotic societies are said to show severe forms of aggression at relatively high frequencies, aggression is invariably unidirectional, and reconciliations are rare following agonistic interactions. In some cases, however, high rates of aggression are reported in tolerant and even near-egalitarian macaque societies. It is explained that the presence of ready mechanisms to promote reconciliation may actually permit more frequent agonistic encounters without threatening the social fabric of the group. One may ask, however, how groups, characterized as despotic, maintain cohesion if reconciliation is rare and aggression is severe and not uncommon. Of course a despot might coerce group members into staying with the group but, although some males do herd females, such coercion does not seem to account for sociality. A second possibility is that the formation and maintenance of social bonds is so effective in despotic groups that aggressive episodes do not threaten social relationships (as has been argued to account for the low frequency of reconciliation within some kin groups). There is no evidence, however, that despotic groups form more secure social bonds; in fact, the frequency of affiliative behavior (presumably important in forming and maintaining social bonds) is low in despotic groups. Another possibility, of course, is that despotic groups do reconcile but that the means used are not readily detected using the standard means of measuring reconciliation and conciliatory tendencies currently available. The most revolutionary explanation would be that reconciliation does not exist as a special mechanism, and that primates maintain social bonds and relationships despite occasional conflicts and competitions leading to agonistic conflicts. Social bonds are maintained regularly regardless of the occurrence of agonistic encounters, but severe or frequent agonistic encounters do have the potential to cause irreparable harm such that one or more individuals might leave the group. In fact, some authors do describe group fissions and expulsions in terms of animals leaving following prolonged or severe fights. Investigators have, therefore, searched for mechanisms that function to reconcile opponents following agonistic encounters. The differential frequency of identified reconciliations in different groups may be accounted for by different time scales for reconciliation in different species, or by different mechanisms not yet identified by investigators.

Some investigators have noted that, in at least some despotic groups, it is the aggressor that is more likely to initiate reconciliations. It is possible that in despotic groups, where the subordinate yields immediately and rarely shows counteraggression, the dominant reconciles because the subordinate has never challenged, and therefore the dominant is not in any state of tension such as the subordinate has already experienced. In groups characterized as showing a tolerant dominance style, the dominant has expressed and the subordinate has received aggression; both have experienced tension and either may initiate reconciliation. In egalitarian groups there may be the greatest need to reconcile because both have aggressed against the other. This tension may be the impetus for reconciliation. Aureli and van Schaik (1991) have suggested that "tension" may be measured by such behavioral indicators as rough scratching, body shaking, and yawning. Demonstrating the existence of behavioral mechanisms, elicited when animals show signs of tension, may be used to demonstrate that mechanisms specific to reconciliation exist. These mechanisms, rather than the general skills used in establishing and maintaining social bonds, may restore relationships following social disruptions, such as fighting between members of a social unit.

The control of the expression of aggression in a social group is achieved by multiple means. These include mechanisms that influence the frequency with which aggression is used as an instrumental act, mechanisms that limit the severity of aggressive expression, social interventions in agonistic encounters that moderate or terminate aggressive expression, and the means by which disruptions due to such expression are repaired. Multiple mechanisms exist to accomplish each of these functions. Many of these mechanisms may be expected to develop during ontogeny, and it may be useful to consider the ontogenetic processes that operate in primates in considering how these mechanisms come into being.

The order Primates can be described as K-selected. Extensive investment, in the relatively small number of young produced, is characteristic. K selection requires prolonged association of the young with at least one caregiver. Mammalian lactation makes association with, at least, a lactating female mandatory during this period. When the young are relatively altricial, a bond between one or more older individuals and the infant is crucial. If the infants are mobile, then they too must bond to the parent. By "bonding" I mean that they repeatedly seek proximity to and interact selectively with one another. The process by which bonding is achieved may vary across taxa. In some animals, a special type of learning called imprinting has been described. There is no requirement for a cognitive concept of kinship or a genetic mechanism to bond a mother to her specific child. The infant, and/or at least the mother, may be primed to become attached to anything within a certain stimulus range (for example, the angle that a stimulus figure subtends on the subject's retina) during a specified time. This usually results in bonding between mothers and infants, but experimental manipulations indicate that geese can be imprinted on humans and chicks imprinted on wooden models of male mallard ducks; imprinting is not based on genetic similarity or concepts of kinship. Imprinting, nonetheless, functions to bond mothers and their biological young under species-typical conditions. The genetic contribution limits the stimuli that the infant can imprint on; this ensures that infants do not bond to rocks, trees, clouds, or objects swirling in the wind. Those that fail to imprint likely wander off and fail to reap the benefits of parental care. The infant is not motivated to bond because it knows that this will enable it to survive. The infant is not motivated to "find its mother" in order to achieve the benefits that mother can provide. The infant is motivated to follow an object within a certain class of stimulus parameters. Attachment to a particular individual results. Motivation and proximal cause are clearly distinct from function.

Mammalian mothers may bond to infants as a function of licking the birth fluids on the newborn baby (Klopfer 1971) or through other activities associated with the parturition process. The process ensures bonding of parent and young under natural conditions, but is not universally successful. If there were active processes ensuring recognition of genes, we would expect near 100 percent success. Humans certainly have to learn to identify their offspring, and this appears to be true of most nonhuman primates as well.

The infant begins life seeking "maternal" proximity; indeed most non-human primates maintain active contact with their attachment object, usually the mother, continually. The infant may cling to the mother with hands and feet and engage in nearly continuous nonnutritive sucking, as in some Old World monkeys. The infant is also attracted to novel stimuli and may explore them. If the infant experiences pain, or some other noxious consequence, as a consequence of exploring an object, the infant withdraws. Most primates, when hurt, also seek proximity to their attachment figure and establish physical contact. "Contact comfort" following any level of distress is typical and functions to put the infant in a position such that mother (or another caregiver) can respond to the source of distress, or carry the infant away from it. Perhaps having mother carry you away when you are distressed relates to the widespread phenomenon of infants being soothed by rocking motions and seeking movement stimulation when upset. Being soothed by movement allows the infant to quietly cling to its mother while being transported. The function and the selective advantages seem to be clear.

The growing infant soon learns that any stimulus that reliably precedes the experience of pain can be responded to by going to mother immediately. Such stimuli can include maternal alarm signals, and other signs, that mother is distressed. A distress call from mother also may signal that mother is likely to leave the area, thereby reducing proximity to the infant if the infant does not also move in the same direction or establish contact with mother. The infant develops a generalized response to all such stimuli; it approaches mother and makes contact. The infant will, therefore, approach mother when mother is under attack, even if the infant would probably be better off staying away. The infant approaches its distressed mother, not to escape a particular stressor, or to aid its mother, but because the infant is distressed. Such approaches can sometimes place the infant in situations where it is likely to get injured, but if most such approaches are beneficial to the infant, natural selection will favor infants who run to their mothers when their mothers are in distress, even though occasionally such approaches are maladaptive.

Infants key on mother's signals of emotional response, sometimes referred to as social referencing. The infant responds to a novel stimulus or person by looking at mother and slowly exploring the novel thing or person, checking with mother repeatedly. The infant will explore whatever mother is attracted to. This may lead to opportunities to learn what is being eaten, or to learn what objects to manipulate, and in what contexts, to achieve particular consequences (local site enhancement). The infant will also be most likely to establish and maintain social relationships with those individuals that associate with mother. Mother does not affiliate with every possible partner, and her infant will be selectively exposed to members of the group. The infant directs some of the same behavior, usually directed to the mother, to these new partners, thereby forming new social bonds. The infant's social network expands in this manner to include other attachment figures that the infant will seek out and associate with. These are likely to include siblings, mother's mother and mother's siblings, and the usual associates of any of these, regardless of genetic relationship. These bonds are not based on genes in common, but selection may favor such bonding because the association is beneficial. The usual associates are often kin, but need not be for the association to be beneficial.

Mother's associates accept the infant because mother accepts it. Mother may associate with a male for any number of reasons, in addition to reproduction. If mother associates with a male, for any reason, the infant may also form a bond to that male. It need not be the infant's father, but it may be. Even an association between a fully adult male and an infant may be a case of mutualism. The infant can benefit by the male's protection, and the male can benefit by being better tolerated by adult males, mothers, or other females when in association with the infant.

As long as the association is beneficial, selection will favor individuals that form social bonds. It may not be possible to find beneficial consequences to both parties in each and every interaction. It is the overall relationship that is beneficial and not each and every interaction that occurs in a relationship. Natural selection favors behavior that was beneficial in the past and cannot be expected to program appropriate behavior for every possible future circumstance. Behavior can be maintained by natural selection if the net results are beneficial, even if the behavior is sometimes costly.

Unusual events, such as mother mating with a male with whom she does not normally interact, or at least not in that particular way, may stimulate the infant to run to mother to make proximity and contact. Mother may ignore or discourage the infant at this time, and the male may push the infant away. The infant is certainly interfering with mating, and this may function to reduce breeding success and the production of a rival sibling, but the infant cannot be described as "trying to" interfere in the mating, or the production of a sibling. Even if that were the functional consequence, and even if selection favored infants who succeeded in achieving that outcome, that would not be the proximal cause for the infant's interference. The infant's interference may be an artifact of the fact that the infant goes to mother any time that anything unusual occurs and the infant is not certain what to do. One should not assume that natural selection has selected for all of the future consequences of typical activities, even though natural selection does act based on the past consequences of activities. After all, the past is not a perfect predictor of the future.

In very young infants we should expect the infant to avoid whatever is painful and to approach its attachment figure when in distress. A curious sequence occurs when mother punishes her child and is thus simultaneously the source of pain and the source of solace. If mother hurts you, and you have no other attachment figures, then mother is the only one that you can go to when in distress. Clinging to her will reduce the distress that her behavior produced! Seeking comfort from mother immediately after being punished by mother may seem paradoxical but will be predicted in these circumstances. If mother persists in attacking and kills her own infant, then her behavior is pathological; she has reduced her own genetic fitness.

The infant seeking its mother, when attacked by its mother, is not showing pathological behavior; it may be maladaptive in a specific instance but it is generally advantageous to seek out mother whenever in distress and no alternatives are available. If the older infant has alternative attachment figures, it may express a new preference when punished by mother, but for the very young infant there are no alternatives to the abusive mother. The distressed infant, running to and contacting its mother after receiving maternal punishment, is not necessarily seeking reconciliation, although such behavior may often result in restoring the usual close association between mother and infant.

An older child who approaches when its mother signals distress may arrive to discover that mother is under attack. Mother may reject the child's attempt to cling or the child may flee the attackers. The child is not necessarily rushing to aid its mother. If mother is winning, or if the child is not afraid of whoever is attacking mother, the child may actively join in the attack and assist mother. This is a particular instance of joint action on the social environment. If mother is already winning, or the child can defeat the opponent, there is little risk involved. This is not exactly what is implied when the sequence is described as the child coming to the aid of its mother. This is very similar to what happens in any sequence when an individual hears agonistic signals from an individual with whom it is socially bonded. The individual makes proximity with the partner and may join in an attack, or not, depending on the identity of the opponent, the relative risk involved in interfering, and the strength of the bond to the partner. If, on arrival, it turns out that the fight involves two individuals that the subject is bonded to, the individual may redirect aggression to an innocent third party, or it may displace its aggression into some other activity, such as mounting one of the combatants. Such behavior certainly does function to interfere in the ongoing fight. It may well restore peace, but the motivation may not have been to restore peace, but rather to join the activity of a bonded partner in response to a challenge, this time from the social environment. To describe such behavior as "peacemaking" describes the function, but when this is used as an example of "peacemaking skills," motivation seems to be implied, i.e., the individual utilizes a skill to achieve a goal.

When two socially bonded individuals get into a fight, each is also a source of comfort for the other. If they are not socially bonded then they may go their separate ways after the fight, but friends often affiliate after a fight. The baseline comparison is not the affiliative rate with the average group member. The matched control should be the specific partner with whom the fight occurred. But note, data for the matched control should not be collected when there is no disturbance in the group, but should be collected when the subject is equally distressed, but due to an interaction with someone other than the current partner. It may well be that, when in distress, individuals seek out specific friends, as when a distressed infant seeks its mother. These friends will also be distressed after a fight in which they fought each other. After the fight each will seek the individuals to whom they are socially bonded. Much as the infant seeks out its own mother, friends may seek each other out when they are distressed. They may seek each other out regardless of the original cause of their distress; two socially bonded individuals may thus approach one another after just having fought with each other. This will produce an opportunity for reconciliation of friends, but the motivation is not to "restore valuable" relationships. As Silk (1997) has stated, an affiliative signal after a fight may be an honest signal of present motivation rather than a signal specific to the initiation of reconciliation. It may nonetheless function to do exactly that (initiate a reconciliation), but the two descriptions imply vastly different cognitive processes and proximal causes.

Some have argued that the proximal cause of the behavior is irrelevant since natural selection acts on the functional consequences and will favor beneficial behavior regardless of the proximal cause. True enough, but it must be remembered that genes cannot code for the function. You cannot have genes for "high genetic fitness," genes for self-preservation or survival, genes for good maternal care, or genes for any functional outcome. You cannot have genes for infanticide or genes for reconciliation. What is selected is the specific structure, or behavior emitted in a specific situation, that produced the outcomes on which natural selection operates. It is the structure that must be identified and the proximal circumstances that elicit the expression of that structure. Whereas natural selection operates on the consequences of behavior in a situation, what is selected is the structure of the behavior that is elicited by specific proximal stimuli. There must be stimuli that trigger behavior in a situation where that behavior will be adaptive, if natural selection is to operate. Whereas the function is what drives natural selection, it is structure, including specific responses to specific proximal stimuli, that is transmitted to future generations. Natural selection may ensure that individuals blink when a puff of air strikes the eyeball. This usually functions to protect the eye and so is selected for, but natural selection cannot produce the functional consequences directly. You cannot have "genes for eyeball protection."

If a fight starts and a socially bonded ally arrives, the ally may aid by aggressively attacking the opponent, if the opponent is subordinate. If the opponent is dominant to the individual interfering, then the third party, having just charged up to the scene, may counter its charge by directing submissive signals to the opponent. This display may distract the attacker and the submissive signals may reduce the likelihood that the third party will be attacked, but should this be called "appeasement"? It may or may not function in that fashion but the label should not indicate a motive. Third-party involvement in fights, no matter what its form, may benefit one or more of the original opponents. Interventions may end the fight, but they also have the potential to prolong the fight when multiple animals join in the original attack, or third parties target bystanders and redirect aggression to others. Fight interference may usually reduce aggressive consequences, or benefit some of the parties involved, but it also may serve to prolong and spread aggression within the group. Exactly the same mechanisms may be involved but the outcome can vary based on the circumstances. Selection may have favored responses to proximal cues that result in ending fights in most circumstances, even if the same structures sometimes have the opposite effect. The proximal causes and motivation for interference, however, may have nothing to do with assessments of the value of partners, or the likelihood that peace will be restored. Fight interference may be a result of bonded individuals acting jointly on the social and physical environment. Interference on behalf of group members, fighting members of other groups, or extraspecifics would not be described as "peacemaking." If the explanation for fight interference when joining allies to defend against intragroup disturbance is the same as the explanation for fight interference when members act jointly in response to extragroup challenges, then both should be described in the same terms. If the structures are identical, then the label should be the same even if different functional consequences result. Calling one "group defense" and the other "peacemaking" seems unwarranted, even if the consequence of such interference functions to protect the group in one case, and usually restores peace among group members in the other. Even if the behavior protects the group or restores the peace, it is not necessarily performed in order to protect the group or restore the peace.

Socially bonded individuals act jointly on the environment. They groom one another, stay in proximity, huddle together in the rain or cold, and each responds to the communication signals of the others. They do not all have equal abilities. The bonds may not be equally valuable, but the bonds may be valuable to all. Examining each interaction separately we can ask who it is that is profiting, or profiting more. We can ask if this bond is mutually beneficial because of reciprocation, because one individual provides one kind of service and the other a second kind, or if there seems to be a difference in the times when one party benefits and when the other benefits. This can be described as trading services for favors in different currencies, as investments, as reciprocation, etc. When we describe individual relationships in these terms, analyzing the costs and benefits of each interaction, we assume a great deal of behavioral micromanagement. Assuming that natural selection acts on each interaction, rather than the total relationship, implies that animals do some kind of mental bookkeeping to make sure that things are "fair." One assumes that a failure to recip-rocate would have to be punished; individuals would coerce one another to maximize benefits to themselves, etc. Only if the two were related to one another would we be able to explain why selection would favor behavior where one individual benefits another at any cost to itself. But if both indi-viduals benefit, this is mutualism and will be selected for regardless of whether one benefits more than the other. To refuse to cooperate because the partner benefits more than you is "spite," since your noncooperation harms the partner as well as yourself. Natural selection would select against such individuals in favor of those that cooperated when it was to their advantage, regardless of the relative advantage their partners received. It may be "fairer" when both benefit equally, but it will be advan-tageous to both when both benefit even if one benefits much more than the other. Before an infant can groom, its mother does all of the grooming. Of course, mother may be improving her classical genetic fitness in doing so. When the infant becomes an adult female, it may aid and groom its now feeble mother. Although the degree of relatedness is the same, the child has little to gain in inclusive fitness benefits from a mother whose repro-ductive value is near zero but . . . Certainly the mother did not defend and groom her child in order to reap grooming benefits in her old age, but the establishment and maintenance of her relationship with her daughter did indeed benefit both. Even in her old age the female can still be of some small benefit to her daughter. Nonetheless it would not be proper to say that the daughter grooms her mother in order to guarantee those benefits.

Social bonds are not permanent. If social partners do not maintain bonds, they will weaken. When bonds weaken and aiding, grooming, and proximity diminish, is it revenge, retribution, or punishment for past fail-ures to reciprocate? It can be described that way. It takes relatively little to find some reason why an animal that breaks a social relationship "should" do so. It is no more difficult to explain why an animal should form a social relationship. There are enough variables and assumptions such that you can always "explain" what you already know. E. O. Wilson (1975) was keenly aware of this problem and, in the unabridged version of his book, correctly pointed to the philosophical error of "affirming the consequent" as one of the pitfalls in sociobiology. He warned that when you explain the function of an event, after you have observed it, you cannot cite the event as proof of the function, even though it does "explain" why it occurs. Many "causes" are neither necessary nor sufficient so that the appearance of the behavior is never proof that the "cause" has occurred. There are many alternative ways to "explain" an event after it has occurred. Science is based on predictions, not post hoc explanations. A post hoc explanation is useful only if it accurately predicts what will happen in the future. Sci-ence involves the testing of such predictions.

It is exciting to think of animals that understand genetic relationships, calculate costs and benefits of services in different currencies, and keep books on friends. It is exciting to think that they scheme like Machiavellian princes, carefully weighing all of the ramifications of every activity. It is very satisfying to think that evolution has selected for the human traits that we treasure most and that these are "natural" products of evolution. Killjoy explanations are unwanted and scorned. There is no fun left if we demystify animals, and people. How much better it is to describe our noblest virtues as products of natural selection. Morality is treasured. Surely there must be a biological basis for the incest taboo, for scorning criminality, and for the highest forms of human behavior where we set aside self-interest to act in ways that support universal justice and a set of ethical ideals. The problem of course is that when individuals put aside self-interest, selection operates against those individuals and in favor of the less moral. It is the tragedy of the commons. I may aspire to morality but often pessimistically deny its existence in most humans, let alone animals. Maybe some apes are ethical and moral beings. Maybe some do have political skills that exceed my own. Maybe some do have the complex cognitive abilities implied by many explanations for their behavior. The burden of proof is on those hypothesizing the existence of the phenomenon, and I will be so delighted to be proven wrong in my gloomy pessimism.

REFERENCES

Aureli, F. and C. P. van Schaik. 1991. "Post-Conflict Behaviour in Lion-Tailed Macaques (*Macaca fascicularis*): Coping with the Uncertainty." *Ethology* 89: 101–14.

Bernstein, I. S. and C. L. Ehardt. 1986. "The Influence of Kinship and Socialization on Aggressive Behaviour in Rhesus Monkeys (*Macaca mulatta*)." *Animal Behaviour* 34:739–47.

Bernstein, I. S., L. Williams, and M. Ramsey. 1983. "The Expression of Aggression in Old World Monkeys." *International Journal of Primatology* 4:113–25.

de Waal, F. B. M. 1989. "Dominance 'Style' and Primate Social Organization." Pp. 243–63 in *Comparative Socioecology: The Behavioral Ecology of Humans and Other Mammals*, edited by V. Standen and R. A. Foley. Oxford: Blackwell Scientific.

Klopfer, P. H. 1971. "Mother Love: What Turns It On?" *American Scientist* 59:404–7.

Kurland, J. A. 1977. "Kin Selection in the Japanese Monkey." *Contributions to Primatology* 12:1–145.

Preushoft, S. and C. P. van Schaik. 2000. "Dominance and Communication: Conflict Management in Various Social Settings." Pp. 77–105 in *Natural Conflict Resolution*, edited by F. Aureli and F. B. M. de Waal. Berkeley: University of California Press.

Schultz, A. H. 1969. *The Life of Primates.* London: Weidenfeld and Nicolson Natural History.

Silk, J. B. 1997. "The Function of Peaceful Post-Conflict Contacts among Primates." *Primates* 38:265–79.

Thierry, B. 1997. "Adaptation and Self-Organization in Primate Societies." *Diogenes* 45:39–71.

Thierry, B. 2000. "Covariation of Conflict Management Patterns across Macaque Species." Pp. 106–25 in *Natural Conflict Resolution,* edited by F. Aureli and F. B. M. de Waal. Berkeley: University of California Press.

Tinbergen, N. 1951. *The Study of Instinct.* Oxford: Clarendon.

Wilson, E. O. 1975. *Sociobiology: The New Synthesis.* Cambridge, MA: Belknap Press of Harvard University Press.

III

Mechanisms of Sociality

5

Proximate Mechanisms Regulating Sociality and Social Monogamy, in the Context of Evolution

C. Sue Carter and Bruce S. Cushing

The proximate mechanisms of sociality are best understood in the context of their adaptive functions and when possible the ultimate/evolutionary origins of such behaviors. Toward this end, it is useful to make comparisons of the mechanisms regulating specific behavioral traits or patterns in different species, including animals that are highly social, as well as less social species. Recent research, conducted primarily in socially monogamous mammals (capable of high levels of sociality, pair bond formation, and biparental care) has revealed that specific hormones, and especially oxytocin (OT) and arginine vasopressin (AVP), influence general sociality, the capacity to form social bonds, and parental behavior. These neurochemicals and their receptors are regulated by genetics as well as epigenetic factors (such as early social experiences and hormonal changes), producing in some cases lifelong adaptations, which may help to account for individual and species-typical variations in social behavior. Based on these findings, the purpose of this review is to summarize current understanding of the physiological mechanisms underlying mammalian sociality and social monogamy, in the context of evolution.

SOCIAL BEHAVIOR IN AN ADAPTIVE AND EVOLUTIONARY CONTEXT

What Is Social Behavior and Why Is It Important?

Social behaviors by definition involve interactions between two or more individuals. These are sometimes classified as either agonistic (aggressive or defensive) or affiliative (prosocial). The causes of agonistic or aggressive behaviors have been the subject of intense study for many decades, although the origins of aggression as well as sociality still remain open to debate (Fuentes 1999; Bernstein, this volume; Sussman and Garber, this volume). Based on frequency or duration, positive social behaviors may be, under most conditions, more common than agonistic behaviors. However, positive social behaviors have only recently become the focus of physiological analyses (reviewed in Carter 1998; Carter and Keverne 2002).

Positive social behaviors require at a minimum mutual willingness to congregate, remain together, and in some cases engage in selective social behaviors. Social behaviors, including social bonds and other forms of social support, facilitate both the survival of the individual, as well as reproduction—necessary in turn for genetic survival. Reproductive behaviors, such as male mounting or female reproductive postures, have direct consequences for fitness; such behaviors are often stereotypical and conserved even across phyla. However, social behaviors that are not directly involved in sexual behavior or direct care of offspring tend to be more variable, not only across species but also between individuals. Both genetic and epigenetic (postgenomic or experiential) processes, mediated by physiological changes and acting throughout the lifespan, can have immediate and long-lasting consequences for social behaviors. Therefore, while these behaviors may still be adaptive, in an evolutionary sense, they are more flexible and often more difficult to study because of their complexity and potential for individual expression.

For example, the complexity of mammalian and especially primate social groups must be understood in terms of the propensity of individuals to accept or reject social interactions with other members of their species (Wickler and Seibt 1983; Fuentes 1999). Social familiarity or novelty is a particularly important determinant of social organization in many mammalian species. However, the reaction to familiarity or novelty can be situation- and gender-specific and may change across the life cycle, usually in directions that favor successful reproduction.

In humans and most other species, positive social relationships and perceived social support are protective (reviewed in Carter 1998; Singer and Ryff 2001). For example, social bonds can provide a sense of safety, reduce anxiety, and may influence physical and mental health. In addition, selective social behaviors and social bonds are often critical to long-lasting relationships. In mammals, the best studied, enduring relationships (defined by selective social behaviors and in some cases by emotional responses) are between mothers and infants (Hrdy 1999), or between breeding pairs, sometimes termed pair bonds. These bonds may play a major role in the structure of larger groups, including families (Fuentes 1999).

Several recent reviews have examined social bonds and attachment from a biobehavioral perspective, and there is little doubt that selective sociality has neurobiological underpinnings (reviewed in Carter 1998; Insel and Young 2001; Carter and Keverne 2002). Understanding of the neurobiology of selective social behaviors has been slowed by the inherent inter- and intraspecific variation in these behaviors and by the tendency of physiologists to study socially promiscuous laboratory species. In addition, considerable confusion and controversy also surround the concept of monogamy.

Social Monogamy

One common method for categorizing species is based on mating systems, which are characterized by the species-typical number of sexual partners and often focused on male reproductive behavior. Based on this system the most common mating system in mammals is polygamy (many mates) or more specifically polygyny (many female mates). The less common alternative is monogamy (one mate) (Kleiman 1977), while polyandry (many male mates) is rarest (but see Birkhead 2000).

Attempts to identify mating systems were historically based on field observations; under these conditions copulation was only rarely observed. Therefore, animals that were seen living together in pairs were assumed to be mating only with their social partner, and were classified as "monogamous." Subsequent studies, especially those using DNA fingerprinting to confirm paternity, have left little doubt that members of apparently "monogamous" pairs may engage in extrapair copulations (Gowaty 1997). It is now well established that sexual exclusivity is not necessarily a reliable trait of species that live in pairs, although in the absence of other partners or when mate guarding is successful social monogamy can promote sexual exclusivity.

Even in the absence of absolute "sexual" or "genetic" monogamy, there is no doubt that pair bonds, characterized by selective social behaviors, and other forms of stable social groups do exist. These relationships may endure beyond the mating season, and in some species individuals may remain together for a lifetime. Based primarily on life histories and occasional laboratory work, social monogamy in mammals has been described in various taxa, ranging from primates to canids to rodents (reviewed in Kleiman 1977; Dewsbury 1987; Fuentes 1999; Sussman and Garber, this volume). For example, among the diverse species that share the traits of social monogamy are wolves, several New World primates including tamarins and marmosets, titi monkeys, and even a few rodents, including prairie voles.

Table 5.1   Characteristics of Mammalian Social Monogamy[a]

- Capacity to form social bonds, including selective social behaviors and selective aggression, which may serve as mate guarding
- Biparental care; male parental behavior and alloparenting
- Tendency to form extended families
- Incest avoidance
- Physical monomorphism (reduced sexual dimorphism)[b]

---

[a]Sexual exclusivity is not an absolute feature of socially monogamous species, but may occur if mate guarding is successful.
[b]Monomorphism also can occur in species that do not show other traits of social monogamy.

Among species that tend to form pair bonds it also is common to observe a suite of other behavioral and physiological traits (Table 5.1). These traits are *not absolute* and may vary based on the social and neuroendocrine history of the individual (Carter 2003), as well as the genetics of a given species (Young 1999; Insel and Young 2001). It is also important to understand that the tendency to form pair bonds may extend to more than one partner, allowing the development of polygamous or polyandrous groups. In addition, a willingness to accept social partners may extend to a tendency for offspring to remain with the parents, forming extended family groups.

For the purposes of this review the term "social monogamy" will be used to refer to a constellation of social behaviors including the capacity to form long-term pair bonds, biparental care of the offspring, and a tendency toward monomorphism, i.e., reduced sexual dimorphism (Kleiman 1977; Boonstra, Gilbert, and Krebs 1993). Among socially monogamous species aggression toward strangers may be seen in both sexes and probably serves to protect the mate and offspring. In species that are inclined toward living in pairs, intense selective aggression, usually toward unfamiliar conspecifics, may be triggered by sexual experience, especially in males (Winslow et al. 1993) and/or by prolonged periods of cohabitation, especially in females (Bowler, Cushing, and Carter 2002).

In contrast, members of polygamous species are less likely to form selective social bonds and males rarely care for the young. Intrasexual aggression also is seen in polygamous species. However, such aggression is less selective and may be directed toward both familiar and unfamiliar conspecifics. In polygamous species, aggression often occurs before mating, especially during competition for either a territory or mate or may be used to establish social rank.

Descriptions of the proximate mechanisms responsible for different kinds of social groups are most successful in the context of the responses of individual animals. For example, among socially monogamous species, individuals may show selective behaviors, including those necessary for social bonding. The tendency toward selectivity in social responses probably relies on mechanisms (see below) that promote both general sociality as well as specific neurophysiological processes necessary to reinforce or reward selective sociality (Insel 2003).

Animals in polygamous species also may be capable of individual recognition, but may be less inclined to be selective in their social or reproductive behaviors. We can postulate that less selective species lack (or fail to use) the mechanisms for reinforcing individual preferences.

Evolutionary Factors Leading to Social Monogamy

The reptilian ancestors of modern mammals, like most modern reptiles, were presumably polygamous. The evolution of the social system we call "social monogamy" (Table 5.1) may have been driven in large part by the fact that female mammals bear the burden of gestation and lactation, with male involvement in the process being potentially minimal, i.e., sperm. The capacity of females to provide almost all of the needs of the young also means that males can attempt to increase fitness by finding additional mates. Thus, mammalian social monogamy is relatively rare and represents a derived characteristic that has evolved independently in several taxa.

It is typically assumed that social monogamy evolved in situations where the male could provide direct benefits to his offspring (Trivers 1972; Wingfield 1990; Lott 1991; Fuentes 1999), for example in the form of food or protection for the young (see Table 5.2). Thus, a male that remained with and defended his sexual partners and later helped to care for his own young would increase his individual fitness. This theory is supported by the observation that social monogamy is far more common in birds than in mammals. It has been estimated that over 90 percent of bird species live in pairs and show biparental care. Male birds can provide significant direct care to their offspring by incubating eggs and feeding the offspring. Thus, in birds approximately equivalent parental investment is not only possible, it is also common.

Sexual Competition as an Ultimate Factor
Producing Sexual Differentiation

Male-male competition is often assumed to be a major evolutionary force capable of driving the selection of marked sex differences and the appearance of "masculine" traits such as generally larger body size in males and an enlarged phallus, or "weapons of war" such as antlers. Males from polygynous species also tend to have larger testes and high levels of sperm production (reviewed in Harcourt, Purvis, and Liles 1995; Birkhead

Table 5.2   Ultimate Causation: Examples of Hypotheses Regarding the Evolution of Social Monogamy

---

Increasing the survival of the offspring by male or other caretakers. This could occur under a variety of conditions. For example, the presence of a male may prevent infanticide by other males, as suggested in some nonhuman primates (Palombit 1999). The presence of a male, or a second caretaker, also may buffer the offspring during harsh environmental conditions. For example, the presence of a partner may reduce neonatal water loss or provide thermal regulation for the pups (Scribner and Wynne-Edwards 1994; Gubernick et al. 1993).

Control of limited resources, including food or territory. This is postulated to be a cause of monogamy in some lemurs that feed on underrepresented food items (Thalmann 2001).

Mate guarding—By males, mate guarding is presumably intended to ensure paternity of his mate's offspring. This has been hypothesized as a possible selective force in the evolution of human monogamy (Marlowe 2000) and in cervids (Brotherton and Manser 1997).

Female choice, allowing male presence outside the mating season. Females may allow the male to help raise the offspring. This choice may be due to environmental conditions. For example, female meadow voles appear to permit males to enter nests in the northern extreme of the range where the male can provide thermal benefits to the offspring (Storey et al. 1994).

Female coercion of males. Several arguments have been suggested. Females may have a hidden estrus and/or ovulation, which forces the male to remain with the female to ensure that he fathers the offspring (Sillentullberg and Moller 1993). Female may abort litter if male leaves, reducing his fitness (McGuire et al. 1992)

2000). It has been argued that these features may be advantageous in "sperm competition," especially under conditions when a given female may potentially mate with more than one male. The evolutionary mechanisms responsible for sexual dimorphism are often assumed to result over generations from an "arms race" in which larger or more dimorphic males, possibly with a correlated increase in sperm, are more sexually successful and thus have higher levels of reproductive fitness. The validity of this theory is difficult to test in mammals, but does gain support from other phyla, especially insects (Birkhead 2000; Hosken, Garner, and Ward 2001).

The majority of mammalian species exhibit sexual dimorphism in body size and appearance. However, monomorphism (or a reduction in masculine physical traits) does occur, and while it is often associated with socially monogamous species, monomorphism also can be found in some polygamous mammals, especially among primates. For example, about 80 percent of primate species do not show marked sex differences in body size. The selection factors that lead to the expression of both social monogamy and monomorphism in nonmonogamous species presumably originated under conditions of reduced intrasexual competition, or in the face of positive selection factors favoring within-species cooperation, perhaps without the necessity for male-male competition (Clutton-Brock 2002; Sussman and Garber, this volume; Bernstein, this volume).

The evolution of a monogamous strategy from a polygynous ancestor requires that there be at least two major changes. First, there must be a reduction in attributes that are associated with polygyny, especially in males. For example, sex differences in body size and external appearance tend to be less obvious in monogamous species (Kleiman 1977; Wingfield 1990). Second, there must also be an accompanying increase in selective social behaviors, i.e., formation of long-term pair bonds and paternal care of the offspring. These changes are presumably the product of varying selection factors (see examples in Table 5.2), which have led to the evolution of monogamy. However, the main goal of this chapter is to review proximate mechanisms that permit the expression of sociality and specifically the traits that characterize social monogamy.

THE PROXIMATE CAUSES OF SOCIAL MONOGAMY

Nature Is Conservative

Current evidence suggests that the same hormones, including neuropeptides and steroids, regulate social behavior, regardless of the mating strategy or the degree of sociality of a particular species. Monogamy has evolved independently in various mammalian taxa, yet all mammals share the same basic neurological and neurochemical building blocks. Through variations in the nervous system, it is possible for different species, males and females within a species, and even individuals to show unique patterns of social behavior (Young 1999). Regardless of the species, selection must be acting upon the same set of neurochemicals/hormones, as well as the receptors for these chemicals, to create the patterns of neuroendocrine and behavioral activity that are associated with varying degrees of sociality.

A Model System: Voles

To illustrate the proximate mechanisms responsible for the development and expression of the social traits of monogamy we will use here examples from the rodent genus *Microtus* (voles). Based upon both field and laboratory data, several species from this genus (including prairie and pine voles) are considered to be socially monogamous, while others (including meadow and montane voles) are usually classified as polygamous. Comparisons among these closely related voles have provided an especially useful perspective on the behavioral, neuroendocrine, and neuroanatomical basis of social behavior (Dewsbury 1987; Carter, DeVries, and Getz 1995; Insel and Young 2001).

The prairie vole (*Microtus ochrogaster*) is found in grasslands of central North America. Field data, primarily from central Illinois, suggest that in prairie voles the most common family group in nature consists of one adult male and one adult female and their offspring (Getz et al. 1993; McGuire et al. 1993; Getz and Carter 1996). Within family groups in Illinois about three-fourths of the adult pairs remain together until one member dies. In less than 10 percent of cases is there evidence that the male has abandoned his female partner. After the death of one member of a pair fewer than 20 percent of the survivors acquire a new mate. Thus, in nature a male-female pair bond is at the core of their social organization. Based on fieldwork it has been suggested that the social organization and mating system of prairie voles evolved as an adaptation to a low-resource habitat, possibly more similar to Kansas than the resource-abundant habitat in Illinois (Getz and Carter 1996). The fact that environmental factors can influence both the evolution and expression of monogamy provides a clue to the physiological substrates of the social behaviors that define monogamy.

Prairie voles present a unique opportunity to study monogamy. In addition to having been studied extensively in the field, especially by Lowell Getz and his collaborators, prairie voles also are easily bred in the laboratory, permitting controlled manipulation. The ability to rear prairie voles in captivity is critical because much of what we know about social behavior comes from studies of domesticated species, such as laboratory rats and mice. However, unlike prairie voles only a few wild-caught rats will reproduce in the laboratory. Laboratory animals are therefore the product of both natural and artificial selection; only individuals capable of successfully reproducing in a potentially stressful laboratory environment can be studied. Thus, the influences of natural versus artificial selection, especially on social behaviors, may be difficult or impossible to distinguish in domesticated animals.

Within- and Between-Species Variation

The forces of nature are not absolute, instead producing variations along a theme. Within-species variations in voles, even when animals are reared and studied under apparently constant laboratory conditions, provide insights into the role of the environment in the development of patterns of sociality. It is interesting to note that prairie voles captured in modern Kansas (possibly more like their ancestral home) are less likely to exhibit the communal traits of social monogamy (Table 5.1) and are physically more sexually dimorphic than animals from a stock originating in Illinois— even when both populations are reared and studied under comparable laboratory conditions (Carter and Roberts 1997). Kansas prairie voles are capable of forming pair bonds and show high levels of male parental behavior, but do not readily form the extended families seen in field studies in Illinois (McGuire et al. 1993). Field research supports the notion that prairie voles from modern Kansas may live a variant on the polygynous lifestyle, with a single male moving between the nests of several females (Swihart and Slade 1989). Thus, even within a species (for example, *M. ochrogaster*) it is possible to observe differences in social behavior that may be as great, if not greater, than those seen between species.

In another example of variation, individuals of the predominantly polygynous meadow voles (*Microtus pennsylvanicus*), which normally shun pair bonding, may in fact form selective social bonds and show biparental behavior when reared under winterlike photoperiods (Parker, Kinney, Phillips, and Lee 2001a; Parker, Phillips, Kinney, and Lee 2001b). Nonetheless, under comparable conditions polygynous meadow voles are less likely than prairie voles to engage in social contact and to exhibit selective partner preferences (Dewsbury 1987; Carter and Getz 1993; Carter et al. 1995).

Neuroendocrine Substrates for Social Monogamy

Many aspects of sexual differentiation, including the origins of sex differences in both anatomy and behavior, are regulated by gonadal steroid hormones. Although both sexes produce androgens and estrogens, circulating levels of androgen are higher in males, while estrogen levels are higher in females. These steroid hormones in turn interact with another class of hormones, known as "peptide" hormones. Of particular interest in understanding the proximate causes of mammalian social monogamy are two uniquely mammalian peptides, oxytocin (OT) and arginine vasopressin (AVP) (Table 5.3). AVP and OT are closely related compounds, both consisting of nine amino acids, and are produced in many of the same regions of the brain (Gimpl and Fahrenholz 2001).

Both steroids and neuropeptides play a critical role in regulating social behavior. Therefore, these compounds have become central to our understanding of the proximate mechanisms responsible for social monogamy. Specifically, it appears that production of and the receptors for both steroid and peptide hormones have undergone a series of modifications, which in turn allowed the emergence of the traits of social monogamy. Furthermore, there is evidence that the production and actions of these same hormonal systems are subject to long-lasting modification by social or hormonal experiences (Francis, Young, Meaney, and Insel 2002; Pedersen and Boccia 2002; Yamamoto et al. 2002; Carter 2003), providing a neural substrate suitable to the task of allowing social behaviors to vary among different species and different individuals within a species.

Table 5.3   Features of the Neurobiology of Oxytocin (OT) and Arginine Vaso-
     pressin (AVP)[a]

- Ancient origins prior to the separation between vertebrates and invertebrates (van Kesteren et al. 1992)
- Specific 9-amino-acid structures of OT and AVP are novel to mammals (Gimpl and Fahrenholz 2001)
- OT is the most abundant neuropeptide in the hypothalamus as indexed by mRNA (Gautvik et al. 1996)
- Synthesized in largest cells in the CNS (magnocellular neurons) as well as other smaller cells (Gimpl and Fahrenholz 2001)
- Transported by neurosecretion to posterior pituitary, but also released in CNS (Gimpl and Fahrenholz 2001)
- Sibling hormones, OT and AVP, have consequences for each other's functions (Engelmann et al. 1996; Carter 1998)
- OT has only one known receptor, but may bind to one or more of the three AVP receptors (Ostrowski 1998)

[a]See text and Carter (1998) for details and exceptions.

Neuropeptides and Social Bonding

Keeping in mind that the evolution of monogamy involved both a reduction in polygamous traits and an increase in positive social interactions, we first turn our attention to mechanisms involved in regulating sociality. OT and AVP have been implicated in social bonding and other forms of positive social interactions. While both peptides appear to play a role in many aspects of social behavior and social memory in both sexes, there are indications that OT plays a greater role in females and AVP in males. OT, produced in the hypothalamus of the CNS and released into the general circulation, is classically associated with birth and lactation, and more recently with maternal behavior (Pedersen 1997; Pedersen and Boccia 2002). Interestingly OT appears to be the most abundant peptide in the hypothalamus, as indexed by mRNA (Gautvik et al. 1996). In addition, OT is novel in having only one known receptor, which may allow this peptide to have a particularly integrative role in physiology and behavior (Ostrowski 1998).

The involvement of OT in establishing monogamy provides a logical progression from its functions in polygynous species. In mice, OT is critical for the establishment of social recognition and social memory, both of which are prerequisites for the formation of pair bonds. In rats, OT regulates many of the social interactions involved in maternal behavior and mother-infant bonds. OT also is released during genital stimulation (Carter 1992) and may function to facilitate short-term male and female social interactions that must occur if mating is to be successful.

In addition to influencing social recognition and social memory, OT also plays a role in regulating social/physical contact (Witt, Carter, and Walton 1990; Witt 1997). Exogenous OT facilitates social contact, but endogenous OT also is important, since treatment with a selective OT receptor antagonist (OTA) significantly reduced physical contact in prairie voles (Williams, Carter, Harbaugh, and Insel 1994; Cho, DeVries, Williams, and Carter 1999). Social contact is critical for prairie voles, as females require direct prolonged contact with a novel male for estrus induction and to become sexually receptive; in addition, social contact and especially sexual contact facilitates the formation of selective partner preferences (Williams, Catania, and Carter 1992). Perhaps it is not surprising that prairie voles have higher circulating levels of OT than rats. Of particular relevance to monogamy and pair bonding is the fact that exogenous OT facilitates the formation of partner preferences in prairie voles (Williams et al. 1994; Insel and Hulihan 1995; Cho et al. 1999; Cushing and Carter 2000). The ability to form selective preferences is a necessary step in the formation of a true pair bond.

Like OT, AVP is produced in the CNS and also acts on brain tissue, as well as peripheral sites such as blood vessels and the kidneys. In certain areas of the brain AVP is a sexually dimorphic hormone, with males producing higher levels than females. This sex difference is due to the fact that AVP synthesis, especially within the limbic system, is facilitated by androgens; thus the concentrations of AVP usually are higher in males (De Vries and Villalba 1997). AVP is not only more abundant in males, but may be especially important to male social behavior (Winslow et al. 1993; Cushing, Martin, Young, and Carter 2001). The behavioral effects of AVP tend to be associated with alertness, behavioral reactivity, arousal (Engelmann et al. 1996), and in many cases the defense of both the individual and the family (Carter et al. 1995).

While an androgen-dependent peptide like AVP might be assumed to be primarily associated with behaviors that would be more typical of polygynous males, there also are indications that AVP plays an important role in behaviors associated with monogamy. Even monogamous males produce more androgen than the females of their species; within the CNS this androgen may serve to increase AVP or its effects, thus encouraging the postcopulatory mate guarding that is typical of monogamous species.

Although the method of action may vary, in male voles AVP seems to play some of the same roles that OT does in females. AVP is released during sexual behavior, but AVP is not essential for sexual behavior since blocking AVP receptors does not prevent mating (Winslow et al. 1993). In addition, even males that do not mate can still develop selective partner preferences. However, AVP is necessary for the induction of postcopulatory aggression or mate guarding in male prairie voles. There also is experimental evidence implicating AVP in the high sociality toward familiar animals that is typical of monogamous species. Experimentally increasing AVP receptors (V_{1a}) in the limbic system of male mice resulted in a significant increase in affiliative contact, although this social behavior was not selective. Furthermore, in prairie voles, AVP given as an *exogenous* treatment can increase social contact, as well as the preference for a familiar partner and the subsequent onset of aggression. Further evidence that AVP is working on its own receptors comes from the fact that the effects of AVP on social behavior and aggression are blocked by pretreatment with a selective V_{1a} AVP receptor antagonist (Winslow et al. 1993; Cho et al. 1999).

DEVELOPMENTAL EFFECTS OF HORMONES AND THE ONTOGENY OF THE TRAITS OF SOCIAL MONOGAMY

Sexual Differentiation and Gonadal Hormones

As mentioned above, certain "masculine" or sexually dimorphic traits (such as body size, phallic development, or the occurrence of precopulatory aggression) are less obvious in males of socially monogamous species when compared to nonmonogamous species. It is generally assumed that androgens, such as testosterone, are essential for many features of masculine anatomy. However, this research is traditionally done in polygamous species.

In monogamous mammals, external sex differences tend to be less obvious and testes size and sperm counts tend to be lower than in polygamous species (Kleiman 1977; Birkhead 2000). In evolutionary terms, it has been hypothesized that these differences reflect reduced pressure from male-male competition and reduced sperm competition, associated with the development of a monogamous mating strategy (Harcourt et al. 1995). With regard to gonadal hormones, the nature of the physiological changes that yield the morphological or behavioral traits of monogamy is not well described.

In monogamous species, gonadal hormones continue to be necessary for certain aspects of sexual differentiation, such as testicular function and the internal ducts and phallic development necessary for sperm production and transport to the female reproductive tract. However, it is known that many of the physical consequences of androgens occur following enzymatic conversion of testosterone into a metabolite known as dihydrotestosterone (DHT) (Breedlove and Hampson 2002; Carter 2002). The enzyme necessary for this conversion (known as 5-alpha reductase) is a likely target for creating at least some of the reductions in sexually dimorphic traits that are characteristic of monogamous species. (DHT is the same androgen that is blamed for male-pattern baldness, and blocking the 5-alpha reductase enzyme is one treatment for baldness.)

Furthermore, because many of the behavioral and neural effects of androgens occur following the intracellular conversion of testosterone to an estrogen (by the aromatase enzyme), the aromatization necessary for the conversion of testosterone to estrogen could be another source of variance, allowing within- or between-species differences. These enzymes allow testosterone to affect physical and behavioral traits separately. For example, low levels of testosterone and reductase and thus lower levels of DHT might create a more monomorphic body. However, testosterone converted to estrogen and localized estrogen receptors are believed important for male behaviors, influencing such behaviors as copulatory behavior (Wood 1996; Ogawa et al. 1998; Rissman, Wersinger, Fugger, and Foster 1999), exploration of novel individuals (Roberts, Williams, Wang, and Carter 1998; Wersinger and Rissman 2000), partner preferences (Bakker, van Ophemert, and Slob 1996), aggression (Ogawa et al. 1997), scent marking (Vagell and McGinnis 1998), and pup retrieval (Ogawa et al. 1998). According to this model, if sufficient aromatase and thus estrogen were available to the nervous system, even apparently "demasculinized" males might be able to use steroid hormones to regulate male sexual behavior. Although not well described in monogamous species, processes such as these, allowing adequate sexual and social behaviors and also permitting sexual monomorphism, may be independent targets for evolutionary change. Changes in various components of these systems would offer a likely substrate for between- and within-species variation.

Steroid Receptors

Receptors for steroids provide another possible target for differences between monogamous and nonmonogamous species. In spite of the fact that males of monogamous species may appear superficially monomorphic, the brains of females and males may actually be less similar than those of polygamous species. One of the most striking differences is seen in the occurrence of cellular receptors for estrogen (ER). For example, ER distributions are similar (but not identical) in male and female rats, but are visibly different between the sexes in prairie voles; in prairie voles hypothalamic ER levels are much lower in males than females (Hnatczuk et al. 1994; Cushing, Le, and Hoffman 2002). The presence or absence of steroid hormones during development also can influence the subsequent availability of steroid receptors. For example, when males are castrated in early life, they later exhibit a more femalelike pattern of hypothalamic ERs (Cushing et al. 2003). Furthermore, there is within-species variation in the traits of monogamy in prairie voles. Kansas prairie voles display a more ratlike pattern of ER, with Kansas males having significantly more ER than Illinois males. These results suggest that a reduced responsivity to estrogen (via fewer ERs in males) may facilitate or permit the development of monogamy.

Neuropeptides, Steroids, and Sex Differences

The synthesis of AVP, which has been associated with certain aspects of masculine behavior, including territoriality and vigilance, is markedly higher in males than in females; this sex difference is seen in both rats and prairie voles (De Vries and Villalba 1997). The sex difference in AVP may be part of neural systems that permit monogamous males to show necessary sexually dimorphic behaviors, such as mounting or even aggression, while continuing to exhibit high levels of social behavior within the family.

Selective sociosexual behaviors are another feature of monogamous species that remains to be explained. At present there is no evidence for a direct role for steroids in the regulation of pair bonds; for example, in adult prairie voles gonadal steroids do not appear to be either essential or necessary in the formation of partner preferences (DeVries and Carter 1999). In addition, males and females gonadectomized as adults are still capable of forming partner preferences. Exogenous OT significantly facilitates partner preference formation in ovariectomized female prairie voles (Williams et al. 1994; Cho et al. 1999), while AVP continues to facilitate the formation of partner preferences in males castrated as adults (Cushing et al. 2003). Ovarian hormones also are not necessary for the induction of female aggression in this species (Bowler et al. 2002). The interaction between steroids and OT also differs in prairie voles from rats. Unlike rats, with the exception of the accessory olfactory system, increasing levels of estrogen do not up-regulate hypothalamic OT receptors in prairie voles (Witt, Carter, and Insel 1991), and there is evidence that OT in female prairie voles can enhance the effects of estrogen (Cushing and Carter 1999). (The relationship between OT and estrogen in rats is in the opposite direction; i.e., estrogen primes the CNS to respond to OT.)

In prairie voles, androgens facilitate the synthesis of AVP (De Vries and Villalba 1997), yet androgens may inhibit the actions of OT (Gimpl and Fahrenholz 2001). It is also likely that gonadal hormones have indirect effects of aggression. However, in comparison to polygynous species, in monogamous mammals gonadal hormones seem to play a less critical role in the regulation of reproduction. Female prairie voles have significantly lower circulating levels of estrogen than rats during estrus (Cushing, Marhenke, and McClure 1995), but at the same time have higher blood levels of OT than rats (Kramer et al., no date). Data from primates are rare. However, a recent study comparing bonnet macaques, which are highly social, to less gregarious and more aggressive pigtail macques, revealed that the more social bonnet macaques have higher levels of OT in cerebrospinal fluid (Rosenblum et al. 2002). Whether socially monogamous primates show a similar pattern remains to be determined.

Although generalities are difficult (and admittedly based on a very small number of species), the social environment—possibly mediated by neuropeptides—seems to be especially important in monogamous species. In contrast, in polygamous species gonadal steroids are relatively more powerful, helping to tie reproduction in such species to the physical environment and factors such as seasonality. The downplaying of the role of gonadal hormones may be related to the relative reliance of monogamous species on social stimuli, rather than photoperiodic cues, to regulate both reproduction and the formation of social groups (Carter et al. 1995).

Oxytocin

Stimulation of young animals during the neonatal period can influence the subsequent expression of adult behavior and physiology (Levine 2001; Meaney 2001; Pedersen and Boccia 2002). Early handling, possibly mediated by maternal responses to the pups, increases the synthesis of OT in the CNS in prairie voles (Carter et al. 2003) and the production of OT receptors at least in rats (Champagne, Diorio, Sharma, and Meaney 2001; Francis et al. 2002). There also is evidence that the early effects of OT have permanent behavioral consequences. For example, in male prairie voles a single neonatal treatment with OT facilitated the onset of behaviors associated with pair bonding (Bales and Carter 2002, 2003) as well as the number of cells that produced OT (Yamamoto et al. 2002). In contrast, neonatal treatment with an OT antagonist (OTA) decreased paternal behavior (Bales, Pfeifer, and Carter, in press) and also inhibited the expression of AVP (Carter 2003; Yamamoto et al. 2003).

Possible Interactions between Steroids and Neuropeptides

While steroids may not be necessary for the formation of partner preferences and pair bonds in adults, steroids may play a critical behavioral role during development by altering the adult response to neuropeptides. Two recent studies provide strong evidence for a major role of steroids in the expression of AVP-regulated social behaviors. Sexually naive male prairie voles (from Illinois) display high levels of spontaneous parental behavior, so much so that it is difficult to increase the expression of parental behavior in males (Carter and Roberts 1997; Lonstein and De Vries 2000). One of the few ways to disrupt spontaneous male parental behavior is to remove the primary source of steroids early after birth by neonatal castration (Lonstein et al. 2002). Neonatal castration also inhibits the formation of partner preferences by adult males in response to centrally administered AVP (Cushing et al. 2003). The results from these studies suggest that gonadal steroids may play a critical role in establishing the adult response to neuropeptides. For example, in male prairie voles, castration, especially if done during the early postnatal period, interferes in later life with neuropeptide-regulated social behaviors. Both male and female prairie voles produce comparatively low levels of testosterone (males) and estradiol (females) in adulthood, at least when compared to rats (Cushing et al. 1995; Cushing et al. 2003). These findings suggest that the absence of gonadal steroids (produced by castration) may inhibit the formation of social behaviors associated with monogamy in male prairie voles. However, in males of this species some exposure to gonadal steroids (in amounts normally produced by the testes) during a "critical period" in early life is necessary for the subsequent expression of at least some of the characteristics of monogamy (Table 5.1). Comparable neonatal gonadectomy studies have not been conducted in females; however, it is possible that ovarian hormones also may have developmental consequences. The consequences of sex differences in gonadal steroids as well as neuropeptides, especially during development, may contribute to sex differences in the CNS and behaviors that in turn characterize monogamous or non-monogamous species.

Hormones of the Hypothalamic-Pituitary-Adrenal (HPA) Axis

Hormones of the HPA axis may be of particular importance to understanding social behaviors. At least some species of socially monogamous mammals have exceptionally high levels of adrenal hormones (Taymans et al. 1997), and it is possible that the adrenal steroids have behavioral consequences related to the development and expression of the traits of monogamy (Carter and Roberts 1997; Carter 1998; Carter and Keverne 2002).

It has long been recognized that the HPA axis is sensitive to various forms of environmental challenge. Manipulations of hormones of the HPA axis also may regulate social behaviors, both in adulthood and during development (Roberts et al. 1996; Carter and Roberts 1997; Roberts et al. 1997). Adrenal hormones, and specifically adrenal corticoids (CORT), are sensitive to social experiences. Hormones of the HPA axis also can directly modulate pair bond formation (DeVries et al. 1995; DeVries et al. 1996; DeVries et al. 2002). In addition, these hormones may influence the sensitivity of animals to neuropeptides, such as OT (of either endogenous or exogenous origins); for example, in rats either a stressor or treatment with CORT increases OT receptor binding (Liberzon and Young 1997). In prairie voles, basal or nonstressed levels of CORT tended to be very high, and the HPA axis in prairie voles also is very reactive to stress—especially when compared to domestic rats and nonmonogamous voles (Carter et al. 1995; Taymans et al. 1997). Another CNS peptide, corticotropin-releasing hormone (CRH), which stimulates the HPA axis, has a dose-dependent effect on pair bonding (DeVries et al. 2002), and there is evidence in rats that CRH receptors show lifelong modifications as a function of early experiences, including maternal licking and grooming (Meaney 2001). The regulation of the HPA axis throughout life depends in part on the social history of the individual. In addition, social regulators in early life are probably species specific. For example, young prairie voles, unlike rats and nonmonogamous montane voles, react quickly to maternal separation with increased CORT secretion (Shapiro and Insel 1990). HPA axis hormones can both facilitate and inhibit the actions of sex steroids and neuropeptides (Liberzon and Young 1997).

Thus, hormones of the HPA axis have developmental consequences for various social and reproductive behaviors (Levine 2001; Meaney 2001; Pedersen and Boccia 2002), possibly interacting with both gonadal steroids and neuropeptides to modulate the expression of the characteristics of social monogamy as well as sexual behaviors (Carter and Roberts 1997; Carter 1998). The involvement of early experience and the hormones of the HPA axis in the permanent "wiring" of the CNS may be of critical importance to the molding of both species and individual differences in sociality.

SUMMARY

Social behavior is notably flexible. However, species-typical patterns of social behavior can be identified and are sufficiently divergent among species to be considered species-typical traits (see, for example, Sussman and Garber, this volume). For example, when compared to the more common pattern seen in polygamous mammals, animals that have been identified as socially monogamous tend to exhibit a novel set of behavioral traits, including pair bonding and male parental care. Embedded in our knowledge of the physiology of sociality are clues to the proximate causes of species differences in social behavior. There is now strong evidence that the appearance of these behaviors results from species and individual differences in neuroendocrine processes, involving both steroids and neuropeptide hormones.

The characteristics of adult animals are influenced by genetics, but also emerge as a function of developmental experiences and long-lasting reorganization of the CNS. In nonmonogamous species, gonadal steroids, including androgens and estrogens, play a major role in the development of sexual dimorphisms. The genetic mechanisms for synthesizing steroid hormones and their receptors also may provide substrates for evolution, and in turn play a role in the emergence of the novel traits of social monogamy. In addition, neuropeptides, including OT and AVP, and their receptors also regulate both the development (Carter 2003) and expression (Carter 1998) of the traits of social monogamy (or, alternatively, polygamy). Steroids and peptides, during both early life and adulthood, can reorganize and modulate the functions of the nervous system, producing long-lasting changes in the capacity to respond to social stimuli. The resultant patterns of social behavior in turn may be considered characteristics of a species or an individual.

REFERENCES

Bakker, J., J. van Ophemert, and A. K. Slob. 1996. "Sexual Differentiation of Odor and Partner Preference in the Rat." *Physiol. Behav.* 60:489–94.

Bales, K. L. and C. S. Carter. 2002. "Oxytocin Facilitates Parental Care in Female Prairie Voles (But Not in Males)." *Horm. Behav.* 41:456.

Bales, K. L. and C. S. Carter. 2003. "Developmental Exposure to Oxytocin Facilitates Partner Preferences in Male Prairie Voles." *Behav. Neurosci.* 117:854–59.

Bales, K., L. Pfeifer, and C. S. Carter. In press. "Sex Differences and Developmental Effects of Manipulations of Oxytocin on Anxiety and Alloparenting in Prairie Voles." *Dev. Psychobiol.* 44.

Birkhead, T. 2000. *Promiscuity: An Evolutionary History of Sperm Competition*. Cambridge, MA: Harvard University Press.

Boonstra, R., B. S. Gilbert, and C. J. Krebs. 1993. "Mating Systems and Sexual Dimorphism in Mass in Microtines." *J. Mammal.* 74:224–29.

Bowler, C. M., B. S. Cushing, and C. S. Carter. 2002. "Social Factors Regulate Female-Female Aggression and Affiliation in Prairie Voles." *Physiol. Behav.* 76:559–66.

Breedlove, M. and E. Hampson. 2002. "Sexual Differentiation of the Brain and Behavior." Pp. 75–114 in *Behavioral Endocrinology,* edited by J. B. Becker, S. M. Breedlove, D. Crews, and M. M. McCarthey. Cambridge, MA: MIT Press.

Brotherton, P. N. M. and M. B. Manser. 1997. "Female Dispersion and the Evolution of Monogamy in the Dik-Dik." *Anim. Behav.* 54:1413–24.

Carter, C. S. 1992. "Oxytocin and Sexual Behavior." *Neurosci. Biobehav. Rev.* 19:303–13.

Carter, C. S. 1998. "Neuroendocrine Perspectives on Social Attachment and Love." *Psychoneuroendocrinology* 23:779–818.

Carter, C. S. 2002. "Hormonal Influences on Human Sexual Behavior." Pp. 205–22 in *Behavioral Endocrinology,* edited by J. B. Becker, S. M. Breedlove, D. Crews, and M. M. McCarthey. Cambridge, MA: MIT Press.

Carter, C. S. 2003. "The Developmental Consequences of Oxytocin." *Physiol. Behav.* 79:383–97.

Carter, C. S., A. C. DeVries, and L. L. Getz. 1995. "Physiological Substrates of Mammalian Monogamy: The Prairie Vole Model." *Neurosci. Biobehav. Rev.* 19:303–14.

Carter, C. S. and L. L. Getz. 1993. "Monogamy and the Prairie Vole." *Sci. Amer.* 268: 100–6.

Carter, C. S. and E. B. Keverne. 2002. "The Neurobiology of Social Affiliation and Pair Bonding." Pp. 299–337 in *Hormones, Brain, and Behavior,* edited by D. W. Pfaff. San Diego: Academic Press.

Carter, C. S. and R. L. Roberts. 1997. "The Psychobiological Basis of Cooperative Breeding." Pp. 231–66 in *Cooperative Breeding in Mammals,* edited by N. G. Solomon and J. A. French. New York: Cambridge University Press.

Carter, C. S., Y. Yamamoto, K. M. Kramer, K. Bales, G. E. Hoffman, and B. S. Cushing. 2003. Long-Lasting Effects of Early Handling on Hypothalamic Oxytocin-Immunoreactivity and Responses to Separation. *Soc. Neurosci. Abst.* 191.14.

Champagne, F., J. Diorio, S. Sharma, and M. J. Meaney. 2001. "Naturally Occurring Variations in Maternal Behavior in the Rat Are Associated with Differences in Estrogen-Inducible Central Oxytocin Receptors." *Proc. Natl. Acad. Sci. USA* 122:12736–41.

Cho, M. M., A. C. DeVries, J. R. Williams, and C. S. Carter. 1999. "The Effects of Oxytocin and Vasopressin on Partner Preferences in Male and Female Prairie Voles (*Microtus ochrogaster*)." *Behav. Neurosci.* 113:1071–80.

Clutton-Brock, T. 2002. "Breeding Together: Kin Selection and Mutualism in Cooperative Vertebrates." *Science* 296:66–72.

Cushing, B. S. and C. S. Carter. 1999. "Prior Exposure to Oxytocin Mimics Social Contact and Facilitates Sexual Behaviour in Females." *J. Neuroendocrinol.* 11:765–69.

Cushing, B. S. and C. S. Carter. 2000. "Peripheral Pulses of Oxytocin Increase Pair Bonding in Female, But Not Male Prairie Voles." *Horm. Behav.* 37:49–56.

Cushing, B. S., W. W. Le, and G. E. Hoffman. 2002. "Mating Strategy and Estrogen Receptor Alpha: A Comparison of Two Populations of Prairie Voles." *Soc. Neurosci. Abst.* 957.13

Cushing, B. S., S. Marhenke, and P. A. McClure. 1995. "Estradiol Concentration and the Regulation of Locomotor Activity." *Physiol. Behav.* 58:953–57.

Cushing, B. S., J. O. Martin, L. J. Young, and C. S. Carter. 2001. "The Effects of Peptides on Partner Preference Formation Are Predicted by Habitat in Prairie Voles." *Horm. Behav.* 39:48–58.

Cushing, B. S., U. Okorie, and L. J. Young. 2003. "Neonatal Castration Inhibits Adult Male Response to Vasopressin but Does Not Alter Expression of V_{1a} Receptor." *J. Neuroendocrinol.* 15:1021–26.

De Vries, G. and C. Villalba. 1997. "Brain Sexual Dimorphism and Sex Differences in Parental and Other Social Behaviors." *Ann. N.Y. Acad. Sci.* 807:273–86.

DeVries, A. C. and C. S. Carter. 1999. "Sex Differences in Temporal Parameters of Pair Bonding." *Can. J. Zool.* 77:885–89.

DeVries, A. C., M. B. DeVries, S. E. Taymans, and C. S. Carter. 1995. "The Modulation of Pair Bonding by Corticosterone in Female Prairie Voles." *Proc. Natl. Acad. Sci. USA* 92:7744–48.

DeVries, A. C., M. B. DeVries, S. E. Taymans, and C. S. Carter. 1996. "Stress Has Sexually Dimorphic Effects on Pair Bonding in Prairie Voles." *Proc. Natl. Acad. Sci.* 93:11980–84.

DeVries, A. C., T. Guptaa, S. Cardillo, M. Cho, and C. S. Carter. 2002. "Corticotropin-Releasing Factor Induces Social Preferences in Male Prairie Voles." *Psychoneuroendocrinology* 27:705–14.

Dewsbury, D. A. 1987. "The Comparative Psychology of Monogamy." Pp. 1–50 in *Comparative Perspectives in Modern Psychology, Nebraska Symposium on Motivation,* edited by D. W. Leger. Lincoln: University of Nebraska Press.

Engelmann, M., C. T. Wotjak, I. Neumann, M. Ludwig, and R. Landgraf. 1996. "Behavioral Consequences of Intracerebral Vasopressin and Oxytocin: Focus on Learning and Memory." *Neurosci. Biobehav. Rev.* 20:341–58.

Francis, D., L. J. Young, M. J. Meaney, and T. R. Insel. 2002. "Naturally Occurring Differences in Maternal Care Are Associated with the Expression of Oxytocin and Vasopressin (V1a) Receptors: Gender Differences." *J. Neuroendocrinol.* 14:349–53.

Fuentes, A. 1999. "Re-evaluating Primate Monogamy." *Amer. Anthropol.* 100:890–907.

Gautvik, K. M., L. de Lecea, V. T. Gautvik, P. E. Danielson, P. Tranque, A. Dopazo, F. E. Bloom, and J. G. Sutcliffe. 1996. "Overview of the Most Prevalent Hypothalamus-Specific mRNAs, as Identified by Directional Tag PCR Subtractions." *Proc. Natl. Acad. Sci. USA* 93:8733–38.

Getz, L. L. and C. S. Carter. 1996. "Prairie-Vole Partnerships." *Amer. Sci.* 84:56–62.

Getz, L. L., B. McGuire, T. Pizzuto, J. E. Hofmann, and B. Frase. 1993. "Social Organization of the Prairie Vole (*Microtus ochrogaster*)." *J. Mammal.* 74:44–58.

Gimpl, G. and F. Fahrenholz. 2001. "The Oxytocin Receptor System: Structure, Function and Regulation." *Physiol. Rev.* 81:629–83.

Gowaty, P. A. (Ed.). 1997. *Feminism and Evolutionary Biology.* New York: Chapman & Hall.

Gubernick, D. J., S. L. Wright, and R. E. Brown. 1993. "The Significance of Father's Presence for Offspring Survival in the Monogamous California Mouse, *Peromyscus californicus*." *Anim. Behav.* 46:539–46.

Harcourt, A. H., A. Purvis, and L. Liles. 1995. "Sperm Competition: Mating System, Not Breeding Season Affects Testes Size of Primates." *Functional Ecology* 9:468–76.

Hnatczuk, O. C., C. A. Lisciotto, L. L. DonCarlos, C. S. Carter, and J. I. Morrell. 1994. "Estrogen Receptor Immunoreactivity in Specific Brain Areas of the Prairie Vole (*Microtus ochrogaster*) Is Altered by Sexual Receptivity and Genetic Sex." *J. Neuroendocrinol.* 6:89–100.

Hosken, D. J., T. W. Garner, and P. I. Ward. 2001. "Sexual Conflict Selects for Male and Female Reproductive Characters." *Curr. Biol.* 11:489–93.

Hrdy, S. B. 1999. *Mother Nature.* New York: Pantheon.

Insel, T. R. 2003. "Is Social Attachment an Addictive Disorder?" *Physiol. Behav.* 79:351–57.

Insel, T. R. and T. J. Hulihan. 1995. "A Gender-Specific Mechanism for Pair Bonding: Oxytocin and Partner Preference Formation in Monogamous Voles." *Behav. Neurosci.* 109:782–89.

Insel, T. R. and L. J. Young. 2001. "The Neurobiology of Attachment." *Nat. Rev. Neurosci.* 2:129–36.

Kleiman, D. 1977. "Monogamy in Mammals." *Quart. Rev. Biol.* 52:39–69.

Kramer, K. M., B. S. Cushing, C. S. Carter, J. Wu, and M. A. Ottinger. No date. "Chemical and Biological Validation of an Enzyme Immunoassay for Plasma Oxytocin." [in review].

Levine, S. 2001. "Primary Social Relationships Influence the Development of the Hypothalamic-Pituitary-Adrenal Axis in the Rat." *Physiol. Behav.* 73:255–60.

Liberzon, I. and E. A. Young. 1997. "Effects of Stress and Glucocorticoids on CNS Oxytocin Receptor Binding." *Psychoneuroendocrinology* 22:411–22.

Lonstein, J. S. and G. J. De Vries. 2000. "Sex Differences in the Parental Behavior of Rodents." *Neurosci. Biobehav. Rev.* 24:669–86.

Lonstein, J. S., B. D. Rood, and G. J. De Vries. 2002. "Parental Responsiveness Is Feminized after Neonatal Castration in Virgin Male Prairie Voles, but Not Masculinized by Perinatal Testosterone in Virgin Females." *Horm. Behav.* 41:80–87.

Lott, D. F. 1991. *Intraspecific Variation in the Social Systems of Wild Vertebrates.* Cambridge: Cambridge University Press.

Marlowe, F. 2000. "Paternal Investment and the Human Mating System." *Behav. Proc.* 51:45–61.

McGuire, B., L. L. Getz, J. E. Hofmann, T. Pizzuto, and B. Frase. 1993. "Natal Dispersal and Philopatry in Prairie Voles (*Microtus ochrogaster*) in Relation to Population Density, Season, and Natal Social Environment." *Behav. Ecol. Sociobiol.* 32:293–302.

McGuire, B., K. D. Russell, T. Mahoney, and M. Novak. 1992. "The Effects of Mate Removal on Pregnancy Success in Prairie Voles (*Microtus ochrogaster*) and Meadow Voles (*Microtus pennsylvanicus*)." *Biol. Reprod.* 47:37–42.

Meaney, M. J. 2001. "Maternal Care, Gene Expression, and the Transmission of Individual Differences in Stress Reactivity across Generations." *Ann. Rev. Neurosci.* 24:1161–92.

Ogawa, S., V. Eng, J. Taylor, D. B. Lubahn, K. S. Korach, and D. W. Pfaff. 1998. "Roles of Estrogen Receptor Alpha Gene Expression in Reproduction-Related Behaviors in Female Mice." *Endocrinology* 139:5070–81.

Ogawa, S., T. F. Washburn, D. B. Lubahn, K. S. Korach, and D. W. Pfaff. 1997. "Modifications of Testosterone-Dependent Behaviors by Estrogen Receptor Alpha Gene Disruption in Male Mice." *Endocrinology* 139:5058–69.

Ostrowski, N. L. 1998. "Oxytocin Receptor mRNA Expression in Rat Brain: Implications for Behavioral Integration and Reproductive Success." *Psychoneuroendocrinology* 23:989–1004.

Palombit, R. A. 1999. "Infanticide and the Evolution of Pair Bonds in Nonhuman Primates." *Evol. Anthropol.* 7:117–29.

Parker, K. J., L. F. Kinney, K. M. Phillips, and T. M. Lee. 2001a. "Paternal Behavior Is Associated with Central Neurohormone Receptor Binding Patterns in Meadow Voles (*Microtus pennsylvanicus*)." *Behav. Neurosci.* 5:1341–48.

Parker, K. J., K. M. Phillips, L. F. Kinney, and T. M. Lee. 2001b. "Day Length and Sociosexual Cohabitation Alter Central Oxytocin Receptor Binding in Female Meadow Voles (*Microtus pennsylvanicus*)." *Behav. Neurosci.* 5:1349–56.

Pedersen, C. A. 1997. "Oxytocin Control of Maternal Behavior: Regulation by Sex Steroids and Offspring Stimuli." *Ann. N. Y. Acad. Sci.* 807:126–45.

Pedersen, C. A. and M. Boccia. 2002. "Oxytocin Links Mothering Received, Mothering Bestowed and Adult Stress Responses." *Stress* 5:259–67.

Rissman, E. F., Wersinger, S. R., Fugger, H. N., and Foster, T. C. 1999. "Sex with Knockout Models: Behavioral Studies of Estrogen Receptor Alpha." *Brain Res.* 835:80–90.

Roberts, R. L., J. R. Williams, A. K. Wang, and C. S. Carter. 1998. "Cooperative Breeding and Monogamy in Prairie Voles: Influence of the Sire and Geographical Variation." *Anim. Behav.* 55:1131–40.

Roberts, R. L., A. S. Zullo, and C. S. Carter. 1997. "Sexual Differentiation in Prairie Voles: The Effects of Corticosterone and Testosterone." *Physiol. Behav.* 62:1379–83.

Roberts, R. L., A. Zullo, E. A. Gustafson, and C. S. Carter. 1996. "Perinatal Steroid Treatments Alter Alloparental Affiliative Behavior in Prairie Voles." *Horm. Behav.* 30:576–82.

Rosenblum, L. A., E. L. P. Smith, M. Altemus, B. A. Scharf, M. J. Owens, C. B. Nemeroff, J. M. Gorman, and J. D. Coplan. 2002. "Differing Concentrations of Corticotropin-Releasing Factor and Oxytocin in the Cerebrospinal Fluid of Bonnet and Pigtail Macaques." *Psychoneuroendocrinology* 27:651–60.

Scribner, S. J. and K. E. Wynne-Edwards. 1994. "Moderate Water Restriction Differentially Constrains Reproduction in 2 Species of Dwarf Hamster (*Phodopus*)." *Can. J. Zool.* 72:1589–96.

Shapiro, L. E. and T. R. Insel. 1990. "Infant's Response to Social Separation Reflects Adult Differences in Affiliative Behavior: A Comparative Developmental Study in Prairie and Montane Voles." *Devel. Psychobiol.* 23:375–95.

Sillentullberg, B. and A. P. Moller. 1993. "The Relationship between Concealed Ovulation and Mating Systems in Anthropoid Primates: A Phylogenetic Analysis." *Amer. Nat.* 141:1–25.

Singer, B. H. and C. D. Ryff (Eds.). 2001. *New Horizons in Health: An Integrative Approach.* Washington, DC: National Academy Press.

Storey, A. E., C. G. Bradbury, and T. L. Joyce. 1994. "Nest Attendance in Male Meadow Voles: The Role of the Female in Regulating Male Interactions with Pups." *Anim. Behav.* 49:1–10.

Swihart, R. and N. Slade. 1989. "Differences in Home-Range Size between Sexes of *Microtus ochrogaster.*" *J. Mammal.* 70:816–20.

Taymans, S. E., A. C. DeVries, M. B. DeVries, R. J. Nelson, T. C. Friedman, S. Detera-Wadleigh, C. S. Carter, and G. P. Chrousos. 1997. "The Hypothalamic-Pituitary-Adrenal Axis of Prairie Voles (*Microtus ochrogaster*): Evidence for Target Tissue Glucocorticoid Resistance." *Gen. Comp. Endocrinol.* 106:48–61.

Thalmann, U. 2001. "Food Resource Characteristics in Two Nocturnal Lemurs with Different Social Behavior: *Avahi occidentalis* and *Lepilemur edwardsi.*" *Int. J. Primatol.* 22:287–324.

Trivers, R. L. 1972. "Parental Investment and Sexual Selection." Pp. 136–79 in *Sexual Selection and the Descent of Man (1871–1971)*, edited by B. G. Campbell. Chicago: Aldine-Atherton.

Vagell, M. E. and M. Y. McGinnis. 1998. "The Role of Gonadal Steroid Receptors Activation in the Restoration of Sociosexual Behavior in Adult Male Rats." *Horm. Behav.* 33:163–79.

van Kesteren, R. E., A. B. Smit, R. W. Dirkds, N. D. Dewith, W. P. M. Deraerts, and J. Joosse. 1992. "Evolution of the Vasopressin/Oxytocin Superfamily: Characterization of a cDNA Encoding a Vasopressin-Related Precursor, Preproconopressin, from the Mollusc *Lymnaea stagnalis.*" *Proc. Natl. Acad. Sci. USA* 89:4593–97.

Wersinger, S. R. and E. F. Rissman. 2000. "Oestrogen Receptor Alpha Is Essential for Female-Directed Chemo-Investigatory Behavior but Not Required for the Pheromone-Induced Luteinizing Hormone Surge in Male Mice." *J. Neuroendocrinol.* 12:103–10.

Wickler, W. and U. Seibt. 1983. "Monogamy: An Ambiguous Concept." Pp. 33–52 in *Mate Choice,* edited by P. Bateson. Cambridge: Cambridge University Press.

Williams, J. R., C. S. Carter, C. R. Harbaugh, and T. R. Insel. 1994. "Oxytocin Centrally Administered Facilitates Formation of a Partner Preference in Female Prairie Voles (*Microtus ochrogaster*)." *J. Neuroendocrinol.* 6:247–50.

Williams, J. R., K. C. Catania, and C. S. Carter. 1992. "Development of Partner Preferences in Female Prairie Voles (*Microtus ochrogaster*): The Role of Social and Sexual Experience." *Horm. Behav.* 26:339–49.

Wingfield, J. C. 1990. "Interrelationship of Androgens, Aggression and Mating Systems." Pp. 187–205 in *Endocrinology of Birds: Molecular to Behavioral,* edited by M. Wada, S. Ishii, and C. G. Scanes. Berlin: Springer-Verlag.

Winslow, J. T., N. Hastings, C. S. Carter, C. R. Harbaugh, and T. R. Insel. 1993. "A Role for Central Vasopressin in Pair Bonding in Monogamous Prairie Voles." *Nature* 365:545–48.

Witt, D. M. 1997. "Mechanisms of Oxytocin-Mediated Sociosexual Behavior." *Ann. N. Y. Acad. Sci.* 807:287–301.

Witt, D. M., C. S. Carter, and T. R. Insel. 1991. "Oxytocin Receptor Binding in Female Prairie Voles: Endogenous and Exogenous Oestradiol Stimulation." *J. Neuroendocrinol.* 3:155–61.

Witt, D. M., C. S. Carter, and D. Walton. 1990. "Central and Peripheral Effects of Oxytocin Administration in Prairie Voles (*Microtus ochrogaster*)." *Pharm. Biochem. Behav.* 37:63–69.

Wood, R. I. 1996. "Estradiol, But Not Dihydrotestosterone, in the Medial Amygdala Facilitates Male Hamster Sex Behavior." *Physiol. Behav.* 59:833–41.

Yamamoto, Y., B. S. Cushing, G. E. Hoffman, P. D. Epperson, K. M. Kramer, and C. S. Carter. 2002. "Neonatal Manipulations of Oxytocin Produce Lasting Effects on Oxytocin Immunoreactivity in the Prairie Vole PVN." *Soc. Neurosci. Abst.* 878.7.

Yamamoto, Y., B. S. Cushing, K. M. Kramer, P. D. Epperson, G. E. Hoffman, and C. S. Carter. 2003. "Neonatal Manipulations of Oxytocin Alter Expression of Oxytocin and Vasopressin Immunoreactive Cells in the PVN in a Gender-Specific Manner." *Soc. Neurosci. Abst.* 191.16.

Young, L. J. 1999. "Oxytocin and Vasopressin Receptors and Species-Typical Social Behaviors." *Horm. Behav.* 36:212–21.

6

Signals, Symbols, and Human Cooperation

T. K. Ahn, Marco A. Janssen, and Elinor Ostrom

Human sociality differs from that of other mammals in that only humans have generated societies whose complexity approaches and eventually surpasses that of social insects and colonial invertebrates (Wilson [1975] 2000). Within complex human societies individuals engage in a wide diversity of cooperative actions leading to joint outcomes. Many have studied how this level of cooperation has emerged in an evolutionary process based on competition.

Various contributions in this book argue that the direct benefits of cooperation may be sufficient to maintain cooperative relationships (see also Clutton-Brock 2002). Direct benefits of cooperation, however, while relevant in explaining cooperative hunting or cooperative breeding, are less powerful in explaining cooperation in complex societies that have evolved during the last several millennia. In many instances, cooperation produces indirect benefits over time rather than immediate returns essential for physical survival.

In fact, we can distinguish two classes of cooperation: (1) where a temptation to defect exists because the individual contributes more than it gains from its own contribution, and (2) where no temptation to defect exists. The second class of cooperation exists in processes where the sheer number of organisms acting together generates fitness advantages. In the terminology of modern economics, the second type of cooperation problem that individual organisms face is called a *coordination* problem. The need for coordination to take advantage of group size is an important source for the evolution of sociality among humans and animals. In a coordination process everyone benefits more from cooperating with others than they contribute.

While the second type of cooperation is largely explained by the principle of evolutionary adaptation between individuals and their evolutionary environment, the evolution of the first kind of cooperation is harder to explain. The Darwinian evolutionary principle assumes that organisms that do not maximize fitness will be weeded out by the force of evolution. At least on the surface, however, the kind of behavior that does not seem to maximize individual fitness is exactly the behavior that is needed for cooperation of the first kind to exist. The core question is why would an individual who contributes more to others than it receives survive in a competitive process? Even though the individual is better off when in a group of cooperating individuals than when in a group of noncooperating individuals, the individual maximizes short-term returns when others cooperate and the individual defects.

This chapter concerns the first type of cooperation among humans in situations in which the temptation to defect exists. In particular, we address how effective signals and symbols evolve to facilitate cooperation. One of the main puzzles in human societies is why costly cooperation is frequent among genetically unrelated people, in nonrepeated interactions, and in the contexts in which gains from reputation are small or absent (Fehr and Gächter 2002). We argue that the ability of humans to use signals and craft symbolic systems facilitates cooperation in nonrepeated interactions and stimulates the development of complex social organizations. This symbolic capability of humans is the key that differentiates them from nonhuman animals. Over time, the use of artificial symbols to establish, to convey, and to detect reputation has brought forth the possibility of human cooperation on unprecedented scales.

Throughout the rest of this chapter, we focus on cooperation of the first type—on behavior that, at least in the short term, does not appear to be fitness maximizing or incentive-compatible. Cooperation can be viewed as a subcategory of altruism (as it is typically understood among evolutionary biologists as a kind of behavior). We use cooperation in a multilateral context. That is, in a two-organism interaction, for example, both participants should have the opportunity to confer payoff (fitness or welfare) to each other. The "prisoner's dilemma" is the most famous example of the situation in which cooperation of this type may or may not exist. In a prisoner's dilemma, two players can cooperate or defect. The payoff matrix of their choices provides the individual highest payoff when a player defects while the opponent cooperates. Selfish rational players will therefore defect, while both will be better off when they both cooperate.

Among animals, cooperation among nonkin has been mainly understood in terms of Trivers's (1971) theory of reciprocal altruism. Reciprocal altruism requires repeated interactions among individuals with the capability of recognizing each other's genetic programming or past behavior. Examples of cooperation based on type-detection and memory of past behavior are observed among fish, vampire bats, and chimpanzees (de Waal 1989, 1997; Kurzban 2003). Evolutionary game models of repeated prisoner's dilemma (Axelrod 1981; Axelrod and Hamilton 1981) provide formal theories that support the evolution of cooperation generated by organisms using tit-for-tat-type strategies. Later studies by Boyd and Lorberbaum (1987), Lorberbaum (1994), and Bendor and Swistak (1997) show that tit for tat, or any other pure or mixed strategy, is not by itself ultimately stable. That is, there are always possible combinations of strategies that can invade a population composed of a single strategy. However, the researchers also note that, in terms of relative stability, tit-for-tat style strategies have the best chance of evolving.

Among animals, it should be noted, the evidence of the supposed reciprocal cooperation is not as strong as the theories suggest (Stephens, McLinn, and Stevens 2002). The main reason seems to be the high discount rates of animals for whom surviving today is frequently what counts the most. In an ingenious repeated prisoner's dilemma experiment using birds as experimental subjects, Stephens et al. (2002) find that only when a low discount rate is artificially induced do blue jays respond in a manner consistent with the tit-for-tat strategy. In sum, cooperation among unrelated animals is rare even in repeated situations when substantial fitness benefits from defection exist.

Humans, on the other hand, show remarkably different cooperative patterns. Cooperation based on the reciprocity principle in repeated situations is ubiquitous, although not universal. Even when a prisoner's dilemma is repeated in an experimental laboratory for a clearly preannounced number of rounds, human subjects sustain cooperation until near the end of the repeated game (for example, Selten and Stoeker 1986; Isaac and Walker 1988; Andreoni and Miller 1993; Schmidt et al. 2001). While these experiments show more cooperation than theories based on rational and egoistic individuals would predict, there is also evidence that infinite repetition does not necessarily guarantee universal cooperation.

One possible explanation for more than predicted levels of cooperation in finitely repeated prisoner's dilemmas but less than full, though still substantial, cooperation in infinitely repeated prisoner's dilemmas is Frank's (1988) account of the role of emotions. Frank argues that a substantial proportion of humans are emotionally committed to reciprocal cooperation. That is, they feel good when mutual cooperation is achieved and feel bad when defecting on cooperative partners or when others defect on them. This hypothesis is recently supported by neuroscientific research. Rilling et al. (2002) performed iterated prisoner's dilemma games, while one of the subjects was connected to an MRI machine. They found that mutual cooperation was associated with consistent activation in brain areas that have been linked with reward processes. On the other hand, humans, similar to some extent to the blue jays in Stephen et al.'s experiments, do not necessarily have the level of prudence required to resist temptation to defect even when defection is not a rational payoff-maximizing behavior. Therefore, the distinction between finite repetition and infinite repetition may be less useful in practice than it is viewed in economic theories. The preference for fair outcomes supported by emotions provides a consistent explanation for cooperation in single-shot or finitely repeated prisoner's dilemma as well as the less-than-full cooperation in potentially long-term relationships.

Another characteristic of human cooperation that differs from animal cooperation is its scope, complexity, and flexibility. While eusocial insects also show fairly complicated social structures, their cooperation is mainly limited to genetically related individuals and to predictable patterns. Kinship still defines an important pattern of cooperation in humans. Human cooperation in modern times is, however, in spite of apparent similarities, qualitatively different from cooperation based on kinship.

First of all, at the individual level, the *scope* of partners with whom an individual cooperates expands far beyond kin or any genetically defined boundaries. Humans cooperate with strangers and build long-term relationships from scratch with nonkin. Even more distinctive, many humans cooperate with strangers with whom there is not much prospect for building lasting relationships. For example, during holidays or business trips, people give the waitress a tip in a restaurant to which they will never return. There is evidence that group-level differences in economic organization and the structure of social interactions explain a substantial portion of the behavioral variation across societies. In a study of fifteen small-scale societies, economists and anthropologists found that the higher the degree of market integration and the higher the payoffs to cooperation in everyday life, the greater the level of prosociality expressed in experimental games (Henrich et al. 2001).

Large *scale* is another uniquely human characteristic of cooperation, except for in kin-based insect colonies. While among nonhuman primates, cooperation is limited to dyadic relationships or small groups, humans, via complex social organizations, have achieved scales of cooperation reaching thousands and even millions of individuals. How do humans achieve cooperation of such scope and scale? That is the central question of this chapter. We answer the question by examining the roles of signals and symbols in human cooperation. The scope and scale of human cooperation have been made possible by the fact that a large proportion of, though not all, humans do not maximize fitness (biology) or material wealth (economics). In the presence of a substantial number of other humans, who are fitness (wealth) maximizers, how these nonmaximizers survive the evolutionary processes is a major puzzle.

Our answer to this puzzle derives from two sources: signals and symbols. For humans, the evolved biological signals are grounded on neurological processes and, thus, are not easy to fake. Further, the ability of humans to detect those signals and to behave contingently based on the detected signals is the fundamental biological mechanism that supports human cooperation at a remarkable scope and scale. In addition, cultural and social development creates secondary signals, some of which are properly called symbols.

For those sharing a culture, cultural symbols frequently define what constitutes cooperation when there are uncertainties regarding the meaning of actions and inform people how to coordinate when there are many different ways of cooperating. Reputational symbols help individuals to detect one another and, thus, potential behavior during nonfrequent encounters in situations where biological signals are not available or are unreliable.

SIGNALS

Signals are a means of communication. Wilson defines biological communication as "the action on the part of one organism (or cell) that alters the probability pattern of behavior in another organism" ([1975] 2000:176). In this definition of communication, signals are the bits of information that emanate, voluntarily or involuntarily, from the sender and reach the receiver in ways that affect the receiver's behavior. The role of biological signals in the evolution of human cooperation is discussed intensively by Frank (1988), who draws on Darwin ([1872] 1873) and modern research on facial expression.

Imagine an individual in a one-shot *sequential* prisoner's dilemma situation. Game-theoretic analysis of the situation based on an assumption that individuals are all selfish maximizers predicts that any individual will defect. Regardless of whether one is a first or second mover, the second obtains a larger payoff by defecting. The individual, knowing this, is predicted to defect.

The available evidence does not strongly support this prediction. For example, in a set of three single-shot experiments using the same experimental protocol in three countries, Japan, Korea, and the United States (see Ahn et al., 2003), between one-half and two-thirds of second movers cooperate in single-shot sequential prisoner's dilemma when the first movers cooperate. On the other hand, when first movers defect, the rates of cooperation are 0 percent among the Korean and U.S. second movers and only 12 percent among the Japanese.

Suppose that there are two types of players: egoists and reciprocators. While egoists are the selfish maximizers whose motivation and behavior fit the standard game-theoretic assumptions, the reciprocators are intrinsically motivated to reciprocate cooperation with cooperation. The presence of reciprocators changes the game-theoretic analysis. Again using backward induction, it can be shown that a first mover is better off by choosing cooperation when the second mover is a known reciprocator.

The problem is, however, how does the first mover know that the second mover is a reciprocator? It would be unreasonable from the evidence available from sequential prisoner's dilemma games to assume that everyone is a reciprocator. If the two individuals can talk to each other face to face, the second mover will probably try to convince the first mover that he or she would reciprocate if and only if the first mover cooperates. Can the first mover trust the second mover's promises? Economists use the term "cheaptalk" to describe a first mover's promise and advise the second mover not to trust what is sent. In other words, talking to or looking at others is not viewed as a solution to a social dilemma.

What if the physical symptoms accompanying the promise send reliable signals of the true intentions of those who make the promises? Biological signals, such as facial expressions, body language, eye movement, and tone of voice, can be reasonably reliable under certain conditions. At a fundamental level, the signals should be related to the true intention of the signal sender and not under his voluntary (willful) control. Other neurologically based spontaneous signals can serve the purposes as well.

Frank (1988) and, more recently, Schmidt and Cohn (2001) provide the details of such mechanisms. Take facial expression as an example. Several categories of facial muscles are not subject to perfect, conscious control. They do respond to emotions and corresponding neurological processes. Suppose a smile is expected to accompany promises of reciprocation. Those who promise to reciprocate without actually intending to do so may generate smiles, but only by conscious efforts. The muscles used in a conscious effort to generate a smile are different from those muscles that are at work to create genuine smiles. A perceptibly different kind of smile is likely to be produced when an individual is self-consciously trying to smile. Or, if the sender of a smiling signal knows that "artificial smiles" are not good signals and thus is aware that the signal receiver is likely to detect the difference, those who are not intending to reciprocate may not even try to send such artificial signals. If this is the case, a spontaneous smile as a cooperative signal can be quite an effective means of achieving cooperation.

The most observable difference between genuine and artificial smiles is symmetry [see Brown and Moore (2002) for an extensive review of the literature on true and false smiles], but the signal receivers' responses to genuine and artificial smiles are not under complete control of rational thinking. Brown and Moore (2000, 2002) conducted experiments in which symmetric and asymmetric smile icons were shown to experimental subjects. They found that subjects allocate more of their endowed resources to partners to whom symmetric smile icons were associated. The evidence is not, however, definitive. Eckel and Wilson (2003) report a trust game experiment in which subjects were shown an image of their counterparts projected on a screen. Four different images were shown, including either smiling or neutral faces of male or female models. Eckel and Wilson report that even though the subjects revealed in surveys that they trusted smiling faces more than neutral faces, these responses did not correlate with the actual trusting behavior of the subjects.

The theory of mind advocates (Baron-Cohen 1995; Byrne 1995; O'Connell 1998) present a similar account of the ways in which intentions of one person can be revealed to another person. The ability to reason about others' ways of thinking, their intentions, and thus their likely behavior is a distinguishing characteristic of human cognition. A lack of the capacity for mind reading is the key symptom of autism that makes social life almost impossible. Among primates, chimpanzees show some level of mind reading, compared to monkeys' simple behavior reading (Cheney and Seyfarth 1990). A substantial difference exists, however, between chimpanzees' level of mind reading intentionality and that of ordinary adult humans. Using Cheney and Seyfarth's scale, chimpanzees rate at maximum a 1.5 level while humans rate at level 4 (Schmidt and Cohn 2001:188, Box 2). The theory of mind proposes, in addition to intentionality, eye movements and shared attention as key mind-reading mechanisms.

The existence of individuals who are internally motivated to reciprocate and the potential capacity to detect others' types from physical signals jointly explain the most consistent finding of experimental research on social dilemmas—that communication enhances cooperation (Ostrom and Walker 1991; Sally 1995). The problem that conditional cooperators face in social dilemmas is the uncertainty regarding whether or not a sufficient number of individuals would also cooperate. This is more than a problem of belief about others' motivations. A conditional cooperator must also be confident, in addition to believing that there are many conditional cooperators in a particular situation, that other conditional cooperators also believe that there are many conditional cooperators, etc.

When people talk to each other and their intrinsic motivations and intentions are reliably revealed to one another, the problem of gaining common knowledge regarding the proportion of reciprocators present can be reduced but not completely eliminated. By making commitments to cooperate, seeing that others also make such commitments, and observing that many of those who make such commitments appear to be trustworthy, a conditional cooperator can be convinced to do his or her share in a collective endeavor. Reliable signaling serves to facilitate cooperation, along with other cooperation-enhancing functions of communication, such as sharing of information about proper ways to cooperate and developing group identity.

Frank, Gilovich, and Regan (1993) report a single-shot prisoner's dilemma experiment where subjects were asked to predict their partners' behavior. A group of three subjects were given thirty minutes to talk to each other. The topic of the talk was not imposed, so the subjects could chat about anything they chose. In all of the taped discussions, everyone made promises that he or she would cooperate in the forthcoming game. After the subjects finished their thirty-minute face-to-face talks, each of them was led to a separate room and asked to predict the likely choices of the other two individuals in a prisoner's dilemma game. Frank reports that 73 of the 97 subjects (68 percent) who were predicted to cooperate actually cooperated in the game. Further, 15 of 25 subjects (60 percent) who were predicted to defect actually defected. While this capability is obviously not perfect, it is better than chance.

Kikuchi, Watanabe, and Yamagishi (1997) classify subjects in a one-shot prisoner's dilemma into high-, medium-, and low-trusters after administering a preexperimental survey. They test a hypothesis that high-trusters maintain higher levels of social intelligence and, thus, can more accurately predict the behavior of their partners. In their experiments, groups of six subjects participated in a thirty-minute discussion about garbage collection before they made decisions in single-shot prisoner's dilemma games. After the subjects made decisions, they were informed of the identities of their partners and asked to predict the partners' decisions. Kikuchi et al. report that the high-trusters predicted 12 of 16 (75 percent) cooperators and 10 of 16 (62 percent) defectors accurately. The accuracy of prediction among the medium- and low-trusters was significantly lower.

Scharlemann, Eckel, Kacelnik, and Wilson (2001) performed a laboratory experiment consisting of a simple two-person, one-shot sequential trust game with monetary payoffs. Each person is shown a photo of his/her partner prior to the game. The photos were chosen from a collection that included those smiling and those not smiling. They find that smiles can increase the level of trust between strangers significantly, although other facial expressions are also likely to contribute to the cooperative outcomes.

Mealey, Daood, and Krage (1996) find that human subjects have an enhanced memory of faces of cheaters. Black-and-white reproductions of photos of faces of Caucasian males were presented to the subjects together with a fictional descriptive sentence giving information about the depicted individual's status (high or low) and character (related to trustworthiness). One week later they were shown a larger set of photos without descriptions and the subjects tended to recognize nontrustworthy agents more frequently. Oda (1997) conducted a similar study where photos represented partners in one-shot prisoner's dilemma games. One week after the experiment, the subjects were biased to remember those faces that had been portrayed as defectors in the game. DeBruine (2002) performed sequential trust games where the subjects were shown faces of playing partners manipulated to resemble either themselves or an unknown person. Resemblance to the subject's own face raised the expressed trust as a first player in the partners, but had no effect on being a trustworthy or reciprocating second player.

Cosmides (1989) identified biased cognitive processes for identifying cheaters among a cooperating group. Cognitive neuroscience shows that social information is distinct from the processing of other kinds of information, and Stone et al. (2002) describe neurological evidence indicating that social exchange reasoning can be selectively impaired while reasoning about other domains is left intact.

Another recent finding is that the amygdala, which lies within the cerebrum of the brain, is required for accurate social judgment of the facial appearance of others. Individuals with complete bilateral amygdala damage were not able to judge unfamiliar individuals by visual cues, although this did not hold for verbal descriptions about unknown others (Adolphs, Tranel, and Damasio 1998). Winston, Strange, O'Doherty, and Dolan (2002) found a neural basis for trustworthiness judgments using event-related functional magnetic resonance imaging. The neural activities used in trustworthiness judgment may relate to structures that process emotions, although it is not known what cues of facial expressions are important in the process of making trustworthiness judgments (Adolphs 2002).

The existence of reliable signals and their roles in the evolution of intrinsic motives can and have been formalized by economists, generating models of cooperation that are significantly different from Axelrod's earlier evolutionary models. The latter rely on the assumption that the prisoner's dilemma is infinitely repeated. As the title of Axelrod's paper—"The Emergence of Cooperation among *Egoists*" (italics added)—indicates, cooperation does not necessarily require that individuals are motivated by intrinsic preferences. One of the implications of the human ability to detect others' types is that among egoists cooperation cannot really evolve under one-shot or finitely repeated dilemma settings.

An indirect evolutionary approach, foretold by Frank (1987) and fully developed by Werner Güth and his colleagues (Güth and Yaari 1992; Güth 1995; Güth and Kliemt 1998; Güth, Kliemt, and Peleg 2000), provides ways to examine the evolutionary consequences in the presence of the players' ability to detect others' types with more than random accuracy. Güth and Yaari, for example, show that, in the context of a simple sequential trust game, trustworthy types can evolve to be a significant proportion of a population when players can detect others' types with more than random accuracy. Ahn (2002) extends the model to a one-shot sequential prisoner's dilemma setting with three preference types. Populations with egoists and conditional cooperators can be stable under a reasonable parameter range of information and the temptation to defect. Janssen (no date) shows that when simulated agents have the ability to learn to estimate the trustworthiness of others, cooperation with strangers in one-shot games can emerge.

The indirect evolutionary approach is widely utilized by social scientists to provide the logic of viability of the preferences that are different from the rational, selfish preference typically assumed in standard economic modeling. In addition to the evolution of the intrinsic motivation for conditional cooperation, the indirect evolutionary approach can also model costly punishment, observed in a wide range of experiments (Ostrom, Walker, and Gardner 1992; Ostrom, Gardner, and Walker 1994; Fehr and Gächter 2002).

SYMBOLS

While signals have immediate, biological attachment to their senders, symbols are secondary, abstract signals constructed socioculturally. The word *symbol* is derived from the Greek word *symbolon.* In ancient Greece it was a custom to break a slate of burned clay into several pieces and distribute them within a group. When the group reunited the pieces were fitted together (Greek *symbollein*). This confirmed that the individuals were members belonging to the group. Two kinds of symbols are important in facilitating human cooperation: cultural and reputational symbols. We will focus on analyzing the construction of reputational symbols.

Symbols in our usage constitute a subset of signals that are broadly understood in a population; we may call them symbolic signals. Others have used "signs" (Bacharach and Gambetta 2001) or signals (Feldman and March 1981) for the same purpose. Some socioculturally constructed symbols closely parallel biological signals in that they are immediately observable characteristics of their carriers. They differ from biological signals in that their meanings are socioculturally constructed: tattoos and ties are immediately observable characteristics but they represent different meanings among Hell's Angels and businessmen.

Effective symbolic signals share the characteristics of the effective biological signals. That is, they are hard to fake. Bacharach and Gambetta (2001:173–74) discuss an incident that one of them experienced at an Oxford college. A group of youngsters outside the college building claimed that they were to have a seminar in the building, but were locked out. Since the college building hosted many valuable paintings and furniture, the author had to be careful about whether the youngsters' claim was trustworthy. Bacharach and Gambetta proudly report that it took only a split second for the author to assess the trustworthiness of the youngsters' claim using their manifest symbols. The likely symbols—glasses, books, clothes, etc., that Oxford graduate students usually carry—are familiar to an Oxford professor who knew that it would be very costly for an intending group of robbers to coordinate so that all of them manifested such signals. Those signals were effective in that specific sociocultural context in which the senders and receivers of the signals shared a common understanding of the meaning of those signals. The symbols that the professor detected from the group of young people would not have been effective in front of an entrance to a seventeenth-century Chinese imperial library.

Reputational symbols are often artificially devised summary information about past behavior and/or other qualifications of a person or a group. Reputational symbols are essentially information-sharing devices, a solution to the collective action problem that a set of potential transaction partners of an individual or a group faces. Viewed as an information-sharing mechanism, reputational symbols represent highly abstract, systemized gossip.

Public documents, such as, for example, the information that merchants of the Middle Ages could obtain from the law merchant (Milgrom, North, and Wengast 1990), are a bridge between gossip and reputational symbols. Various types of public and private documentation that are available to the general public upon request or to a qualified set of individuals, such as visas, drivers' licenses, and credit cards, are repositories of information with varying levels of abstraction on the past behavior of certain actors.

Reputational symbols are the most abstract and artifactually condensed information. Let us illustrate the meaning and roles of reputational symbols, using the eBay feedback system as an example. The Internet creates a new kind of marketplace with greatly improved information capability that can overcome the limits of conventional markets. The problem of trust among buyers and sellers, however, has become a key obstacle in expanding the scope of online transactions. How can I, as a buyer, be sure that my credit card and other personal information will not be misused? Or, will the seller send me the merchandise at all?

eBay maintains a "feedback profile" that it describes as an "official reputation" for each of its users. The immediate form of the profile is a username/score pair, for example, in the form of "John (125)." This deceptively simple reputation symbol, when reflected upon, reveals how far humans have come to devise systems of mutually beneficial cooperation, which at the same time protect the system itself and the trustworthy users of it from the potential invasion of untrustworthy exploiters.

First of all, the username, as any name would, assigns symbols to individuals so that each can be identified to any number of others. Animals above a certain evolutionary stage are known to identify others as individuals and store memories specific to each of them. This can be done without language or any cultural devices. Humans, even without taking into account sociocultural mechanisms, have superb capabilities of recognizing the individuality of others through stored memories of appearances and voices. Naming is a step forward. Having each individual named provides further means to share the stored individual-specific information with others who have not had firsthand experience with the named person.

Scores in parentheses in the eBay feedback profile are an even higher-level symbolic representation of the reputational information. They utilize the number system, which by itself is a very modern achievement in the long history of human cultural evolution. The eBay scores are constructed by aggregating comments from the transaction partners of an individual. Comments are coded so that a positive feedback is counted as +1, a negative comment as –1, and a neutral comment as 0. Thus, John (125) is a symbolic condensation of the information that the person using John as the username has received 125 more positive comments than negative ones from the individuals with whom he has done transactions.

There are potential shortcomings of any specific reputation symbols. For example, Malaga (2001) argues that eBay's reputation management is problematic in that it aggregates all positive and negative feedback, leading to overly aggregated information. There is a barrier to entry for new users as many avoid those sellers with a low reputation score. There is also a potential problem in the accuracy of one's reputation since only half of the participants provide feedback on reputation (Resnick and Zeckhauser 2002). Despite these potential problems, the eBay reputation system works very well in practice. Resnick et al. (2002) show that a high reputation of a seller leads to about 8 percent higher prices in a controlled experiment with high- and low-reputation sellers selling the same products.

While eBay-type reputational symbols can be rather simple to construct, other reputational symbols evolve over time by trial and error. Certificates and licenses also serve as reputational symbols that signify, in addition to trustworthiness, the qualification for certain performances. When established and trusted, the certificates and licenses expand the possibilities of mutually beneficial transactions. In the early years of the automobile industry in the United States, the considerable ambiguity due to the lack of standards was an obstacle for the industry's further development. Far more manufacturers existed than there are today, and each firm produced by its own standard. The production quality differed, but consumers could not assess quality before they made the purchases and for some time thereafter. Rao (1994), in his study of the earlier times of the American automobile industry, reports that a series of racing contests organized by national and local newspapers served as a reputation-establishment mechanism as a result of the firms' performance in those contests. Manufacturers' recorded performances, certified by the newspapers as quasi-public institutions, were symbols of the performance of their products and affected the purchase choices of consumers. Eventually, the low performing manufacturer exited from the automobile market and high performers remained in and developed more reliable standards among them.

The presence of socioculturally constructed reputational symbols, and systems of reputation in general, dramatically alter the behavioral incentives that social actors face in cooperative endeavors. Economists have traditionally assumed that all individuals are engaged in the rational pursuit of their own self-interest and have studied how various reputational mechanisms affect the behavior of such rational actors (Rubinstein 1979; Kreps, Milgrom, Roberts, and Wilson 1982; Fudenberg and Maskin 1986; Weigelt and Camerer 1988; Kandori 1992; Tirole 1996).

Reliable reputational symbols, as economists have shown in various ways, induce cooperation by self-interested individuals. One of the implications of these studies is that with reasonably well-functioning reputational symbols, the *behavioral* difference between intrinsically motivated and extrinsically motivated types may disappear: with reliable symbols, self-interested individuals will cooperate out of their own selfish concerns, i.e., to reap the long-term benefit of sustained cooperation. Reliable symbols further assure the intrinsically motivated conditional cooperators that a large proportion of others will also cooperate. In terms of behavior and relative payoffs, therefore, the logical conclusion is that different types of players behave in a similar manner and obtain the same payoffs in the presence of a reliable reputation system.

If we consider the possibility that many humans do not possess the level of prudence needed to behave rationally, a great deal of which involves resisting immediate gratification to ensure future gains, it is not too unreasonable to hypothesize that the reputation systems do have different effects on different types. That is, while the intrinsically motivated cooperators may find it easier to resist the temptation to defect, the level of self-restraint that self-interested individuals need may be significantly higher.

Devising and maintaining reliable systems of reputational symbols involve collective action problems among many individuals. The eBay reputation profile depends on the users' taking time to provide accurate inputs on their transaction partners' trustworthiness. Symbols are in that sense a public good, like all language, which once provided benefits to everyone whether or not they contributed their time and resources for the establishment of the reputation symbols.

Symbols evolve for diverse reasons. Over time, people learn to mimic the symbols and, thus, nullify their effectiveness. As symbols lose their effectiveness, the incentives to behave in a trustworthy manner decrease. Thus, the trustworthy users of a system of symbols find it necessary to constantly evolve the symbolic systems of reputation. The evolutionary arms races that are observed in the biological world also occur in the socio-cultural world. And the quality of any reputational symbol as a public good may erode over time if substantial investment is not made in "policing" the use of the system. Recent events in American corporate industry (Enron, among others) have called the reputational symbols awarded by auditing firms into serious question.

CONCLUSION

The great potential of human cooperation is rooted in the evolved human biological capacity to use signals. This potential that is the foundation for all society is further developed by socioculturally constructed reputational symbols. Both signals and symbols are not just devices for cooperation but essentially serve as the mechanisms by which intrinsically motivated conditional cooperators can evolve to compose a large proportion of a population. At the same time, neither biological signals nor sociocultural symbols work perfectly. They are both prone to errors and abuses.

The scope and scale at which a society can maintain cooperative endeavors greatly affect the society's destiny. In a world where the types of social interactions change so rapidly, it is important to craft sophisticated systems of rules and symbols so that society will not enter a Hobbesian world in which predation replaces cooperation. A world that becomes highly affected by the Internet, terrorist networks, and a global market for commodities faces immense challenges in developing reliable symbolic systems that allow trustworthy actors to detect each other to maintain human sociality.

ACKNOWLEDGMENTS

We gratefully acknowledge support from the Center for the Study of Institutions, Population, and Environmental Change at Indiana University through National Science Foundation grants SBR9521918 and SES0083511.

REFERENCES

Adolphs, R. 2002. "Trust in the Brain." *Nature Neuroscience* 5(3):192–93.
Adolphs, R., D. Tranel, and A. Damasio. 1998. "The Human Amygdala in Social Judgement." *Nature* 393:470–74.

Ahn, T. K. 2002. "Information and the Evolution of Preferences in One-Shot Prisoner's Dilemma." Working paper, Workshop in Political Theory and Policy Analysis, Indiana University, Bloomington.

Ahn, T. K., E. Ostrom, and J. Walker. 2003. "Incorporating Motivational Heterogeneity into Game-Theoretic Models of Collective Action." *Public Choice*.

Andreoni, J. and J. H. Miller. 1993. "Rational Cooperation in the Finitely Repeated Prisoner's Dilemma: Experimental Evidence." *Economic Journal* 103:57–85.

Axelrod, R. 1981. "The Emergence of Cooperation among Egoists." *American Political Science Review* 75:306–18.

Axelrod, R. and W. D. Hamilton. 1981. "The Evolution of Cooperation." *Science* 211:1390–96.

Bacharach, M. and D. Gambetta. 2001. "Trust in Signs." Pp. 148–84 in *Trust in Society*, edited by C. Cook. New York: Russell Sage Foundation.

Baron-Cohen, S. 1995. *Mindblindness: An Essay on Autism and Theory of Mind.* Boston: MIT Press.

Bendor, J. and P. Swistak. 1997. "The Evolutionary Stability of Cooperation." *American Political Science Review* 91:290–307.

Boyd, R. and J. Lorberbaum. 1987. "No Pure Strategy Is Evolutionarily Stable in the Repeated Prisoner's Dilemma Game." *Nature* 327:58–59.

Brown, W. M. and C. Moore. 2000. "Is Prospective Altruist-detection an Evolved Solution to the Adaptive Problem of Subtle Cheating in Cooperative Ventures? Supportive Evidence Using the Wason Selection Task." *Evolution and Human Behavior* 21:25–37.

Brown, W. M. and C. Moore. 2002. "Smile Asymmetry and Reputation as Reliable Indicators of Likelihood to Cooperate: An Evolutionary Analysis." *Advances in Psychology Research* 11:59–78.

Byrne, R. B. 1995. *Thinking Primates.* Oxford: Oxford University Press.

Cheney, D. L. and R. M. Seyfarth. 1990. *How Monkeys See the World.* Chicago: Chicago University Press.

Clutton-Brock, T. 2002. "Breeding Together: Kin Selection and Mutualism in Cooperative Vertebrates." *Science* 296:69–72.

Cosmides, L. 1989. "The Logic of Social Exchange: Has Selection Shaped How Humans Reason? Studies with the Wason Selection Task." *Cognition* 31:187–276.

Darwin, C. [1872] 1873. *The Expressions of the Emotions in Man and Animals.* New York: D. Appleton.

de Waal, F. B. M. 1989. "Food Sharing and Reciprocal Obligations among Chimpanzees." *Journal of Human Evolution* 18:433–59.

de Waal, F. B. M. 1997. *Bonobo: The Forgotten Ape.* Berkeley: University of California Press.

DeBruine, L. M. 2002. "Facial Resemblance Enhances Trust." *Proceedings of the Royal Society London B* 269:1307–12.

Eckel, C. C. and R. K. Wilson. 2003. "The Human Face of Game Theory: Trust and Reciprocity in Sequential Games." Pp. 245–74 in *Trust, Reciprocity, and Gains from Association: Interdisciplinary Lessons from Experimental Research*, edited by E. Ostrom and J. Walker. New York: Russell Sage Foundation.

Fehr, E. and S. Gächter. 2002. "Altruistic Punishment in Humans." *Nature* 415: 137–40.

Feldman, M. S. and J. G. March. 1981. "Information in Organizations as Signals and Symbols." *Administrative Science Quarterly* 26:171–86.

Frank, R. H. 1987. "If Homo Economicus Could Choose His Own Utility Function, Would He Want One with a Conscience?" *American Economic Review* 77(4): 593–604.

Frank, R. H. 1988. *Passions within Reason: The Strategic Role of the Emotions.* New York: Norton.

Frank, R. H., T. Gilovich, and D. Regan. 1993. "The Evolution of One-Shot Cooperation: An Experiment." *Ethology and Sociobiology* 14:247–56.

Fudenberg, D. and E. Maskin. 1986. "The Folk Theorem in Repeated Games with Discounting or with Incomplete Information." *Econometrica* 54:533–54.

Güth, W. 1995. "An Evolutionary Approach to Explaining Cooperative Behavior by Reciprocal Incentives." *International Journal of Game Theory* 24:323–44.

Güth, W. and H. Kliemt. 1998. "The Indirect Evolutionary Approach: Bridging the Gap Between Rationality and Adaptation." *Rationality and Society* 10(3):377–99.

Güth, W., H. Kliemt, and B. Peleg. 2000. "Co-evolution of Preferences and Information in Simple Games of Trust." *German Economic Review* 1(1):83–110.

Güth, W. and M. Yaari. 1992. "An Evolutionary Approach to Explaining Reciprocal Behaviour in a Simple Strategic Game." Pp. 23–34 in *Explaining Process and Change,* edited by H. Kliemt. Ann Arbor: University of Michigan Press.

Henrich, J., R. Boyd, S. Bowles, C. Camerer, E. Fehr, H. Gintis, R. McElreath, M. Alvard, A. Barr, J. Ensminger, K. Hill, F. Gil-White, M. Gurven, F. Marlowe, J. Q. Patton, N. Smith, and D. Tracer. 2001. *Economic Man in Cross-Cultural Perspective: Behavioral Experiments in Fifteen Small-Scale Societies.* Working paper 063, Santa Fe Institute.

Isaac, R. M. and J. M. Walker. 1988. "Group Size Effects in Public Goods Provision: The Voluntary Contribution Mechanism." *Quarterly Journal of Economics* 103:179–200.

Janssen, M. A. No date. "Evolution of Cooperation in a One-Shot Prisoner's Dilemma Based on Recognition of Trustworthy and Untrustworthy Agents." Submitted.

Kandori, M. 1992. "Social Norms and Community Enforcement." *Review of Economic Studies* 59:63–80.

Kikuchi, M., Y. Watanabe, and T. Yamagishi. 1997. "Judgment Accuracy of Others' Trustworthiness and General Trust: An Experimental Study." *Japanese Journal of Experimental Social Psychology* 37:23–36 (in Japanese with an English abstract).

Kreps, D. M., P. Milgrom, J. Roberts, and R. Wilson. 1982. "Rational Cooperation in the Finitely Repeated Prisoner's Dilemma." *Journal of Economic Theory* 27: 245–52.

Kurzban, R. 2003. "Biological Foundations of Reciprocity." Pp. 105–27 in *Trust, Reciprocity, and Gains from Association: Interdisciplinary Lessons from Experimental Research,* edited by E. Ostrom and J. Walker. New York: Russell Sage Foundation.

Lorberbaum, J. 1994. "No Strategy Is Evolutionarily Stable in the Repeated Prisoner's Dilemma." *Journal of Theoretical Biology* 168(May):117–30.

Malaga, R. A. 2001. "Web-based Reputation Management Systems: Problems and Suggested Solutions." *Electronic Commerce Research* 1:403–17.

Mealey, L., C. Daood, and M. Krage. 1996. "Enhanced Memory for Faces of Cheaters." *Ethology and Sociobiology* 17:119–28.

Milgrom, P., D. North, and B. Wengast. 1990. "The Role of Institutions in the Revival of Trade: The Law Merchant, Private Judges, and the Champaign Fairs." *Economics and Politics* 2:1–23.

O'Connell, S. 1998. *Mindreading: An Investigation into How We Learn to Love and Lie.* New York: Doubleday.

Oda, R. 1997. "Biased Face Recognition in the Prisoner's Dilemma." *Evolution and Human Behavior* 18:309–15.

Ostrom, E., R. Gardner, and J. Walker. 1994. *Rules, Games, & Common-Pool Resources.* Ann Arbor: University of Michigan Press.

Ostrom, E. and J. Walker. 1991. "Communication in a Commons: Cooperation without External Enforcement." Pp. 287–322 in *Laboratory Research in Political Economy*, edited by T. R. Palfrey. Ann Arbor: University of Michigan Press.

Ostrom, E., J. Walker, and R. Gardner. 1992. "Covenants with and without a Sword: Self-Governance Is Possible." *American Political Science Review* 86(2):404–17.

Rao, H. 1994. "The Social Construction of Reputation: Certification Contests, Legitimation, and the Survival of Organizations in the American Automobile Industry: 1895–1912." *Strategic Management Journal* 15:29–44.

Resnick, P. and R. Zeckhauser. 2002. "Trust Among Strangers in Internet Transactions: Empirical Analysis of eBay's Reputation System." Pp. 127–57 in *The Economics of the Internet and E-Commerce*, edited by M. R. Baye. Amsterdam: Elsevier Science.

Resnick, P., R. Zeckhauser, J. Swanson, and K. Lockwood. 2002. *The Value of Reputation on eBay: A Controlled Experiment.* Unpublished manuscript, University of Michigan. http://www.si.umich.edu/~presnick/papers/postcards/

Rilling, J. K., D. A. Gutman, T. R. Zeh, G. Pagnoni, G. S. Berns, and C. D. Kilts. 2002. "A Neural Basis for Social Cooperation." *Neuron* 35:395–405.

Rubinstein, A. 1979. "Equilibrium in Supergames with the Overtaking Criterion." *Journal of Economic Theory* 21:1–9.

Sally, D. 1995. "Conversation and Cooperation in Social Dilemmas: A Meta-Analysis of Experiments from 1958 to 1992." *Rationality and Society* 7:58–92.

Scharlemann, J. P. W., C. C. Eckel, A. Kacelnik, and R. K. Wilson. 2001. "The Value of a Smile: Game Theory with a Human Face." *Journal of Economic Psychology* 22:617–40.

Schmidt, D., R. Shupp, J. Walker, T. K. Ahn, and E. Ostrom. 2001. "Dilemma Games: Game Parameters and Matching Protocols." *Journal of Economic Behavior and Organization* 46:357–77.

Schmidt, K. L. and J. F. Cohn. 2001. "Human Facial Expressions as Adaptations: Evolutionary Questions in Facial Expression Research." *Yearbook of Physical Anthropology* 44:3–24.

Selten, R. and R. Stoecker. 1986. "End Behavior in Sequences of Finite Prisoner's Dilemma Supergames: A Learning Theory Approach." *Journal of Economic Behavior and Organization* 7:47–70.

Stephens, D. W., C. M. McLinn, and J. R. Stevens. 2002. "Discounting and Reciprocity in an Iterated Prisoner's Dilemma." *Science* 298:2216–18.

Stone, V. E., L. Cosmides, J. Tooby, N. Kroll, and R. T. Knight. 2002. "Selective Impairment of Reasoning about Social Exchange in a Patient with Bilateral Limbic System Damage." *Proceedings of the National Academy of Science* 99:11531–36.

Tirole, J. 1996. "A Theory of Collective Reputations (with Applications to the Persistence of Corruption and to Firm Quality)." *Review of Economic Studies* 63(1):1–22.

Trivers, R. L. 1971. "The Evolution of Reciprocal Altruism." *Quarterly Review of Biology* 45(4):35–57.

Weigelt, K. and C. Camerer. 1988. "Reputation and Corporate Strategy: A Review of Recent Theory and Applications." *Strategic Management Journal* 9(5):443–54.

Wilson, E. O. [1975] 2000. *Sociobiology: The New Synthesis.* Cambridge, MA: Harvard University Press.

Winston, J. S., B. A. Strange, J. O'Doherty, and R. J. Dolan. 2002. "Automatic and Intentional Brain Responses during Evaluation of Trustworthiness of Faces." *Nature Neuroscience* 5(3):277–83.

7

Darwinian Evolution by the Natural Selection of Heritable Variation

Definition of Parameters and Application to Social Behaviors

James M. Cheverud

One of the major advances in studies of sociality over the last forty years has been the introduction of Darwinian evolutionary theory (Hamilton 1964; Wilson 1975). The evolutionary concepts of altruism, kin selection, and sexual selection have dominated interpretations of social behavior and interindividual interactions. However, at times there has been confusion among researchers about the meaning of terms used in evolutionary theory and a consequent lack of empirical work measuring the crucial theoretical parameters. Here, I will first review the concepts of quantitative evolutionary theory (Falconer and Mackay 1996; Lynch and Walsh 1998), their application to situations in which social interactions play an important role in determining variation in individual characteristics (Wolf et al. 1998), and, briefly, empirical studies on cooperative breeding that have successfully measured the parameters specified in evolutionary theory.

Darwinian evolution occurs through the natural selection of heritable variation. There are thus two main aspects to this theory: heritable variation, concerned with how variations are passed from parent to offspring, and natural selection, concerned with the relationship between the characteristics of an organism and their survival and reproduction in a specific environment. At the turn of the last century, the conundrum of heritable variation was resolved by the rediscovery of Mendel's laws and their quantitative application to patterns of heritable variation (Provine 1971). Likewise, a quantitative theory of selection was developed so that Darwinian evolution could be defined as the change in the population mean from one generation to the next (Δz) equal to the amount of heritable variation (G) times the amount of selection (β),

$$\Delta z = G\beta \tag{1}$$

The terms of this classic response to selection equation have specific meanings in evolutionary biology, and this specificity directs the kinds of measurements that need to be taken in evolutionary studies and the kinds of assumptions made in evolutionary interpretation. We will consider selection and heritable variation in turn.

SELECTION

Selection is the relationship between the characteristics of organisms and their survival and reproduction in relation to others in the same breeding population. The selection gradient (β) measures this relationship as the slope of the least squares linear regression of relative fitness (w_r) on the trait in question (see Figure 7.1). Thus,

$$\beta = \text{cov}(w_r, P)/V_P \tag{2}$$

where $\text{cov}(w_r, P)$ is the covariance between relative fitness and the phenotypic value (P) and V_P is the variance in the phenotypic value. The phenotypic value is any measurement taken on an individual (Falconer and Mackay 1996). For example, we may measure the level of aggression an individual displays and the number of offspring he or she produces in a lifetime as a measure of fitness. In this instance, relative fitness is the number of offspring produced divided by the average number of offspring produced per individual.

To come to an accurate understanding of a trait's evolution one must consider it as part of a larger complex of related traits rather than as an atomistic, independent property of an animal. This is because direct selection on any one trait will indirectly cause apparent selection on other, correlated traits (Lande and Arnold 1983). For example, we should probably consider aggression as a complex set of traits differing depending on social contexts, such as male-male interactions, male-female interactions, adult-infant interactions, and interactions with strangers. The effects of intertrait correlation on selection are illustrated in Figure 7.2a. There are two correlated phenotypes, X and Y, with their joint distribution in the population represented by an ellipse.

Threshold selection is applied to trait X alone so that everyone to the right of the threshold survives and reproduces while those with X values lower than the threshold do not. This is a simple and rather artificial form of selection that probably does not occur so starkly in nature but serves to

James M. Cheverud

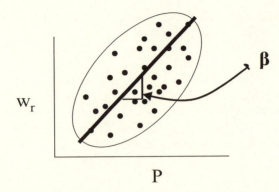

W_r

P

Figure 7.1   Selection is the differential survival and reproduction of individuals with different phenotypes. It is measured by the slope of the linear
regression of relative fitness (w_r) on phenotypic value (P), = $\text{cov}(w_r,P)/V_P$.

illustrate the concepts of direct and indirect selection on traits. The $\text{cov}(w_r,$
$P)$ is equivalent to the difference in population means before and after
selection. As shown in Figure 7.2a, because phenotypes X and Y are correlated, direct selection on X results in apparent selection on Y. If we examined Y alone, we would gain the false impression that increased values of
Y were being selected for because of their direct relationship to relative fitness. For example, direct selection may be actually present for aggression
directed at strangers. This may lead to apparent selection on correlated
features, such as aggression toward infants. If we only scored aggression
directed toward infants ignoring related traits, we would falsely conclude
that selection favors aggression toward infants. It is important to consider
related features jointly when studying selection (ibid.).

In evolutionary theory, indirect selection is accounted for by defining
the selection gradient (β) as the partial regression of relative fitness on the
phenotype. The partial regression measures the relationship between relative fitness and the phenotype of interest independent of all other traits
(ibid.). This is illustrated in Figure 7.2b, where phenotypes X' and Y' refer
to values of X independent of Y and values of Y independent of X, respectively. By definition, X' and Y' are uncorrelated so that when the selection
threshold is applied to X', there is no indirect selection on Y'. This illustrates the point that it is dangerous to consider traits separately rather than
as part of integrated behavioral or morphological systems.

As an example of selection on multiple characters, Table 7.1 presents
selection gradients on adult body size (tarsus length) and bill dimensions
in Darwin's finches due to drought (Grant and Grant 1989). Selection coefficients are presented both before (S) and after (β) correction for correla-

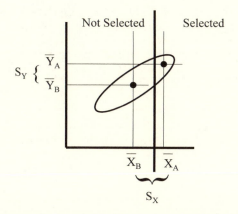

Figure 7.2a   Apparent selection on Y with direct selection on correlated trait X.

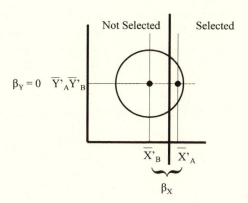

Figure 7.2b   Direct selection on Y independent of direct selection on correlated trait X.

tions between traits. Inspection of the uncorrected selection coefficients (S) indicates apparent positive selection on body size and both bill dimensions. However, after correction for intertrait correlations (β), it is apparent that selection has been primarily on bill depth alone and not on body size at all. Thus ecological interpretations of the cause of this selection should concentrate on the advantages of a deep bill in drought conditions. The apparent indirect selection on body size was actually due to direct selection on bill depth. Grant and Grant (1989) interpret these selection gradients as indicating that drought had changed the relative abundance of different sized seeds, making large seeds relatively more abundant, and that a deep beak would more effectively process these seeds. This is

Table 7.1   Selection on Body Size and Bill Dimensions in
              Adult Darwin's Finches during Drought (Grant and
              Grant 1989)[a]

Character	β	S
Bill length	0.33	0.45*
Bill depth	0.62*	0.58**
Tarsus length	−0.14	0.27

[a]β is the selection gradient corrected for intertrait correlation and S is
selection prior to adjustment.
*, significant at the .05 level; **, significant at the .01 level.

biomechanically reasonable and fits with observations made in the field on
seed abundance and bird feeding during the drought.

When measuring selection it is also important to keep in mind that relative fitness is measured over an animal's entire lifetime, not just at any
one stage, and that selection forces may differ at different life stages. Again
using the example of selection on Darwin's finches due to drought, Table
7.2 presents the corrected (β) and uncorrected (S) selection gradients for
body size and bill dimensions in juvenile birds. Unlike the situation in
adults, direct selection here is on body size, not on bill dimensions. Thus
the same environmental factor has different selective effects at different
life stages.

Selection forces may also be different for different aspects of relative fitness. Preziosi and Fairbairn (2000) studied body size in adult female
waterstriders in terms of reproductive longevity and fecundity. They
found selection for smaller body size in relation to reproductive longevity
and selection for greater body size for fecundity (see Figure 7.3). When
these two aspects of relative fitness are combined in an overall lifetime

Table 7.2   Selection on Body Size and Bill Dimensions in
              Young Darwin's Finches during Drought (Grant and
              Grant 1989)[a]

Character	β	S
Bill length	0.19	0.29*
Bill depth	0.08	0.30*
Tarsus length	0.47*	0.38*

[a]β is the selection gradient corrected for intertrait correlation and S is
selection prior to adjustment.
*, significant at the .05 level.

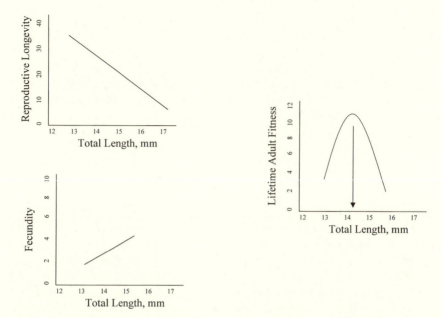

Figure 7.3 Selection on waterstrider female total body length due to reproductive longevity, fecundity and total adult fitness (based on Preziosi and
Fairbairn, 2000). Arrow indicates optimal length.

measure, there is no directional selection at all on female body size.
Instead, there is stabilizing selection for an intermediate optimum. This
form of selection does not cause evolution of the population mean when it
is already placed at the optimum.

Sometimes researchers mistakenly conflate selection and relative fitness. They discover variations in relative fitness in a population. Some
individuals have more mates than others, some survive and others die, or
some are more fecund than others. These variations are then taken as evidence of selection among individuals. After all, the individual who is more
fecund has more offspring and contributes more of his or her genes to the
next generation. However, as we have seen, this is not the definition of
selection. Selection is the relationship between relative fitness and a phenotype and thus is a property of a population, not of an individual.
Variations in relative fitness that are not related to the phenotype in question cause genetic drift, random genetic change in a population. The distinction between selection and relative fitness is clarified by the concept of
opportunity for selection (Crow and Kimura 1967). Variance in relative fitness provides the opportunity for selection. Whether that opportunity is

realized or not depends on causal correlations with phenotypes in a population. Thus, whenever interpretations of behaviors involving adaptation by natural selection are used, we are assuming a direct causal connection between the trait in question and variations in relative fitness. This is a hypothesis that needs to be tested by measuring natural selection. As we have seen from the examples given above and from literature surveys, selection can, indeed, be measured in natural populations (Kingsolver et al. 2001).

HERITABLE VARIATION

Selection, by itself, cannot cause evolution unless there is heritable variation for the trait of interest [see equation (1)]. Selection on phenotypic variation may be strong but the effects of that selection will not be passed on to the next generation unless part of that variation is heritable. R. A. Fisher (1918) introduced a mathematical and measurement theory for heritable variation early in the twentieth century. This theory divided phenotypic values and variation into heritable and nonheritable parts. The heritable parts are due to the Mendelian inheritance of genes in a population. He also assumed random mating. Figure 7.4 gives the definitions of genotypic values at a single locus along a number line measuring the trait of interest (Falconer and Mackay 1996). Genotypic values are the average phenotypic values of individuals carrying the specified genotype. The genotypic values (SS, SL, LL) are plotted along the number line of phenotypic values. The scale of the line has been transposed so that the origin (= 0) of the line is midway between the two homozygous genotypic values. Then the rescaled additive genotypic value (a) is defined as half the difference between the two homozygotes ($-a$ for SS and $+a$ for LL). The rescaled heterozygous genotypic value is the deviation of the heterozygote from the midpoint of the homozygotes (d). Fisher then derived the average, or heritable, effect of an allele (α_i) as the average phenotypic deviation from the population mean of individuals receiving this allele from one parent, the other allele from the other parent, coming at random from the population

$$\alpha_i = p_j[a + d(p_j - p_i)] \tag{3}$$

where p_i is the frequency of the allele in question and p_j is the frequency of the alternate allele ($p_i + p_j = 1$). This is the heritable value of the gene. Fisher determined that the breeding value (A), or heritable value, of an individual is equal to the sum of the average effects of the alleles it carries and that the variance of this breeding value (V_A) is

$$V_A = 2p_1p_2\alpha^2 \tag{4}$$

GENOTYPIC VALUES AT A SINGLE LOCUS

a = additive genotypic value
d = dominance genotypic value

Figure 7.4   Genotypic values at a single locus. a, additive genotypic value; d, dominance genotypic value.

summed over all loci, where p_1 and p_2 are the allele frequencies and α is the average effect of a gene substitution ($\alpha_1 - \alpha_2$) (Falconer and Mackay 1996).

While this model adequately describes heritable variation, in order to use it for measurement in a population one would have to consider and measure all the genes contributing to variation in any given trait. This is a Herculean task only recently approached in model systems with special breeding designs (Lynch and Walsh 1998). However, Fisher also determined that the level of heritable variance is related to the phenotypic similarity among relatives [cov(relatives)]. He showed that

$$\text{cov(relatives)} = rV_A \tag{5}$$

where r is the genetic correlation among relatives (Falconer and Mackay 1996). The simplest and most direct way to estimate the level of heritable variance is to obtain the covariance between offspring and their parents. Since these share (1/2) of their genes, their genetic correlation is (1/2) and

$$V_A = 2\text{cov(offspring, parent)} \tag{6}$$

The parent-offspring covariance is usually estimated from the slope of the least squares linear regression of offspring values on parental values (see Figure 7.5). This slope is referred to as the heritability of the trait and denotes the amount of heritable variation relative to the total phenotypic variation.

As with selection, it is often important to consider a system of related traits together in an analysis of heritable variation because functionally and developmentally related traits tend to be inherited together. There are two reasons for traits being inherited together. The first is pleiotropy, in which one gene affects many traits. The second is linkage disequilibrium,

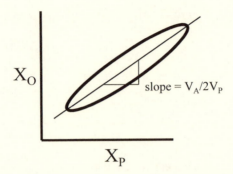

Figure 7.5 Parent-offspring regression slope estimates heritable variance, where X_P is parent's phenotype, X_O is the offspring phenotype, and heritable variance $(V_A) = 2V_P(\text{slope})$.

in which two separate, usually linked, genes are inherited together because of deviations from random mating. The effect of linkage disequilibrium can be transitory because random mating for approximately seven to ten generations degrades linkage disequilibrium to near zero. Genetic correlations and covariances between traits measure the magnitude of trait coinheritance (Falconer and Mackay 1996). For example, the inheritance of aggression in various social contexts is not likely to be independent; some genes may affect aggression in many social contexts. Note that this genetic correlation among traits is different from the genetic correlation among individuals referred to in equation (5). Intertrait genetic covariances can also be measured using the covariance among relatives. With parents and offspring, the genetic covariance between traits is the sum of the phenotypic covariance of trait X in the offspring with trait Y in the parent and the covariance between trait Y in the offspring with trait X in the parent. These heritable variances and covariances are compiled in a symmetric heritable variance/covariance matrix usually symbolized as G with the number of rows and columns corresponding to the number of traits (Lande 1979).

The role of heritable variation in evolution is often dismissed on the grounds that most traits in most populations show heritable variation (Mousseau and Roff 1987). However, heritable covariances can have a major effect on evolutionary processes, limiting evolutionary response to selection and causing substantial deviations from the optimal, adaptive evolutionary path (Figure 7.6; Cheverud 1984). This is because when traits are inherited together they will tend to evolve together regardless of which trait is being selected (Lande 1979). Genetic correlations between traits result in correlated responses to selection (Falconer and Mackay 1996). So

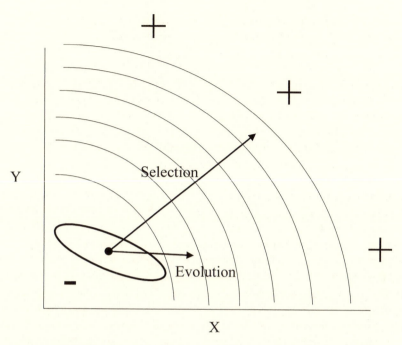

Figure 7.6   Isoclines of equal fitness for traits X and Y running from low (–) to high (+) fitness values.

if traits X and Y are genetically correlated and jointly inherited, direct selection on trait X (as in Figure 7.2) results in a correlated response in trait Y, even with no direct selection on Y.

EVOLUTION IN A SOCIAL ENVIRONMENT

The model described above considers situations in which a phenotype is affected by its own genes and random environmental factors. However, when those environmental factors include the effects of other conspecifics, a more complex and interesting situation arises. In many animals the environment provided by conspecifics influences the morphological, physiological, and behavioral development of the young (Wolf et al. 1998). The most obvious instance of this is the developmental environment a mammalian mother provides for her offspring. Indeed, most of the variation among newborn mammal phenotypes is due to variations in the environment provided by the mother. The environment provided by different mothers for their offspring will vary from mother to mother both because of the varying genes mothers carry and because of the varying external environments the mothers have experienced. Thus the phenotype of an offspring is affected by two sets of genes, those carried by its mother, responsible for the environment she provides her offspring and having only an indirect effect on offspring characteristics, and those carried by the offspring itself and having a direct effect on the animal's own phenotype. This maternal-offspring interaction may be the most primitive form of social interaction in mammals and can be taken as a paradigm for the evolution of social interactions generally (Zihlman, this volume).

A common means of modeling this situation is to consider maternal-offspring interactions as a two-trait system composed of heritable maternal performance, measured as the effect of maternal environment on the offspring phenotype, and heritable direct genetic effects of the offspring's genes on its own phenotype. As discussed above, when dealing with a system of traits we need to consider their coinheritance, in this case the heritable covariance of the direct genetic effects (A_0) and the indirect maternal genetic effects (A_i), written as $\mathrm{cov}(A_0, A_i)$. This covariance measures the degree of coinheritance between the maternal effect on an offspring character and the direct genetic effect on the same character. Evolution of offspring traits under this model is complex, with unexpected outcomes (Cheverud 1984; Kirkpatrick and Lande 1989). For example, when the heritable correlation between the direct and indirect genetic effects is strongly negative, direct selection on the offspring trait, although favoring an increase in trait value, will result in an evolutionary decrease. Likewise, with a negative heritable covariance, direct selection for an increase in the offspring trait can lead to a correlated decrease in maternal performance, thereby limiting selection response. These unusual consequences occur when there is a negative heritable correlation between direct and indirect effects. While one might think, on the surface, that this would be a rare occurrence, strongly negative heritable correlations are the rule in mammals for weanling weight (Cheverud 1984), a time of life at which there is an exceptionally large opportunity for selection.

Evolutionary models of traits affected by the social environment are still being developed (Wolf 2003), and many exciting results are likely to be obtained in the future. These models also have begun to direct empirical research into the evolution of social behaviors (Hunt and Simmons 2000; Agrawal, Brodie, and Brown 2001; Rauter and Moore 2001; Moore, Haynes, Preziosi, and Moore 2002; Peripato and Cheverud 2002; Wade and Shuster 2002).

The new models of evolution in a social environment relate directly to the most commonly applied paradigm in behavior ecology and sociobiology, the evolution of altruism by kin selection (Hamilton 1964; Wilson

1975). The central concept in sociobiology is that we can understand the social behavior of animals living in groups by using the concept of altruism. Altruism has a very specific meaning in evolutionary studies as it refers to the relative value of selection gradients on multiple traits. Hamilton (1964) showed that the frequency of an "altruistic" behavior is expected to increase in a population when

$$r > |c|/b \tag{7}$$

where r is the genetic correlation between two interacting individuals, $|c|$ is the absolute value of the fitness costs of performing the "altruistic" behavior, and b is the fitness benefit obtained by the recipient of that behavior because of the behavior's effect on the recipient's phenotype. In order for a trait to be considered "altruistic" in an evolutionary sense, it must entail a cost of performing the behavior and a benefit for the recipient. Thus "altruism" is a statement about relative selection gradients and not about the behavior of an individual. The same behavior may be altruistic in one context but not in another because selection gradients change with changing ecological circumstances. For a trait to be considered as under altruistic selection, the covariance between relative fitness and that trait itself must be negative, while the covariance of relative fitness and the trait affected in the recipient must be positive. It would seem then that a primary research strategy in sociobiology would be to measure these selection gradients and show that, indeed, a behavior is under an altruistic pattern of selection. Even though the kin selection model has been used widely in interpreting behavioral evolution, these measurements have rarely been made.

Clutton-Brock (2002) presents a general taxonomy of selective causes for social evolution that can be defined by the relationships between relative fitness and the two phenotypes involved, the behavior itself and the phenotype it affects in the recipient. As noted above, altruistic selection is defined by a negative relationship between relative fitness and the trait itself (–) combined with a positive relationship between relative fitness and the affected phenotype in the recipient (+) or a (–/+) selection pattern. Parasitic selection is the mirror image of altruism, being defined as having a (+/–) selection pattern with positive selection on the source phenotype and negative selection on the recipient phenotype. "By-product mutualism" and "pseudo-reciprocity" are specified by a (+/0) selection pattern, with the recipient's phenotype not under direct selection. Its evolution occurs as a correlated response to selection favoring the source phenotype. Finally, mutualism, reciprocity, and group augmentation are all specified by a (+/+) selection pattern, in which both the source and recipient phenotypes are under the same form of selection. Both (+/0) and (+/+) forms of selection lead to the evolution of cooperation without the involvement of evolutionary altruism. We must consider these as viable alternative evolutionary models for social behaviors.

Researchers have been more active in measuring the genetic correlation among pairs or sets of interacting individuals than in measuring selection. This can be done either using genealogies obtained from long-term studies or from genetic markers (Moore and Kukuk 2002). However, the left-hand side of Hamilton's rule [equation (7)] involves an unusual, unspoken assumption. In deriving his rule, Hamilton assumed that the only genetic factor affecting the recipient's phenotype was indirect effect of the "altruist." He did not allow for direct genetic effects of genes carried by the recipient on its own phenotype. This seems a rather restrictive and unrealistic assumption. Cheverud (1984), drawing on forty years of research in the agricultural sciences, rederived Hamilton's rule allowing for direct genetic effects on the recipient's phenotype. He obtained the result that altruism would increase in the population when

$$r + \mathrm{cov}(A_0, A_i)/V_{Ai} > |c|/b \tag{8}$$

where $\mathrm{cov}(A_0, A_i)$ is the heritable covariance between the direct (A_0) and indirect (A_i) effects on the recipient's phenotype and V_{Ai} is the heritable variance in the indirect effect. The heritable covariance can dominate this inequality so that altruism cannot evolve regardless of minimal costs and massive benefits when the heritable covariance is negative. Alternatively, altruism can evolve despite massive costs and minimal benefits if the covariance is relatively large and positive (ibid.). Studies of the evolution of social interactions require that the patterns of heritable variation be considered realistically and that selection be directly measured.

The concept of inclusive fitness has played an important role in the development of sociobiological theory. An individual's inclusive fitness is the combination of its own fitness and the fitness of those individuals it interacts with, weighted by the genetic correlation between interactants. The "altruistic" trait will increase in frequency in a population when it is positively associated with inclusive fitness, although it must still be, by definition, negatively associated with individual fitness. This concept is valid but can be difficult to operationalize in empirical studies. For example, the fitness benefits of a behavior may be double-counted, included in the fitness of the recipient and again in the fitness of the altruist (Clutton-Brock 2002). In models of evolution in a social environment this double counting is avoided by assigning fitness directly to the individual who survives and/or reproduces. The effects of the interaction among individuals on their phenotypes are accounted for as a correlated response to selection rather than conflating the altruistic behavior and its effects (Wolf 2003).

EVOLUTION OF COOPERATIVE BREEDING

Cooperative breeding occurs when animals other than the parents help care and provide for newborns. Kin selection has long been assumed to be the evolutionary basis for helping behavior in cooperatively breeding species, including birds, primates, and carnivores. In such breeding systems there are reproductives and helpers. Helpers expend energy taking care of the reproductives' offspring and thus are assumed to have fitness costs. However, recent literature reviews have cast doubt on the ability of kin selection to explain aspects of cooperative breeding because careful study of selection pressures involved with these behaviors indicates that while the behavior certainly benefits the offspring of reproductives, it does not seem to entail fitness costs to the helper (Clutton-Brock 2002; Griffin and West 2002).

It is commonly observed that helpers in cooperative breeding systems expend energy in caring for young. Achenbach and Snowdon (2002) showed that cotton-top tamarin helper males lose about 10 percent of their body weight when caring for the young and that more helpers in a group leads to smaller weight loss by individual tamarins. Russell, Sharpe, Brotherton, and Clutton-Brock (2003) also show decrements in helper growth incurred by care of young in meerkats. However, in meerkats helper weight loss due to cooperative breeding does not lead to longer-term selective costs because of behavioral compensating mechanisms, such as increased foraging during the nonbreeding period and limitations on helping behavior in later reproductive periods (Russell et al. 2003). Thus, careful long-term studies have found that helping behavior may not entail fitness costs. Instead, there may be direct positive selection for the helping behavior by increasing the survival, mating success, rearing success, and dispersal success of helpers, as found in the meerkats (Clutton-Brock 2002) and in callitrichid monkeys (Epple 1978; Sussman and Garber 1987; Tardif 1990). In the cooperative breeding Florida scrub jays, helpers benefit by having better access to resources in a saturated environment, enhancing their chances of obtaining either their natal territory or budding off a new territory and becoming breeders themselves, and having higher survival on their natal territory than dispersing animals (Woolfenden and Fitzpatrick 1985). Thus, altruistic selection does not seem to be involved in helping behavior; instead the selection pressures indicate mutualism. Cooperative behavior evolves because it has direct benefits for the helper and for the individuals it interacts with.

Given the contrary evidence of careful long-term studies (Clutton-Brock 2002), we may consider the reasons for the widespread acceptance of kin selection models for cooperative breeding and other social behaviors. Kin selection models are quite complex and compelling. They seem to account for evolutionary processes in terms of individual behavioral acts that are relatively simple to understand. However, as seen above, evolutionary processes occur in populations, not individuals. "Altruistic" acts are not the same thing as "altruistic" selection. Furthermore, the conflict portrayed in these models is perhaps more culturally appealing to some of us than are cooperation and mutualism. Finally, kin selection models and inclusive fitness provide a framework for explaining apparent true altruism in terms of self-interest, and self-interest is seen as the guiding light of our culture and value systems (Fehr and Rockenbach 2003).

SUMMARY

It is necessary to understand modern Darwinian theory in order to better understand the evolution of behavior. The parameters of this theory are very specifically defined in quantitative terms. It is important to keep these definitions in mind when studying or interpreting evolutionary processes. Darwinian evolution occurs by the natural selection of heritable variation.

Natural selection is the differential survival and/or reproduction of phenotypes in a population. It is measured by the slope of the regression of relative fitness on a phenotype. A common misconception is that selection is the differential survival and/or reproduction of individuals. However, differential survival and/or reproduction are the sources of evolutionary change by both genetic drift and selection. Differential survival and/or reproduction of individuals must be causally correlated with a phenotype in order for it to be considered as a source of selection, otherwise it produces genetic drift. The same individual births and deaths that contribute to selection on one character or gene produce genetic drift in other characters or genes. It is also important to consider complexes of related traits in analyses and interpretations of selection because intertrait correlations may produce the appearance of selection on a trait that is actually due to direct selection on other, correlated traits.

Heritable variation is the other component of the Darwinian evolution equation. It measures the extent to which differences in individuals can be passed on from parent to offspring. While it is often thought that heritable variation is ubiquitous for traits and thereby only plays a passive, permissive role in Darwinian evolution, quantitative evolutionary theory shows that the coinheritance of traits can play a critical role in determining both the rate and direction of evolutionary change, modifying the impact of selection. This is especially true in situations where an individual's development is impacted by a heritable social environment. In this case the heritable correlation between the direct effects of genes carried by an individual and the indirect effects of genes carried by others, which affect an individual's phenotype through their effects on the social developmental environment, can dominate evolutionary outcomes. For example, a strong negative heritable correlation between direct and indirect effects can block the evolution of altruistic behaviors despite high fitness benefits to the recipient and low fitness costs to the actor. Alternatively, a strong positive correlation facilitates the evolution of altruistic behavior despite low benefits and high costs.

A consideration of recent fieldwork on cooperative breeding behavior in which selection has been measured indicates that behaviors previously assumed to have evolved under altruistic selection are actually under mutualistic selection with direct benefits for both interacting parties. In order to better understand the evolution of social behavior, researchers need to actually measure the fitness consequences for both the actor and recipient involved in social interactions rather than assuming altruistic selection based on incomplete information. They also need to consider the potential effects of patterns of heritable variation on the evolutionary outcome of this selection.

REFERENCES

Achenbach, G. and C. T. Snowdon. 2002. "Costs of Caregiving: Weight Loss in Captive Adult Male Cotton-Top Tamarins (*Saguinus oedipus*) Following the Birth of Infants." *Int. J. Primatol.* 23:179–89.

Agrawal, A. F., E. D. Brodie III, and J. Brown. 2001. "Parent-Offspring Coadaptation and the Dual Genetic Control of Maternal Care." *Science* 292:1710–12.

Cheverud, J. 1984 "Evolution by Kin Selection: A Quantitative Genetic Model Illustrated by Maternal Performance in Mice." *Evolution* 38:766–77.

Clutton-Brock, T. 2002. "Breeding Together: Kin Selection and Mutualism in Cooperative Vertebrates." *Science* 296:69–72.

Crow, J. and M. Kimura. 1967. *An Introduction to Population Genetics Theory.* New York: Burgess.

Epple, G. 1978. "Reproductive and Social Behavior of Marmosets with Special Reference to Captive Breeding." *Primates Med.* 10:50–62.

Falconer, D. S. and T. Mackay. 1996. *An Introduction to Quantitative Genetics.* New York: Longman.

Fehr, E. and B. Rockenbach. 2003. "Detrimental Effects of Sanctions on Human Altruism." *Nature* 422:137–40.

Fisher, R. A. 1918. "The Correlation between Relatives on the Supposition of Mendelian Inheritance." *Trans. Roy. Soc. Edinb.* 52:399–433.

Grant, B. R. and P. R. Grant. 1989. *Evolutionary Dynamics of a Natural Population: The Large Cactus Finch of the Galápagos.* Chicago: University of Chicago Press.

Griffin, A. S. and S. A. West. 2002. "Kin Selection: Fact and Fiction." *TREE* 17:15–21.

Hamilton, W. D. 1964. "The Genetical Evolution of Social Behaviour I, II." *J. Theor. Biol.* 7:1–52.

Hunt, J. and L. W. Simmons. 2000. "Maternal and Paternal Effects on Offspring Phenotype in the Dung Beetle *Onthophagus taurus.*" *Evolution* 54:936–41.

Kingsolver, J. G., H. E. Hoekstra, J. M. Hoekstra, D. Berrigan, S. N. Vignieri, C. E. Hill, A. Hoang, P. Gibert, and P. Beerli. 2001. "The Strength of Phenotypic Selection in Natural Populations." *Am. Nat.* 157:245–61.

Kirkpatrick, M. and R. Lande. 1989. "The Evolution of Maternal Characters." *Evolution* 43:485–503.

Lande, R. 1979. "Quantitative Genetic Analysis of Multivariate Evolution, Applied to Brain: Body Size Allometry." *Evolution* 33:402–16.

Lande, R. and S. J. Arnold. 1983. "The Measurement of Selection on Correlated Characters." *Evolution* 37:1210–26.

Lynch, M. and B. Walsh. 1998. *Genetics and Analysis of Quantitative Traits.* Sunderland, MA: Sinauer Associates.

Moore, A. J. and P. F. Kukuk. 2002. "Quantitative Genetic Analysis of Natural Populations." *Nat. Rev. Genet.* 3:971–78.

Moore, A. J., K. F. Haynes, R. F. Preziosi, and P. J. Moore. 2002. "The Evolution of Interacting Phenotypes: Genetics and Evolution of Social Dominance." *Am. Natur.* 160:S186–97.

Mousseau, T. A. and D. A. Roff. 1987. "Natural Selection and the Heritability of Fitness Components." *Heredity* 59:181–97.

Peripato, A. C. and J. M. Cheverud. 2002. "Genetic Influences on Maternal Care." *Am. Natur.* 160:S173–85.

Preziosi, R. F. and D. J. Fairbairn. 2000. "Lifetime Selection on Adult Body Size and Components of Body Size in a Waterstrider: Opposing Selection and Maintenance of Sexual Size Dimorphism." *Evolution* 54:558–66.

Provine, W. 1971. *The Origins of Theoretical Population Genetics.* Chicago: University of Chicago Press.

Rauter, C. M. and A. J. Moore. 2001. "Quantitative Genetics of Growth and Development Time in the Burying Beetle *Nicrophorus pustulatus* in the Presence and Absence of Post-Hatching Parental Care." *Evolution* 56:96–110.

Russell, A. F., L. L. Sharpe, P. N. M. Brotherton, and T. H. Clutton-Brock. 2003. "Cost Minimization by Helpers in Cooperative Vertebrates." *Proc. Nat. Acad. Sci.* 100:3333–38.

Sussman, R. W. and P. A. Garber. 1987. "A New Interpretation of the Social Organization and Mating System of the Callitrichidae." *Int. J. Primatol.* 8:73–92.

Tardif, S. 1990. "Rearing Experience and Maternal Performance in Captive Cotton-Top Tamarins." *Am. J. Primatol.* 20:237.

Wade, M. J. and S. M. Shuster. 2002. "The Evolution of Parental Care in the Context of Sexual Selection: A Critical Reassessment of Parental Investment Theory." *Am. Natur.* 160:285–92.

Wilson, E. O. 1975. *Sociobiology: The New Synthesis.* Cambridge, MA: Harvard University Press.

Wolf, J. B. 2003. "Genetic Architecture and Evolutionary Constraint When the Environment Contains Genes." *Proc. Nat. Acad. Sci.* 100:4655–60.

Wolf, J. B., E. D. Brodie III, J. M. Cheverud, A. J. Moore, and M. J. Wade. 1998. "Evolutionary Consequences of Indirect Maternal Effects." *Trends in Ecology and Evolution* 13:64–70.

Woolfenden, G. E. and J. W. Fitzpatrick. 1985. *The Florida Scrub Jay: Demography of a Cooperative-Breeding Bird*. Princeton, NJ: Princeton University Press.

IV
Primate Sociality

8

Rethinking Sociality

Cooperation and Aggression among Primates

Robert W. Sussman and Paul A. Garber

Feeding competition is considered to be the driving force behind group-life.

— Wrangham 1980:288

Within-group competition for food and safe positions is a virtually inevitable and universal cost of group living.

— Janson 1988, cited in Sterck et al. 1997:293

Agonistic relationships are an especially important organizing feature in primate groups.

— Sterck et al. 1997:291

Among the characteristics passed down was a fairly typical mammalian set of aggressive behavior patterns. Many primate species defend their territories ferociously . . . territorial fights can be frequent and fierce. . . . The same applies to fights inside groups, where the most frequent aggression is between rival males.

— Wrangham and Peterson 1996:130–31

Sexual selection, the evolutionary process that produces sex differences, has a lot to answer for. Without it, males wouldn't possess dangerous bodily weapons and a mindset that sanctions violence. But males who are better fighters can stop other males from mating, and they mate more successfully themselves. Better fighters tend to have more babies.

— ibid.:173

In some monkey populations males kill other males
and then, with the competition thus muted, proceed to
copulate prolifically with females.
　　　　　　— (Goodwin 1992, quoted in Wright 1999:286)

THEORETICAL INTRODUCTION

The idea that competition and aggression, mainly over access to food and
sexual partners, are central to an understanding of the origins of group living and sociality in human and nonhuman primates remains a dominant
theory in primatology and behavioral ecology today. Using this paradigm,
competitive and aggressive interactions are expected to be widespread
among conspecifics residing in neighboring groups, as well as among
members of the same group (e.g., Wrangham 1980; van Schaik and van
Hooff 1983; Janson 1988; Wrangham and Peterson 1996; Sterck et al. 1997;
Cowlishaw and Dunbar 2000). Competition theory also is a fundamental
tenet of sociobiology or neo-Darwinian theories. As explained in sociobiological terms, driven by their selfish genes, individuals seek to maximize
their genetic contributions to the next generation (Wrangham and Peterson 1996). In order to do this, they compete with one another, resulting in
individual differences in reproductive success. Moreover, because females
are required to devote more energy than males in the production and care
of offspring, it is argued that they compete principally over resources such
as food and water. Males, on the other hand, invest little energy in reproduction directly and devote more of their efforts to maintaining priority or
exclusive access to sexual partners (e.g., Wilson 1975; Wrangham 1980,
1999; Cowlishaw and Dunbar 2000; Silk 2002a). Overall, within-species
competition (along with predation risk) is assumed to be a primary influence in the evolution of primate mating and social systems, and sociality
in general (van Schaik 1989; van Hooff and van Schaik 1994; Sterck et al.
1997).

Sociobiologists describe two types of competition: contest and scramble
(Nicholson 1954; Wilson 1975; Wrangham 1980; van Schaik 1989; Isbell
1991; Cowlishaw and Dunbar 2000). Contest competition occurs when
individuals compete directly over resources—it is measurable. The result
of contest competition is that subordinate individuals are excluded from
exploiting monopolizable resources in the presence of more dominant
individuals. Scramble competition, on the other hand, is difficult to measure directly. It is based on the assumption that individuals lose access to
resources because other group members have already used them. This is
more likely to occur at small, ephemeral, or highly dispersed feeding sites
at which animals that arrive first, regardless of social status, are able to
rapidly deplete the resource. In most instances, however, there is no

empirical basis for predicting if a later arriving animal would have obtained full, partial, or no access to the resource, or in fact whether it would have visited the feeding site at all. Given that scramble competition can only be inferred, its role in the evolution of individual foraging and social strategies remains unclear.

In a review of theories of competition, Janson (2000) outlines cases that appear to fit predictions derived from models based on scramble and contest competition. For example, species with scramble competition are said to have weak and undeveloped social bonds, while those that face strong contest competition have linear matrilineal dominance hierarchies and strong bonds. However, Pruetz evaluated the accuracy of the above models in a study of vervet and patas monkeys, and her findings ran counter "to the expectations of theories of feeding competition, which use broad patterns of food availability to predict patterns of contest competition and dominance among female primates" (1999:201). Her findings indicated that "models of female primate social behavior are too broad in the terminology used to describe conditions of food availability predicted to lead to contest competition" (ibid.:249). "Additionally, not all of the relevant variables are quantified by researchers who set out to test hypotheses related to questions of female primate social behavior and food availability" (ibid.:268). Nonetheless, according to sociobiological theory, competition is pervasive, a constant occurrence in most group-living primates, and a major factor influencing social structure (Janson 1988, 2000; van Schaik 1989; Isbell 1991; Janson and Goldsmith 1995; Sterck et al. 1997; Koenig, Beise, Chalise, and Ganzhorn 1998; Boinski et al. 2002).

In virtually all primate species, however, cooperative and affiliative interactions are considerably more common than agonistic interactions. For example, in white-faced capuchins (*Cebus capucinus*), Fedigan (1993) observed 1078 affiliative interactions compared to only 136 agonistic interactions. Similarly, in moustached tamarins (*Saguinus mystax*) the ratio of cooperative to aggressive interactions was reported to be 52:1 (Garber 1997). In ringtailed lemurs (*Lemur catta*), adult females spent approximately twenty-five minutes per day in active affiliative interactions and less than a minute per day in agonistic interactions (Sussman, Andrianasolondraibe, Soma, and Ichino 2003). After fifteen years of studying baboons, Strum observed that "aggression was not as pervasive or important an influence in evolution as had been thought, and that social strategies and social reciprocity were extremely important" (2001:158). We will develop this point further below. However, given a theoretical perspective of competition, how can we understand the context and prevalence of affiliative interactions among members of a social group?

Over the past two decades, many primatologists have described evidence of affiliation, alliance formation, and cooperation as a reaction or behavioral response designed to secure resources against other groups or group mates, or to counteract high levels of within-group aggression. Wrangham (1980) has argued that among female-bonded primate species, social groups evolved essentially to allow females, mainly kin, to fend off other groups in competition over resources (Wrangham 1980, 1999; van Schaik 1983; Dunbar 1988; Cowlishaw and Dunbar 2000). He states: "groups have evolved as a result of the benefits of cooperation, between allies competing against others of the same species" (Wrangham 1980:291). Within-group cooperation is seen in much the same way. Wrangham states: "The model suggests that cooperative behavior arises ultimately because it pays two subordinate animals to form alliances at the expense of a dominant. . . . Ecological pressures favour cooperation, while genetic considerations favour kin as partners" (ibid.:290–291). However, as the number of individual group members increases, so does the potential for feeding competition among individuals. As summarized in a recent textbook on evolutionary psychology: "The major cost of sociality is increased competition" (Gaulin and McBurney 2000:340).

Using a similar theoretical approach, the new field of research on reconciliation behavior initiated by Frans de Waal assumes that many friendly or affiliative behaviors are the result of competition and aggression. "The reconciliation hypothesis predicts that individuals try to 'undo' the social damage inflicted by aggression, hence, they will actively seek contact, specifically with former opponents" (de Waal 2000:587). "Reconciliation ensures the continuation of cooperation among parties with partially conflicting interests" (ibid.:589). In studies of reconciliation, however, it has been difficult to differentiate "friendly and affiliative" interactions from those that are considered "reconciliatory." This has led to problems in identifying and comparing the social function of cooperative behavior within and among primate species (Fuentes, Ray, and Dolhinow 1996; Fuentes et al. 2002; Silk 1997, 2002b; Arnold and Whiten 2001; Sanz, Weghorst, and Sussman 2001; Bernstein this volume; Fuentes this volume).

In summary, it is commonly argued that a consequence of social group living is an expected increase in feeding competition and in the rate of aggressive interactions among conspecifics (Wrangham 1980). Males are expected to compete for access to reproductive partners, and females are expected to compete for access to important feeding sites. A major tenet of sociobiology is that there are strong evolutionary advantages in forming affiliative bonds with group members, especially kin, if this helps your group outcompete individuals in other groups for access to resources. However, individual group members also must compete for resources and sexual partners within their social group. Opportunities for such competitive interactions may occur on an hourly or daily basis and threaten to weaken social bonds required for coordinated and cooperative group-level behavior. It is reasoned, therefore, that reconciliatory behaviors evolved to help reestablish social bonds fractured and refractured by within-group aggression and competition.

The framework described above has been used to interpret the social and mating systems of many primate species. However, sufficient data required to substantiate the basic assumptions of the "competition-aggression/affiliation-reconciliation" model have not been collected, and alternative theories on the causes of aggression and cooperation have not been adequately investigated. After over thirty years of studying baboons, Altmann states that

> some apparently competitive interactions may not be what they seem. When a baboon is feeding, it is sometimes interrupted. . . . The timing of these interruptions was not what would be expected if interruptions were a form of food competition. . . . Indeed, for most of the foods examined, interruption times were indistinguishable from random (time-independent) distributions. (1998:368–69)

Considering contest and scramble competition, Chapman and Chapman state: "The relative frequency of occurrence of these two types of competition has rarely been quantified" (2000:28). Smuts emphasizes that "aggression and affiliative behaviors of male and female primates vary depending on the species, the social context, and the individual" (1987:411). An understanding of this variation "awaits a clearer appreciation and investigation of the complex social environments in which these differences find their varied articulations" (ibid.).

We are concerned that some authors have accepted the competition-aggression/affiliation-reconciliation paradigm as a default explanation without critically evaluating its assumptions or appropriately testing alternative hypotheses. In particular, there can be considerable advantages to both kin and nonkin group members in developing dyadic, polyadic, and group-level affiliative and cooperative behaviors in which partners receive collective benefits (Dugatkin 1997; Clutton-Brock et al. 2001, 2002; Clutton-Brock 2002; Johnson, Kays, Blackwell, and Macdonald 2002; Korst-jens, Sterck, and Noe 2002; Stephens, McLinn, and Stevens 2002; Bernstein this volume; Cheverud this volume; Strier this volume). Theories on the importance of cooperative mutualism (see below) and other low-cost forms of social cooperation are generally lacking from the discussion of primate sociality.

A Call for a New Approach to the Study of Primate Sociality

We believe that there are two major problems with the competition-aggression/affiliation-reconciliation model of primate sociality as currently conceived. The first problem is theoretical and the second is empirical. The current paradigm assumes that competition is the main driving force behind both affiliative and agonistic social behavior. Certainly there is no question that affiliative, agonistic, and competitive behaviors are a consequence of social life and that agonism and competition can have a major effect on the life of individuals. However, there are reasons to believe that competition is not the main driving force of primate sociality and affiliative behavior.

We argue that primate sociality and affiliative, cooperative, as well as agonistic behaviors are best understood in terms of the mutual benefits and collective advantages that individuals obtain as members of a functioning social unit.

Evolution, generally, proceeds at an extremely slow pace, and, therefore, there is no justification to assume that we are observing dramatic evolutionary events in every population under study. For example, a number of experiments with thousands of generations of bacteria and fruit flies have demonstrated that rapid directional selection occurs only after critical environmental change. Little genetic change occurs once a population has adjusted to the environment (Lenski and Travisano 1994; Appenzeller 1999; Ehrlich 2000). As Ehrlich states: "The secret of biological evolution also is vast amounts of time" (2000:15). "The average levels of selection implied by the fossil record are almost impossible to detect in what is called 'ecological time'" (ibid.:34). If, instead, we assume that interactions between individuals are neutral or that the outcomes of such interactions are highly variable or random in relation to evolutionary phenomena at any particular point in time, then competition over food and mates may not be directly responsible for driving various aspects of sociality.

By neutral we mean that, although some of these events may be extremely important to a particular individual's fitness, they may have no direct effect on evolutionary changes in the population at large. After all, evolution occurs at the level of the population and in evolutionary and not ecological time. Both selection and random genetic drift occur through differential reproductive success of members of a population. Selection is differential reproductive success causally correlated with a phenotype, through interaction with the environment. Genetic drift occurs when differential success is random with respect to the phenotype. The life or death of an individual can contribute to both selection and genetic drift, depending on the characters considered. While variant alleles at one locus may be causally correlated with differential reproductive success and thus be under selection, variant alleles at unlinked loci will evolve under genetic drift. In fact, strong selection results in extreme genetic drift in loci not causally correlated to the phenotype. We cannot classify an individual death as a selective death or a random one without reference to a phenotype of interest and without comparison to others in the population with other phenotypes. The relationship between individual interactions and fitness or evolution must be understood in terms of specific phenotypes, where the phenotypes can be related to variance in related genotypes and to differential environmental influences on these phenotypes. The results of social interactions normally have not been examined at this level.

Under this set of assumptions, the current sociobiological paradigm fails to explain adequately the context, function, and social tactics underlying affiliative and agonistic behavior. In other words, if the purpose of behavioral interactions is proximate, and not solely to pass on one's own genes at the expense of others, there must be other factors driving social behavior and sociality among primates. Moreover, as indicated in recent models of resource distribution, except at very large group size, within-group feeding competition appears to play a much smaller role in social interactions than previously expected (Johnson et al. 2002). We will make some suggestions in the final section of the chapter.

Our second problem involves the database currently available to test theories of primate sociality. Data on the contexts, functions, and effectiveness of affiliative and agonistic behaviors in wild primates are extremely limited, especially among species that are arboreal or live in rainforests. In this chapter, we focus on the following three questions: How much time do different primate species actually spend in social behavior? How much of this behavior is friendly and how much is agonistic? How do these numbers vary among populations of the same species and among different species? This is simply a first step. Other questions should drive future research on primate sociality:

- Are sexually dimorphic species less affiliative and more aggressive than those that are not?
- Are terrestrial species more social/aggressive than arboreal ones?
- What are the contexts in which friendly and agonistic behaviors occur, and are the contexts consistent across species?
- Are there differences in the frequency and quality of social behaviors between kin, friends, and nonkin?
- What is the difference in the frequency and severity of agonistic interactions between these classes of individuals?
- When agonism is measured, are distinctions made between mild spats and more violent fights, and what are the patterns of these differences among different group members?
- Can one find a consistent pattern between affiliative and agonistic behavior across species, as well as among different populations of the same species?
- Are friendly and agonistic interactions independent of one another?

- What does this tell us about the concept of reconciliation?
- What are the costs and benefits to the interactants?
- Do cooperative behaviors actually involve a cost to the actor or do both the actor and the interactant benefit?
- And, finally, do the existing data strongly support one theoretical perspective over another?

We believe that, at the present time, none of these questions can be answered fully. Further, we will not attempt to answer them here. However, in order to illustrate the problem, we present and compare data on a basic question concerning primate sociality, namely, how much time do diurnal, social-living prosimians, New World monkeys, Old World monkeys, and apes spend in social behavior and how much of this time is affiliative versus agonistic?

METHODS

We have reviewed much, but certainly not all, of the literature on the socioecology of wild diurnal primates in order to identify the percentage of time group members spend in active social interactions, the size of the social group, the length of the study, and rates of agonism, affiliation, and aggression (events per individual per hour). Not all information was available in all studies. However, we included all of the studies in our sample unless the published data were transformed mathematically in such a way that it was not possible to reconstruct the basic information or presented in the form of histograms with no indication of sample size.

Our data set includes information on twenty-five genera, forty-nine species, and seventy-eight studies (Table 8.1). In these studies, generally, affiliative behaviors include grooming, playing, food sharing, huddling, and coalition or alliance formation of two or more individuals. Agonistic behaviors include fighting, visual or vocal threats, submissive gestures, and evidence of displacement. When collecting data on activity cycles, investigators normally include only "active" social interactions. Behaviors associated with what might be considered "passive" social interactions (such as resting in contact or coordinated activity) or social communication (such as vocal behavior or marking) are not included in these data. In some studies, mild agonistic interactions (which for clarity we refer to as "agonism" here) such as instantaneous spats and displacements are distinguished from more serious interactions (a subset of agonism and referred to as "aggression" here) such as biting, fighting, and extended chases. Where relevant, we will be making this distinction.

It is important to highlight several limitations in our data set. In general, researchers have used different definitions of common behavioral cate-

Table 8.1 Activity Budget and Rates of Agonism in Diurnal Primates

Species	Time Social (%)	Affiliative (%)	Agonistic	Reference
Diurnal Prosimians				
Varecia variegata	<1.0	rare	rare	1
Eulemur fulvus	<1.0	rare	rare	1
	1.0	1.0	rare	2
Eulemur rubriventer	2.0	2.0	rare	2
Lemur catta	2.6	2.1	0.5%	3
	2.6	1.9	0.7%	4
Eulemur fulvus	2.8	2.5	0.3%	4
Varecia variegata	3.1	3.1	0.02/hr male	5
			0.17/hr female	
Eulemur fulvus	3.5	3.5	rare	6
Propithecus verreauxi	3.8	no data	0.35/hr	7
Eulemur mongoz	4.0	4.0	ND	8
Eulemur coronatus	4.3	4.0	0.3%	9
Propithecus diadema	5.0	4.5	0.12/hr	10
Eulemur fulvus	5.8	5.4	0.4%	9
	8.5	7.7	0.8%	4
Lemur catta	8.6	7.1	1.4%	4
Overall Mean Prosimian	3.7 (± 2.3) (mean weighted by species = 3.68)			
New World Monkeys				
Alouatta palliata	0.8			11
	0.9			12
Brachyteles arachnoides	0.9			13
Callicebus torquatus	0.9			14
Ateles paniscus	0.9			15
Alouatta palliata	1.0			16
Cebus olivaceus	1.3			17
Alouatta seniculus	1.7			18
Saguinus mystax	1.8	1.4	0.41%	19
Alouatta palliata	1.9		0.003/ind/hr aggrs	20
Callicebus torquatus	1.9		0.0006/ind/hr aggrs	21
Alouatta palliata	2.0			16
Cebus olivaceus	2.1			17
Saguinus fuscicollis	2.2	1.9	0.35%	19
Brachyteles arachnoides	2.7		0.0006/ind/hr agon	22
Saguinus fuscicollis	2.8			23
Callithrix geoffroyi	2.8			24
Saimiri sciureus	2.9		0.0047/ind/hr aggrs	25
Leontopithecus rosalia	3.1			26
Saguinus mystax	3.1			23
	3.5		0.20%	27
			0.0066/ind/hr aggrs	
Alouatta pigra	3.9		0.10%	28
Lagothrix lagotricha	4.0	3.1	0.60%	29
	4.8			30
Alouatta caraya	4.9		0.38%	31
			0.019/ind/hr-aggrs	
Callithrix humeralifer	5.0	4.2		32

(continued)

Table 8.1 Activity Budget and Rates of Agonism in Diurnal Primates (*continued*)

Species	Time Social (%)	Affiliative (%)	Agonistic	Reference
Leontopithecus rosalia	6.1			33
	7.0			26
Leontopithecus chrysomelas	9.1	8.5		32
Cebus capucinus	9.9		0.016/ind/hr aggrs	34
Cebus apella	9.9			35
Cebus capucinus	12.5	7.7	0.034/ind/hr aggrs	36
Saimiri oerstedii	13.9		0.000009/ind/hr aggrs	37
Callithrix jacchus	14.0	11.1	0.051/ind/hr agon	38
Ateles geoffroyi	17.0		0.18%	39
	22.0		0.92%	39
Alouatta palliata			0.007/ind/hr-aggrs	40
Ateles geoffroyi			0.0043/ind.hr-aggrs	41
Cebus apella			0.042/ind/hr-aggrs	42
			0.007/ind/hr-aggrs	42
Leontopithecus rosalia			0.0012/ind/hr aggrs	43
Saguinus fuscicollis			0.01/ind/hr aggrs	44
Saguinus nigricollis			0.20%	45
Overall				
Mean NWM		5.1% (± 5.1) (mean weighted by species = 5.76)		
Old World Monkeys				
Colobus badius	1.9	1.8		46
Presbytis potenziani	1.9			47
Macaca silenus	2.4			48
Colobus badius	2.6	2.5		46
Macaca silenus	3.4			49
Colobus guereza	5.9	5.7	0 aggrs event in 7793 scans	50
Presbytis entellus	7.4	6.7		50
Colobus badius	8.0	7.6		46
Colobus badius	8.2	7.9		46
Colobus guereza	8.3	8.3	0 aggrs event in 8917 scans	51
Macaca silenus	8.4			49
Colobus badius	8.5	8.5		46
Macaca sylvanus	10.0			52
Papio anubis	10.4	5.2		53
Cercopithecus mitis	10.4			54
Cercopithecus l'hoesti	11.4			54
Macaca sylvanus	11.5			52
Rinopithecus bieti	13.1	9.8		55
Macaca nigra	18.7			56
Macaca fuscata	21.7	18.9		57
Macaca nigra	23.1			56
	23.5			56
Presbytis francoisi	27.9	27.5		58
Presbytis entellus			0.084/ind/hr females agon	59
			0.01/ind/hr males agon	60
Papio cynocephalus			0.14/ind/hr agon males	61
			0.11/ind/hr agon males	62
			0.079/ind/hr agon males	62
			0.14/ind/hr agon males	62

(*continued*)

Table 8.1 Activity Budget and Rates of Agonism in Diurnal Primates (*continued*)

Species	Time Social (%)	Affiliative (%)	Agonistic	Reference
Cercopithecus aethiops			0.0007/ind/hr aggrs	63
Erythrocebus patas			0.0007/ind/hr aggrs	63
Papio anubis			0.084/ind/hr agon males	53
Papio cynocephalus			0.037/ind/hr agon males	64
Overall				
Mean OWM	**10.7% (± 7.1) (mean weighted by species = 11.6)**			
Apes				
Pongo pygmaeus	1.6[a]			65
Hylobates lar	3.0			66
Gorilla gorilla	3.6			67
Hylobates muelleri	4.0			68
Pongo pygmaeus	5.4[a]			69
Gorilla gorilla	7.0	6.7	0.3%	70
Hylobates lar	11.0		0.009/hr	71
Hylobates syndactylus	15.0		0.15/hr	72
Pan troglodytes (includes resting)	22.0	9.0 (groom)	0.067/ind/hr agon males	73
Pan troglodytes	24.9	16.8 (groom)		74
Gorilla gorilla		0.39/hr	0.20/hr	75
Pan troglodytes			0.03/hr	76
			0.05/hr	77
			0.016/hr males	78
			0.009/hr females	78
Overall				
Mean apes	**9.7% (± 8.2) (mean weighted by species = 9.7)**			

[a]Orangutan data include any proximity between adults (76) or all individuals (77), not all of which would include active social interaction.

%, percentage of total activity budget—rare, rarely observed.

References

1: Vasey (1997); 2: Overdorff (1991); 3: Gould (1994); 4: Sussman (unpublished); 5: Morland (1991); 6: Tattersall (1977); 7: Richard (1978); 8: Curtis (1997); 9: Freed (1996); 10: Hemingway (1995); 11: Estrada et al. (1999); 12: Smith (1977); 13: Milton (1984); 14: Kinzey (1981); 15: Symington (1988); 16: Stone (1996); 17: Miller (1992 (1996); 18: Gaulin and Gaulin (1982); 19: Castro (1991); 20: Larose (1996); 21: Easley (1982); 22: Strier (1986 (1987); 23: Peres (1991); 24: Passamani (1998); 25: Mitchell (1990); 26: Dietz et al. (1997); 27: Garber (1997); 28: Silver et al. (1998); 29: Stevenson (1998); 30: Defler (1995); 31: Bicca-Marques (1993); 32: Rylands (1982); 33: Peres (1986); 34: Mitchell (1989); 35: Zhang (1995); 36: Fedigan (1993); 37: Boinski 1986); 38: Digby (1994); 39: Fedigan and Baxter (1984); 40: Jones (1980); 41: Klein and Klein (1977); 42: Janson 1984 (1988); 43: Baker (1991); 44: Goldizen et al. (1996); 45: de la Torre et al. 1995); 46: Decker (1994); 47: Fuentes (1996); 48: Kurup and Kumar (1993); 49: Menon and Poirer (1996); 50: Fashing (2001); 51: Newton (1992); 52: Ménard and Vallet (1997); 53: Harding (1980); 54: Kaplin and Moermond (2000); 55: Kirkpatrick et al. (1998); 56: O'Brien and Kinnaird (1997); 57: Agetsuma (1995); 58: Burton et al. (1995); 59: Borries et al. (1991); 60: Laws and Laws (1984); 61: Strum (1982); 62: Noe and Sluijter (1995); 63: Pruetz (1999); 64: Smuts (1985); 65: Rodman (1973); 66: Gittens and Raemaekers (1980); 67: Watts (1988); 68: Leighton (1987); 69: Galdikas (1985); 70: Olejniczak (in prep.); 71: Bartlett (1999); 72: Chivers (1974); 73: Boesch and Boesch-Achermann (2000); 74: Teleki (1981); 75: Schaller (1963); 76: Bygott (1974); 77: Ghigliari (1984); 78: Goodall (1986).

gories and have recorded their data using different sampling procedures. Moreover, different species and different individuals within the same species are likely to vary considerably in the expression and conspicuousness of social interactions. Therefore, the frequency of social behaviors in certain individuals may be overrepresented in the data, whereas in other individuals those same behaviors are underrepresented. In addition, most individuals spend the vast majority of their day in peaceful and close proximity to conspecifics; however, spatial proximity alone is usually not included in studies of daily activity cycle. With this in mind, we view published percentages and rates of social interactions as general values that are likely to have considerable variance. We also refrained from attempting sophisticated statistics on such varied data sets. Nonetheless, in those species for which we have values from more than a single study group, the percentage of time engaged in social behavior is quite constant. For example, in four studies of activity budgets in mantled howler monkeys (*Alouatta palliata*), the mean percentage of time spent in social behavior was 1.32 with a range of only 0.8–2.0 percent. In two studies of *Lagothrix lagotricha* the percentages ranged from 4.0 to 4.8 percent. Similar evidence of overall consistency in percentage of time engaged in social interactions was found among the prosimian, Old World monkey, and ape samples. We calculated the mean percentage of all studies within major taxonomic groups and a weighted mean for the percentage for each species. There was no significant difference between these means (Table 8.1). Thus, we have confidence that the data represent reasonable approximations of the true activity patterns of these primates and that the simple, descriptive statistics used are justified for our purpose.

RESULTS

Activity Budget

In reviewing the literature we found that diurnal prosimians spent an average of 3.7 percent (±2.3) of their activity budget engaged in direct social interactions (Table 8.1).

Studies in which lemurs were reported to engage in more frequent social behavior either were conducted during the mating season or involved cases in which observations of social interactions, grooming, and "other" were lumped together in a single category. In all studies but one, agonistic behavior accounted for less than 1 percent of the activity budget of diurnal lemurs (Table 8.1).

Similarly, most species of New World monkeys also spent only a small fraction of their day involved in overt social behavior. Over 72 percent of the groups studied in our platyrrhine sample (26/36) devoted 5 percent or less of their activity budget to social interactions (Table 8.1). This included behaviors such as grooming, fighting, playing, visual or vocal threats, copulations, displacement, huddling, active alliance formation, and food sharing. Mean percentage of activity budget devoted to social interaction in New World monkeys was 5.1 percent (±5.1). Only four species of New World monkeys, *Cebus capucinus, Saimiri oerstedii, Callithrix jacchus, and Ateles geoffroyi*, were found to devote more than 10 percent of their daily activity budget to social activities. In several of these species, the primary social activity was grooming. For example, although *C. jacchus* spent 14 percent of its activity budget in social behavior, 80 percent of that involved grooming (of the 14 percent of time spent in social behavior, 11.1 percent is grooming). Similarly, in *Leontopithecus chrysomelas*, grooming accounted for 94 percent of all social interactions (of the 9.1 percent of time spent in social behavior, 8.6 percent is grooming). These data strongly suggest that overt social interactions accounted for only a small part of an individual's daily activity budget, and that the vast majority of these interactions are affiliative rather than agonistic. Overall, there were no significant differences in the frequency of social interactions between diurnal prosimians and New World monkeys ($t = 1.3$, df = 50, $p = .17$).

In general, Old World monkeys are larger in body size and more sexually dimorphic in body mass and canine length than prosimians and New World monkeys. However, excluding allogrooming, the frequency of social interactions was found to be relatively similar. In 15 of the 21 studies reviewed (71 percent), social interactions accounted for between 2 and 12 percent of the daily activity budget. In the remaining 6 studies, social interactions accounted for between 13 and 28 percent (3 of which were of groups of the same species, *Macaca nigra*). The mean value for our Old World monkey sample was 10.7 percent (±7.1). This was significantly higher than the frequency of social interactions among New World monkeys ($t = 3.5$, df = 59, $p = 0.0006$) and prosimians ($t = 3.8$, df = 39, $p = .0004$). Old World monkeys groom more frequently than do lemurs and New World monkeys, and allogrooming alone accounted for most of the differences in the frequency of social interactions among these taxa. For example, in a study of Japanese macaques (Agetsuma 1995), social behavior accounted for 21.7 percent of the activity budget, 87 percent of which was grooming. In these macaques, nongrooming social interactions accounted for only 2.8 percent of the activity budget. Similarly, *Colobus guereza* was observed to engage in within-group social interactions during 8.3 percent of its activity budget, 81 percent of which was devoted to grooming. When grooming is omitted from the analysis, other forms of social interaction accounted for only 4.4 percent (±5.1) of the activity budget of Old World monkeys ($N = 12$ studies). This value is comparable to that found in New

World monkeys (t = 0.98, df = 15, p = .033) and diurnal lemurs (t = 0.50, df = 27, p = .061). Thus, allogrooming appears to take on an added significance among Old World monkeys.

The data on ape social interactions are quite variable because each genus exhibits a very different type of social structure. Whereas gibbons live in small, pair-bonded, family groups and gorillas are found in cohesive multimale/multifemale groups, individual chimpanzees come together intermittently in a fission-fusion type of society, with only mothers and their young usually remaining together throughout the activity cycle. Orangutans do not live in permanent social groupings and meet only rarely in the wild. Social interactions ranged from 3.6 percent of the activity budget of the mountain gorilla, to 4–15 percent in gibbons, to 25 percent in chimpanzees, with much of the latter being between females and their offspring (Table 8.1). As in the Old World monkeys, most ape social interactions take the form of grooming or bouts of play. Among our sample of lesser and great apes, 9.7 percent of the mean daily activity budget was devoted to within-group social interactions.

Agonism and Aggression

In many studies, agonistic and aggressive interactions are presented as a rate—that is, the number of events per observation hour. In some cases, rates are based on data collected using the "all-occurrence" method, in which all observations of these behaviors are recorded (Mitchell 1990; Bicca-Marques 1993; Calegaro-Marques and Bicca-Marques 1997). In other investigations, these interactions are scored only when they involve a focal animal or scan sample of unidentified individuals and occur during an instantaneous or predetermined sampling interval (Fedigan 1993; Garber 1997). The all-occurrence method is likely to show higher rates of agonistic behavior than the focal animal or scan sampling methods because this behavior occurs rarely. In addition, it is common for such data to be reported as a single category—"agonism"—and therefore it is impossible to separate mild spats, displacements, stares, and avoidance from more intense forms of agonistic interactions such as chasing, fighting, and biting that can result in severe injury, death, and social disruption. Below we discuss intense versus mild agonism when these data are available.

In the literature we searched on prosimians, the mean rate of agonism was 0.16 events per hour (N = 4; Table 8.2). Both New World and Old World monkeys averaged approximately 0.6 agonistic events per hour. Among apes, rates of agonism were extremely low, averaging 0.09 events per hour (Table 8.2). Our data on Old World monkeys indicate that rates of agonistic behavior ranged from 0.067 events per hour among male vervet monkeys to 1.19 events per hour among male cynocephalus baboons (N = 13 studies).

Table 8.2 Summary of Agonistic and Affiliative Social Interactions in Primates[a]

	Prosimians	NWM	OWM	Apes
Agonism				
Events per hour	0.16	0.60	0.58	0.09
Events per individual per week	0.002	3.6	6.3	0.0001
Social interactions (%) that are affiliative	93.2 ± 7.3	86.1 ± 10.5	84.8 ± 17.5	95.7[b]

[a]Data are calculated from information presented in Table 8.1. Events/hr represent studies using all occurrence data (recorded any time it is observed) on the number of agonistic social interactions recorded per observation hour. Data on events/ind/week are presented to account for the fact that by chance alone individuals in larger groups are more likely to vie for food or space more frequently than individuals in smaller groups. We assumed that animals are active 12–14 hours per day. Therefore, events/indiv/week represent 14 hours × 7 days or approximately 98 hours of observation. Data on percentage of social interactions that are affiliative were calculated using mean values for each species. In the case of apes, data are available for only one species and one study (*Gorilla gorilla*).
[b]Only one species.

In the five studies of *Papio cynocephalus* we examined, rates of agonism among males ranged from 0.67 to 1.19 events per hour (mean = 0.92).

Based on the expectation that within-group feeding and reproductive competition increase with increasing group size, where possible, we corrected the rate of agonism by the number of potential interactants in the group. Although we acknowledge that not all group members are equally involved in aggressive interactions, there is reason to assume that the number of aggressive events occurring per hour of observation is influenced by the number of potential interactants. In Tables 8.1 and 8.2, we present data on rates of agonism per adult group member per hour and per week (assuming a 12–14 hour active day). The mean values ranged from 6.3 times per individual per week in Old World monkeys to 3.6 times per individual per week in New World monkeys to extremely rare in the apes and prosimians (Table 8.2). The highest frequency of agonism per individual group member per week was 10–11 times in *P. cynocephalus*.

We found a small number of studies in which aggression was recorded separately from agonism. In 12 of 14 studies (85.7 percent) of New World monkeys, the average group member was involved in less than two aggressive interactions per week. Species characterized by the greatest rates of aggressive interactions were *C. capucinus* and *C. apella* (3.5 and 4.2 aggressive interactions per individual per week, respectively). For Old World monkeys there are three studies in which aggression (fighting and chasing) was separated from milder forms of agonistic interactions (two of *C. aethiops* and one of *E. patas*). In these three studies (all focusing on adult females), rates of aggression averaged 0.007 per individual per hour (range = 0.008–0.014), or 1 aggressive event every 142 hours for the entire group. How closely these values reflect levels of aggression in other species remains unclear. However, our values for apes (Table 8.2) support these very low rates of within-group aggressive interactions.

Cooperation and Affiliation

Quantitative data on social cooperation and affiliative behaviors other than grooming, playing, and huddling are not commonly reported in the literature, although qualitative accounts of these behaviors are available. Notable examples of cooperative and affiliative behaviors in primates include cooperative infant care and food sharing (e.g., Sussman and Kinzey 1984; Goldizen 1989; Garber 1997; Mitani and Watts 2001), male vigilance, and protection and defense of neonates (Boinski 1987; Rose and Fedigan 1995; Savage, Snowden, Giraldo, and Soto 1996; Gould, Fedigan, and Rose 1997; Treves 1998, 2000), alliance and friendship formation (e.g., Altmann 1980; Strum 1982; Smuts 1985; Strier 1993; Cords 2002; Silk 2002b), coordinated hunting (e.g., Rose 1997; Boesch and Boesch-Achermann 2000), and coordinated range and resource defense (e.g., see relevant papers in Boinski and Garber 2000).

In our sample, among diurnal prosimians ($N = 7$), 93.2 percent (range 78.5–99 percent) of all social interactions represent affiliative interactions (Table 8.2). In New World monkeys ($N = 10$), affiliative social interactions accounted for 86.1 percent (range 61.6–97.3 percent) of all social interactions. The percentage of affiliative social interactions in our Old World monkey sample ($N = 7$) was almost identical to that found in New World monkeys (84.8 percent, range 50–98.5 percent; Table 8.2). We found only one study on apes in which the percentage of time engaged in affiliative behavior was reported. This was a study of *Gorilla gorilla* by Olijniczak (in prep.). In this research, affiliative behaviors accounted for 95.7 percent of all social interactions in lowland gorillas. Clearly affiliative interactions represent the overwhelming majority of primate social interactions and form the basis of individual social bonds.

DISCUSSION

Within societies all across the planet, be they small nomadic groups of kin wandering through the grasslands or millions of unrelated individuals living in a metropolis, whether modern or prehistoric, cooperation is the glue that binds us together. It is difficult to even imagine a society in which cooperation, at some level or another, has not been integral. (Dugatkin 1999:2–3)

Published data on diurnal prosimians, New World monkeys, Old World monkeys, and apes indicate that most species devote between 3 and 11 percent of their activity budget to active social interactions and are characterized by extremely low rates of agonism and aggression. Langurs, macaques, spider monkeys, and chimpanzees tend to engage in social interactions more frequently than do other primates. This is explained principally by a higher incidence of allogrooming and a more persistent mother-infant bond in some species. Given the relatively low rates of agonistic and aggressive behavior that characterize most species of nonhuman primates, we question whether social affiliation and cooperation are best understood primarily as assets in competition and reconciliation or whether cooperative behaviors serve a greater role in the life of diurnal primates, generally.

We recognize that rarely occurring behaviors may be extremely important to an animal's survival and fully acknowledge that the frequency of an activity or behavior may not accurately measure the importance of that behavior in structuring the social organization of the group or species. Certainly a serious fight causing the injury or death of an individual will affect its life trajectory just as a rare predatory event can be disastrous to an individual or group. However, as stated above, evolution does not operate on individuals but, rather, on populations over vast amounts of time. Antipredator tactics, for example, are already in place when we are making our observations. Certainly, being eaten by a predator affects the fitness of the victim but whether it effects evolutionary change in the population is a much more complex matter. If the predator and prey population has been together for many millennia and there has been no change in their relationship, we are witnessing proximate consequences of the predatory event but evolutionary consequences are likely to be insignificant or nonexistent.

Variance in fitness provides the opportunity for selection but evolution by both selection and genetic drift occurs through differential reproductive success. For evolution to occur, natural selection must act upon underlying genetic variation. The problem is that most measurements of natural selection are limited to phenotypes. The underlying assumption behind many selection analyses is that there is a causal connection between fitness and the trait in question. However, environmental variables can independently affect fitness, and it has been shown that there is often a widespread lack of correspondence between predicted and observed evolutionary trajectories in natural populations (Merilä, Sheldon, and Kruuk 2001; Kruuk, Merilä, and Sheldon 2003).

Diurnal primates live in social groups. Why this is so is an historical, evolutionary question. Most primate social interactions are affiliative. Aggression and affiliation are necessary consequences of social life. That they exist requires no explanation. How they are used by individuals to solve changing social, reproductive, and ecological problems does require an explanation. If an individual's survival is enhanced by the collective advantages of living in a cohesive and socially integrated behavioral unit, then an understanding of the ability to maintain affiliative and coordinated behaviors and to minimize agonistic and eccentric behaviors is likely to provide critical insight into the evolution of sociality and group-living in primates. We hypothesize that affiliation is the major governing principle of primate sociality and that aggression and competition represent important but secondary features of daily primate social interaction.

Clutton-Brock (2002) has recently provided evidence that the benefits of cooperation in vertebrate societies, generally, may show parallels to those in human societies, where cooperation between unrelated individuals is frequent and social institutions are often maintained by generalized cooperation and reciprocity. Cooperation and affiliation represent behavioral tactics that can be used by group members to obtain resources, maintain or enhance their social position, increase reproductive opportunities, and reduce the stress of social isolation (Brown 1983; Sapolsky, Alberts, and Altmann 1997; Taylor et al. 2000; Strum 2001; Clutton-Brock 2002). Many affiliative and cooperative behaviors can be explained by individual actions that may benefit several individuals. In acts of cooperation both participants may receive immediate benefits from the interaction. Coordinated behaviors such as joint resource defense, range defense, cooperative hunting, alliance formation, cooperative food harvesting, and predator vigilance can be explained in terms of immediate benefits to participating individuals. Acts that appear to benefit recipients may also benefit actors. These benefits need not be equal for each individual. If the cost to the actors of affiliative behavior is low, even if the rewards are low and/or variable, we should expect affiliation and cooperation to be common. This intraspecific mutualism may help to explain observations that nonhuman primates live in relatively stable social groups and solve the problems of everyday life in a generally cooperative fashion.

Another type of cooperative behavior has been referred to as "by-product mutualism" or "no cost" cooperation (Dugatkin 1997). Brown described this as occurring when

> each animal must perform a necessary minimum itself that may benefit another individual as a by-product. These are typically behaviors that a solitary individual must do regardless of the presence of others, such as hunting for food. In many species these activities are more profitable in groups than alone. (1983:30)

Furthermore, Dugatkin states:

This category might be thought of as the simplest type of cooperation in that no kinship need be involved, nor are the cognitive mechanisms that require scorekeeping or the population structure required for group-selected cooperation necessary for by-product mutualism to evolve. As such by-product mutualism is "simple" in the sense of what is needed for cooperation to evolve, and this in turn might make it the most common category of cooperation, when all is said and done. (1997:31–32)

Recently, mathematical (e.g., Pepper and Smuts 2000; Hauert, De Monte, Hofbauer, and Sigmund 2002; West, Pen, and Griffin 2002) and genetic models (e.g., Cheverud, this volume) have been developed illustrating how the evolution of cooperation among nonrelated and nonreciprocating individuals can occur. Furthermore, there is growing evidence that endocrinological (e.g., Sapolsky et al. 1997; Carter, Lederhendler, and Kirkpatrick 1999; Taylor et al. 2000; Carter and Cushing, this volume) and neurological (Rilling et al. 2002; Ahn et al., this volume) mechanisms exist that help sustain, maintain, and positively reinforce social cooperation among mammals. In a recent set of experiments, Stephens et al. (2002) have shown that blue jays will cooperate and develop stable cooperative strategies, and give up immediate rewards, if continued cooperation leads to enhanced long-term benefits. Finally, there is evidence that, in general, resources are distributed heterogeneously in time and space and that, under these conditions, feeding competition and group living might be less costly than previously thought (Johnson et al. 2002). This has been referred to as the resource dispersion hypothesis (RDH). The RDH asserts that

resources might be distributed such that the smaller territory that will support a primary pair (or whatever the minimum social unit) might also support additional individuals at minimal cost to the primary occupants and without any requirement for cooperation between them. (Johnson et al. 2002:4)

Given these new findings, we might speculate on an evolutionary scenario that does not necessitate selfish genes, constant competition, complex calculations of kin recognition or relationships, or complicated predictions of future reciprocity. In experiments using MRI scans, mutual cooperation has been associated with consistent activation in two broad brain areas linked with reward processing, the anteroventral striatum and the orbitofrontal cortex. It has been proposed that activation of this neural network positively reinforces cooperative behavior (Rilling et al. 2002). Furthermore, both of these brain areas are rich in neurons able to respond to dopamine, the neurotransmitter known for its role in addictive behaviors. The dopamine system evaluates rewards—both those that that flow from the environment and those conjured up by the mind. When something good happens, dopamine is released, which makes the individual take some action. "The dopamine system works unconsciously and globally providing guidance for making decisions when there is not time to think things through" (Blakeslee 1999:347). In experiments with rats, for example, in which electrodes are placed in the striatum, the animals continue to press a bar to stimulate the electrodes, apparently receiving such pleasurable feedback that they will starve to death rather than stop pressing the bar. With these systems, investigators believe that they have identified a pattern of neural activation "that may be involved in sustaining cooperative social relationships, perhaps by labeling cooperative social interactions as rewarding" (Rilling et al. 2002:403).

Another physiological mechanism associated with affiliation and nurturing is the neurological circuitry related to maternal responses in mammals. Orchestrating the broad suite of maternal responses is the hormone called oxytocin. Oxytocin has been related to every type of animal bonding—parental, fraternal, sexual—and even to the capacity to soothe oneself (Angier 1999; Carter et al. 1999; Taylor et al. 2000). It has been suggested that, although its primary role may have been in forging the mother-infant bond, oxytocin's ability to influence brain circuitry may have been co-opted to serve other affiliative purposes that allowed the formation of alliances and partnerships, thus facilitating the evolution of cooperative behaviors.

If cooperation and spatial proximity among group-living animals are rewarding in a variety of circumstances and if physiological and neurological "feel good" feedback systems reinforce cooperative behavior, this behavior can persist in the absence of any conscious recognition that material gains also might flow from mutual cooperation. Animals appear to be wired to cooperate in certain circumstances (Carter et al. 1999; Taylor et al. 2000; Rilling et al. 2002; Carter and Cushing, this volume). Social affiliation and cooperative behaviors provide psychological, physiological, and ecological benefits that are reinforced by hormonal and neurological systems. These benefits may have their origin in the general benefits of mutual cooperation and in the strong maternal-infant bond and long juvenile period that characterize group-living primates. Oxytocin and endogenous opioid mechanisms may be at the core of basic cooperative social responses (Taylor et al. 2000). Thus, we propose that if cooperation is beneficial among group-living animals and if hormonal and neurological systems exist that reinforce cooperation by making individuals "feel good," cooperative behavior, if not selected against, could become a reward in its own right. This could explain the independent evolution of sociality in many animal lineages, as well as cooperation among nonrelatives and "nonselfish" altruistic behavior.

Again, we acknowledge the important role of aggression and competition in understanding primate social interactions. Our perspective, however, is that affiliation, cooperation, and social tolerance, associated with the long-term benefits of mutualism, form the core of social group living and that, in most instances, aggression and competition are better understood as social tactics and individual adjustments to the immediate and ephemeral conditions of particular social situations.

Finally, we highlight the importance of collecting data on the frequency and context of social behavior in order to better understand the proximate mechanisms that govern everyday interactions within social groups. We must better understand who does what to whom, how often, and when. As stressed by Silk: "we need to pay more attention to methodological details, such as how we should interpret information about the content, frequency, quality and patterns of social interactions" (2002b:440). Furthermore, since active social behavior generally takes up such a small proportion of an individual's time, social interactions must be understood within a wider context, such as the general activity pattern, the life history of the individual, group and population demography, and potential recent perturbations to the ecosystem that may affect the group or population.

Until we have a better understanding of these proximate mechanisms, hypotheses concerning evolutionary explanations of cooperation, agonism, and sociality may be misleading. In any case, we agree with Clutton-Brock that

> if mutualism proves to be important in maintaining cooperative animal societies, the benefits of cooperation in animals may be more similar to those of cooperation in humans than has been previously supposed. In humans, unrelated individuals commonly assist each other . . . [and] generalized reciprocity appears to be important in maintaining many social institutions . . . [these] trends appear to have close parallels in other cooperative animals. (2002:72)

ACKNOWLEDGMENTS

We would like to acknowledge the assistance and encouragement of Audrey Chapman and Jim Miller of the American Association for the Advancement of Science, Program of Dialogue between Ethics, Science and Religion, for their continued assistance and encouragement and for making the symposium that led to this volume possible. We would also like to thank the following for their comments on earlier drafts of this paper: Jim Cheverud, Agustin Fuentes, Lisa Gould, Jane Phillips-Conroy, Katherine Mackinnon, Michelle Sauther, Karen Strier, Linda Sussman, and the participants of the AAAS symposium. We also thank Diana,

Jenni, Katya, and Sara for sharing their insights into factors affecting cooperation and aggression in primate interactions.

REFERENCES

Agetsuma, N. 1995. "Foraging Strategies of Yakushima macaques (*Macaca fuscata yakui*)." *Int. J. Primatol.* 16:595–610.

Altmann, J. 1980. *Baboon Mothers and Infants.* Cambridge, MA: Harvard University Press.

Altmann, S. A. 1998. *Foraging for Survival: Yearling Baboons in Africa.* Chicago: Chicago University Press.

Angier, N. 1999. "Illuminating How Bodies Are Built for Sociality." Pp. 350–52 in *The Biological Basis of Human Behavior: A Critical Review,* edited by R. W. Sussman. Upper Saddle River, NJ: Prentice Hall.

Appenzeller, T. 1999. "Test Tube Evolution Catches Time in a Bottle." *Science* 284:2108–10.

Arnold, K. and A. Whiten. 2001. "Post-Conflict Behaviour in Wild Chimpanzees (*Pan troglodytes schweinfurthii*) in the Budongo Forest, Uganda." *Behaviour* 138: 649–90.

Baker, A. J. 1991. *Evolution of the Social System of the Golden Lion Tamarin* (Leontopithecus rosalia): *Mating System, Group Dynamics and Cooperative Breeding.* Ph.D. thesis, University of Maryland, College Park.

Bartlett, T. Q. 1999. *Feeding and Ranging Behavior of the White-Handed Gibbon* (Hylobates lar) *in Khao Yai National Park, Thailand.* Ph.D. thesis, Washington University, St. Louis, MO.

Bicca-Marques, J. C. 1993. "Padrao de atividades diarias do bugio-preto *Alouatta caraya* (Primates, Cebidae): uma analise temporal e bioenergetica." Pp. 35–49 in *A Primatologia no Brasil—4,* edited by M. E. Yamamoto and M. B. C. Sousa. Natal: Editora Universitaria-UFRN (Universidade Federal do Rio Grande do Norte).

Blakeslee, S. 1999. "How Brain Uses a Simple Dopamine System." Pp. 347–49 in *The Biological Basis of Human Behavior: A Critical Review,* edited by R. W. Sussman. Upper Saddle River, NJ: Prentice Hall.

Boesch, C. and H. Boesch-Achermann. 2000. *The Chimpanzees of the Tai Forest: Behavioural Ecology and Evolution.* New York: Oxford University Press.

Boinski, S. 1986. *The Ecology of Squirrel Monkeys in Costa Rica.* Ph.D. thesis, The University of Texas at Austin.

Boinski, S. 1987. "Mating Patterns in Squirrel Monkeys (*Saimiri oerstedi*)." *Behav. Ecol. Sociobiol.* 21:13–21.

Boinski, S. and P. A. Garber (Eds.). 2000. *On the Move: How and Why Animals Travel in Groups.* Chicago: University of Chicago Press.

Boinski, S., K. Sughrue, L. Selvaggi, R. Quantrone, M. Henry, and S. Cropp. 2002. "An Expanded Test of the Ecological Model of Primate Social Evolution: Competition Regimes and Female Bonding in Three Species of Squirrel Monkeys (*Saimiri oerstedii, S. boliviensis and S. sciureus*)." *Behaviour* 139:227–62.

Borries, C., V. Sommer, and A. Srivastave. 1991. "Dominance, Age, and Reproductive Success in Free-Ranging Female Hanuman Langurs (*Presbytis entellus*)." *Int. J. Primatol.* 12:231–58.

Brown, J. L. 1983. "Cooperation: A Biologist's Dilemma." Pp. 1–37 in *Advances in the Study of Behaviour,* edited by J. S. Rosenblatt. New York: Academic Press.

Burton, F. D., K. A. Snarr, and S. E. Harrison. 1995. "Preliminary Report on *Presbytis francoisi leucocephalus.*" *Int. J. Primatol.* 16:311–27.

Bygott, J. D. 1974. *Agonistic Behaviour and Dominance in Wild Chimpanzees.* Ph.D. thesis, Cambridge University.

Calegaro-Marques, C. and J. C. Bicca-Marques. 1997. "Comportamento agressivo em um grupo de bugios-pretos, *Alouatta caraya* (Primates, Cebidae)." Pp. 29–38 in *A Primatologia no Brasil—5,* edited by S. F. Ferrari and H. Schneider. Belém: SBPr/UFPA (Sociedade Brasileira de Primatologia/Universidade Federal do Para).

Carter, S., I. Lederhendler, and B. Kirkpatrick (Eds.). 1999. *The Integrative Neurobiology of Affiliation.* Boston: MIT Press.

Castro, N. R. 1991. *Behavioral Ecology of Two Coexistent Tamarin Species* (Saguinus fuscicollis nigrifrons *and* Saguinus mystax mystax, *Callitrichidae, Primates*) *in Amazonian Peru.* Ph.D. thesis, Washington University, St. Louis, MO.

Chapman, C. A. and L. J. Chapman. 2000. "Determinants of Group Size in Primates: The Importance of Travel Costs." Pp. 24–42 in *On the Move: How and Why Animals Travel in Groups,* edited by S. Boinski and P. A. Garber. Chicago: University of Chicago Press.

Chivers, D. J. 1974. "The Siamang in Malaya: A Field Study of a Primate in Tropical Rain Forest." *Contrib. Primatol.* 4:1–335.

Clutton-Brock, T. 2002. "Breeding Together: Kin Selection and Mutualism in Cooperative Vertebrates." *Science* 296:69–72.

Clutton-Brock, T., P. N. M. Brotherton, A. F. Russell, M. J. O'riain, D. Gaynor, R. Kansky, A. Griffin, M. Manser, L. Sharpe, G. M. Mcilrath, T. Small, A. Moss, and S. Monfort. 2001. "Cooperation, Control, and Concession in Meerkat Groups." *Science* 291:478–81.

Clutton-Brock, T., A. F. Russell, L. L. Sharpe, A. J. Young, Z. Balmforth, and G. M. Mcilrath. 2002. "Evolution and Development of Sex Differences in Cooperative Behavior in Meerkats." *Science* 297:253–56.

Cords, M. 2002. "Friendship among Adult Female Blue Monkeys (*Cercopithecus mitis*)." *Behaviour* 139:291–314.

Cowlishaw, G. and R. Dunbar. 2000. *Primate Conservation Biology.* Chicago: University of Chicago.

Curtis, D. J. 1997. *The Mongoose Lemur* (Eulemur mongoz)*: A Study in Behavior and Ecology.* Ph.D. thesis, University of Zurich.

de la Torre, S., F. Campos, and T. de Vries. 1995. "Home Range and Birth Seasonality of *Saguinus nigricollis* in Ecuadorian Amazonia." *Am. J. Primatol.* 37:39–56.

de Waal, F. B. M. 2000. "The First Kiss: Foundations of Conflict Resolution Research in Animals." Pp. 15–33 in *Natural Conflict Resolution,* edited by F. Aureli and F. B. M. de Waal. Berkeley: University of California Press.

Decker, B. S. 1994. "Effects of Habitat Disturbance on the Behavioral Ecology and Demographics of the Tana River Red Colobus (*Colobus badius rufomitratus*)." *Int. J. Primatol.* 15:703–37.

Defler, T. R. 1995. "The Time Budget of a Group of Wild Woolly Monkeys (*Lagothrix lagotricha*)." *Int. J. Primatol.* 16:107–20.

Dietz, J. M., C. A. Peres, and L. Pinder. 1997. "Foraging Ecology and Use of Space in Wild Golden Lion Tamarins (*Leontopithecus rosalia*)." *Am. J. Primatol.* 41: 289–305.

Digby, L. J. 1994. *Social Organization and Reproductive Strategies in a Wild Population of Common Marmosets* (Callithrix jacchus). Ph.D. thesis, University of California, Davis.

Dugatkin, L. A. 1997. *Cooperation among Animals: An Evolutionary Perspective.* New York: Oxford University Press.

Dugatkin, L. A. 1999. *Cheating Monkeys and Citizen Bees: The Nature of Cooperation in Animals and Humans.* New York: Free Press.

Dunbar, R. I. M. 1988. *Primate Social Systems.* Ithaca, NY: Cornell University Press.

Easley, S. P. 1982. *Ecology and Behavior of* Callicebus torquatus: *Cebidae, Primates.* Ph.D. thesis, Washington University, St. Louis, MO.

Ehrlich, P. R. 2000. *Human Natures: Genes, Cultures, and the Human Prospect.* Washington, DC: Island Press.

Estrada, A., S. Juan-Solano, T. O. Martinez, and R. Coates-Estrada. 1999. "Feeding and General Activity Patterns of a Howler Monkey (*Alouatta palliata*) Troop Living in a Forest Fragment at Los Tuxtlas, Mexico." *Am. J. Primatol.* 48:167–83.

Fashing, P. J. 2001. "Activity and Ranging Patterns of Guerezas in the Kakamega Forest: Intergroup Variation and Implications for Intragroup Feeding Competition." *Int. J. Primatol.* 22:549–78.

Fedigan, L. 1993. "Sex Differences and Intersexual Relations in Adult White-Faced Capuchins." *Int. J. Primatol.* 14:853–77.

Fedigan, L. M. and M. J. Baxter. 1984. "Sex Differences and Social Organization in Free-Ranging Spider Monkeys (*Ateles geoffroyi*)." *Primates* 25:279–94.

Freed, B. Z. 1996. *Co-occurrence among Crowned Lemurs* (Lemur coronatus) *and Sanford's Lemurs* (Lemur fulvus sanfordi) *of Madagascar.* Ph.D. thesis, Washington University, St. Louis, MO.

Fuentes, A. 1996. "Feeding and Ranging in the Mentawai Island Langur (*Presbytis potenziani*)." *Int. J. Primatol.* 17:525–48.

Fuentes, A., N. Malone, C. Sanz, M. Matheson, and L. Vaughn. 2002. "Conflict and Post-Conflict Behavior in a Small Group of Chimpanzees." *Primates* 43:233–35.

Fuentes, A., E. Ray, and P. Dolhinow. 1996. "Post-Agonistic Interactions in the Hanuman Langur: 'Reconciliation' or Not?" *Am. J. Phys. Anthropol.* 22(Supplement):107.

Galdikas, B. M. F. 1985. "Orangutan Sociality in Tanjung Putting." *Am. J. Primatol.* 9:101–19.

Garber, P. A. 1997. "One for All and Breeding for One: Cooperation and Competition as a Tamarin Reproductive Strategy." *Evol. Anthropol.* 5:187–99.

Gaulin, S. J. C. and C. K. Gaulin. 1982. "Behavioral Ecology of *Alouatta seniculus* in Andean Cloud Forest." *Int. J. Primatol.* 3:1–32.

Gaulin, S. J. C. and McBurney, D. H. 2000. *Psychology: An Evolutionary Approach.* Saddle River, NJ: Prentice Hall.

Ghigliari, M. P. 1984. *The Chimpanzees of Kibale Forest.* New York: Columbia University Press.

Gittens, G. P. and J. J. Raemaekers. 1980. "Siamang, Lar and Agile Gibbons." Pp. 63–105 in *Malayan Forest Primates,* edited by D. J. Chivers. New York: Plenum Press.

Goldizen, A. W. 1989. "Social Relationships in a Cooperatively Polyandrous Group of Saddleback Tamarins (*Saguinus fuscicollis*)." *Behav. Ecol. Sociobiol.* 24:79–89.

Goldizen, A. W., J. Mendelson, M. van Vlaardingen, and J. Terborgh. 1996. "Saddle-Back Tamarin (*Saguinus fuscicollis*) Reproductive Strategies: Evidence from a Thirteen-Year Study of a Marked Population." *Am. J. Primatol.* 38:57–83.

Goodall, J. 1986. *The Chimpanzees of Gombe: Patterns of Behavior.* Cambridge, MA: Harvard University Press.

Goodwin, F. K. 1992. "Conduct Disorder as Precursor to Adult Violence and Substance Abuse: Can the Progression Be Halted?" Address to the American Psychiatric Association, May 5, Washington, DC.

Gould, L. 1994. *Patterns of Affiliative Behavior in Adult Male Ringtailed Lemurs* (Lemur catta) *at the Beza-Mahafaly Reserve, Madagascar.* Ph.D. thesis, Washington University, St. Louis, MO.

Gould, L., L. M. Fedigan, and L. M. Rose. 1997. "Why Be Vigilant? The Case of the Alpha Animal." *Int. J. Primatol.* 18:401–14.

Harding, R. S. O. 1980. "Agonism, Ranking, and the Social Behavior of Adult Male Baboons." *Am. J. Phys. Anthropol.* 53:203–16.

Hauert, C., S. De Monte, J. Hofbauer, and K. Sigmund. 2002. "Volunteering as Red Queen Mechanism for Cooperation and Public Goods Games." *Science* 296:1129–32.

Hemingway, C. A. 1995. *Feeding and Reproductive Strategies of the Milne-Edwards' sifaka,* Propithecus diadema edwardsi. Ph.D. thesis, Duke University, Durham, NC.

Isbell, L. A. 1991. "Contest and Scramble Competition: Patterns of Female Aggression and Ranging Behaviour among Primates." *Behav. Ecol.* 2:143–55.

Janson, C. H. 1984. "Female Choice and Mating System of the Brown Capuchin Monkey, *Cebus apella* (Primates, Cebidea)." *Zeitschrift Tierpsychologie* 65: 177–200.

Janson, C. H. 1988. "Food Competition in Brown Capuchin Monkeys (*Cebus apella*): Quantitative Effects of Group Size and Tree Productivity." *Behaviour* 105: 53–76.

Janson, C. H. 2000. "Primate Socio-Ecology: The End of a Golden Age." *Evolutionary Anthropology* 9:73–86.

Janson, C. H. and M. Goldsmith. 1995. "Predicting Group Size in Primates: Foraging Costs and Predation Risks." *Behav. Ecol* 6:326–36.

Johnson, D. P., R. Kays, P. G. Blackwell, and D. W. Macdonald. 2002. "Does the Resource Dispersion Hypothesis Explain Group Living?" *Trends in Ecol. and Evol.* 17:563–70.

Jones, C. B. 1980. "The Functions of Status in the Mantled Howler Monkey, *Alouatta palliata* Gray: Intraspecific Competition for Group Membership in a Folivorous Neotropical Primate." *Primates* 21:389–405.

Kaplin, B. A. and T. C. Moermond. 2000. "Foraging Ecology of the Mountain Monkey (*Cercopithecus l'hoesti*): Implications for Its Evolutionary History and Use of Disturbed Forest." *Am. J. Primatol.* 50:227–46.

Kinzey, W. G. 1981. "The Titi Monkeys, Genus *Callicebus.*" Pp. 241–76 in *Ecology and Behavior of Neotropical Primates,* Vol. 1, edited by A. F. Coimbra-Filho and R. A. Mittermeier. Rio de Janeiro: Academic Basileira de Ciencias.

Kirkpatrick, R. C., Y. C. Long, T. Zhong, and L. Xiao. 1998. "Social Organization and Range Use in the Yunnan Snub-Nosed." *Int. J. Primatol.* 19:13–52.

Klein, L. L. and D. J. Klein. 1977. "Feeding Behaviour of the Colombian Spider Monkey *Ateles belzebuth.*" Pp. 153–81 in *Primate Ecology,* edited by T. H. Clutton-Brock. London: Academic Press.

Koenig, A., J. Beise, M. K. Chalise, and J. U. Ganzhorn. 1998. "When Females Should Contest for Food—Testing Hypotheses about Resource Density, Distribution, Size, and Quality with Hanuman Langurs (*Presbytis entellus*)." *Behav. Ecol. Sociobiol.* 42:225–37.

Korstjens, A. H., E. H. M. Sterck, and R. Noe. 2002. "How Adaptive or Phylogenetically Inert Is Primate Social Behaviour? A Test with Two Sympatric Colobines." *Behaviour* 139:203–25.

Kruuk, L. E. B., J. Merilä, and B. C. Sheldon. 2003. "When Environmental Variation Short-Circuits Natural Selection." *Trends Ecol. and Evol.* 18:207–9.

Kurup, G. U., and A. Kumar. 1993. "Time Budget and Activity Patterns of the Lion-Tailed Macaque (*Macaca silenus*)." *Int. J. Primatol.* 14:27–39.

Larose, F. 1996. *Foraging Strategies, Group Size, and Food Competition in the Mantled Howler Monkey,* Alouatta palliata. Ph.D. thesis, University of Alberta, Edmonton.

Laws, J. W. and J. V. H. Laws. 1984. "Social Interactions among Adult Male Langurs (*Presbytis entellus*) at Rajaji Wildlife Sanctuary." *Int. J. Primatol.* 5:31–50.

Leighton, D. R. 1987. "Gibbons: Territoriality and Monogamy." Pp. 135–45 in *Primate Societies,* edited by B. Smuts, D. Cheney, R. Seyfarth, R. Wrangham, and T. Struhsaker. Chicago: University of Chicago Press.

Lenski, R. E. and M. Travisano. 1994. "Dynamics of Adaptation and Diversification: A 10,000 Generation Experiment with Bacterial Populations." *Proc. Nat. Acad. Sci. USA* 91:6808–14.

Ménard, N. and D. Vallet. 1997. "Behavioral Responses of Barbary Macaques (*Macaca sylvanus*) to Variations in Environmental Conditions in Algeria." *Am. J. Primatol.* 43:285–304.

Menon, S. and F. E. Poirer. 1996. "Lion-Tailed Macaques (*Macaca silenus*) in a Disturbed Forest Fragment: Activity Patterns and Time Budget." *Int. J. Primatol.* 17:969–85.

Merilä, J., B. C. Sheldon, and L. E. B. Kruuk. 2001. "Explaining Stasis: Microevolutionary Studies of Natural Populations." *Genetica* 112:119–22.

Miller, L. E. 1992. *Socioecology of the Wedge-Capped Capuchin Monkey* (Cebus olivaceus). Ph.D. thesis, University of California, Davis.

Miller, L. E. 1996. "The Behavioral Ecology of Wedge-Capped Capuchin Monkeys (*Cebus olivaceus*)." Pp. 271–88 in *Adaptive Radiations of Neotropical Primates,* edited by M. A. Norconk, A. L. Rosenberger, and P. A. Garber. New York: Plenum.

Milton, K. 1984. "Habitat, Diet, and Activity Patterns of Free-Ranging Woolly Spider Monkeys (*Brachyteles arachnoides,* E. Geoffroy 1806)." *Int. J. Primatol.* 5:491–514.

Mitani, J. and D. Watts. 2001. "Why Do Chimpanzees Hunt and Share Meat?" *Anim. Behav.* 61:915–24.

Mitchell, B. J. 1989. *Resources, Group Behavior, and Infant Development in White-Faced Capuchin Monkeys,* Cebus capucinus. Ph.D. thesis, University of California, Berkeley.

Mitchell, C. L. 1990. *The Ecological Basis for Female Social Dominance: A Behavioral Study of the Squirrel Monkey* (Saimiri sciureus) *in the Wild.* Ph.D. thesis. Princeton University, Rutgers, NJ.

Morland, H. S. 1991. "Preliminary Report on the Social Organization of Ruffed Lemurs (*Varecia variegata variegata*) in a Northeast Madagascar Rainforest." *Folia Primatol.* 56:157–61.

Newton, P. 1992. "Feeding and Ranging Patterns of Forest Hanuman Langurs (*Presbytis entellus*)." *Int. J. Primatol.* 13:245–86.

Nicholson, A. J. 1954. "An Outline of the Dynamics of Animal Populations." *Aust. J. Zool.* 2:9–65.

Noe, R. and A. A. Sluijter. 1995. "Which Adult Male Savanna Baboons Form Coalitions?" *Int. J. Primatol.* 16:77–106.

O'Brien, T. G. and M. G. Kinnaird. 1997. "Behavior, Diet, and Movements of the Sulawesi Crested Black Macaque (*Macaca nigra*)." *Int. J. Primatol.* 18:321–51.

Olejniczak, C. In preparation. *The Social System of Western Lowland Gorillas* (Gorilla gorilla gorilla *Savage and Wyman 1847) at Mbeli Bai in the Nouabale-Ndoki National Park, Northern Republic of Congo and the Value of Bai Vegetation as a Food Resource.* Ph.D. thesis, Washington University, St. Louis, MO.

Overdorff, D. J. 1991. *Ecological Correlates to Social Structure in Two Prosimian Primates:* Eulemur fulvus rufus *and* Eulemur rubriventer *in Madagascar.* Ph.D. thesis, Duke University, Durham, NC.

Passamani, M. 1998. "Activity Budget of the Geoffroy's Marmoset (*Callithrix geoffroyi*) in an Atlantic Forest in Southeastern Brazil." *Am. J. Primatol.* 46:333–40.

Pepper, J. W. and B. B. Smuts. 2000. "The Evolution of Cooperation in an Ecological Context: An Agent-Based Model." Pp. 45–76 in *Dynamics in Human and Primate Societies,* edited by T. A. Kohler and G. J. Gumerman. New York: Oxford University Press.

Peres, C. A. 1986. *Costs and Benefits of Territorial Defense in Golden Lion Tamarins* (Leontopithecus rosalia). M.A. thesis, University of Florida, Gainesville.

Peres, C. A. 1991. *Ecology of Mixed-Species Groups of Tamarins in Amazonian Terra Firme Forests.* Ph.D. thesis, Cambridge University.

Pruetz, J. D. 1999. *Socioecology of Adult Female Vervet* (Cercopithecus aethiops), *and patas monkeys* (Erythrocebus patas) *in Kenya: Food Availability, Feeding Competition, and Dominance Relationships.* Ph.D. thesis, University of Illinois, Urbana.

Richard, A. F. 1978. *Behavioral Variation: Case Study of a Malagasy Lemur.* Lewisburg, PA: Bucknell University Press.

Rilling, J. K., D. A. Gutman, T. R. Zeh, G. Pagnoni, G. S. Berns, and C. D. Kitts. 2002. "A Neural Basis for Social Cooperation." *Neuron* 35:395–405.

Rodman, P. S. 1973. "Population Composition and Adaptive Organization among Orang-utans of the Kutai Reserve." Pp. 171–209 in *Comparative Ecology and Behavior of Primates,* edited by R. P. Michael and J. H. Crook. New York: Academic Press.

Rose, L. M. 1997. "Vertebrate Predation and Food Sharing in *Cebus* and *Pan*." *Int. J. Primatol.* 18:727–65.

Rose, L. M. and L. M. Fedigan. 1995. "Vigilance in White-Faced Capuchins, *Cebus capucinus*, in Costa Rica." *Animal Behaviour* 49:63–70.

Rylands, A. B. 1982. *The Behaviour and Ecology of Three Species of Marmosets and Tamairns (Callitrichidae, Primates) in Brazil.* Ph.D. thesis, University of Cambridge.

Sanz, C. S., J. A. Weghorst, and R. W. Sussman. 2001. "An Examination of the Relationship between Post-Conflict Interaction Method and Theory." *Am. J. Phys. Anthropol. Supplement* 32:130.

Sapolsky, R. M., S. C. Alberts, and J. Altmann. 1997. "Hypercortisolism Associated with Social Subordinance or Social Isolation among Wild Baboons." *Arch. Gen. Psych.* 54:1137–43.

Savage, A., C. T. Snowden, L. H. Giraldo, and L. H. Soto. 1996. "Paternal Care Patterns and Vigilance in Wild Cotton-Top Tamarins (*Saguinus oedipus*)." Pp. 187–99 in *Adaptive Radiations of Neotropical Primates,* edited by M. A. Norconk, A. L. Rosenberger, and P. A. Garber. New York: Plenum Press.

Schaller, G. B. 1963. *The Mountain Gorilla.* Chicago: University of Chicago Press.

Silk, J. 1997. "The Function of Peaceful Post-Conflict Contacts among Primates." *Primates* 38:265–79.

Silk, J. 2002a. "Introduction" to special issue: "What Are Friends For? The Adaptive Value of Social Bonds in Primate Groups." *Behaviour* 139:173–76.

Silk, J. 2002b. "Using the 'F'-Word in Primatology." *Behaviour* 139:421–46.

Silver, S. C., L. E. T. Ostro, C. P. Yeager, and R. Horwich. 1998. "Feeding Ecology of the Black Howler Monkey (*Alouatta pigra*) in Northern Belize." *Am. J. Primatol.* 45:263–79.

Smith, C. C. 1977. "Feeding Behavior and Social Organization in Howling Monkeys." Pp. 97–129 in *Primate Ecology,* edited by T. H. Clutton-Brock. San Francisco, CA: Academic Press.

Smuts, B. B. 1985. *Sex and Friendship in Baboons.* Hawthorne, NY: Aldine de Gruyter.

Smuts, B. B. 1987. "Sexual Competition and Mate Choice." Pp. 385–420 in *Primate Societies,* edited by B. Smuts, D. Cheney, R. Seyfarth, R. Wrangham, and T. Struhsaker. Chicago: University of Chicago Press.

Stephens, D. W., C. M. McLinn, and J. R. Stevens. 2002. "Discounting and Reciprocity in an Iterated Prisoner's Dilemma." *Science* 298:2216–18.

Sterck, E. A., D. P. Watts, and C. P. van Schaik. 1997. "The Evolution of Female Social Relationships in Primates." *Behav. Ecol. Sociobiol.* 41:291–310.

Stevenson, P. R. 1998. "Proximal Spacing between Individuals in a Group of Woolly Monkeys (*Lagothrix lagotricha*) in Tinigua National Park, Colombia." *Int. J. Primatol.* 19:299–311.

Stone, K. E. 1996. "Habitat Selection and Seasonal Patterns of Activity and Foraging of Mantled Howling Monkeys (*Alouatta palliata*) in Northeastern Costa Rica." *Int. J. Primatol.* 17:1–30.

Strier, K. B. 1986. *The Behavior and Ecology of the Woolly Spider Monkey, or Muriqui* (Brachyteles arachnoides *E. Goeffroy 1806*). Ph.D. thesis, Harvard University, Cambridge, MA.

Strier, K. B. 1987. "Ranging Behavior of Woolly Spider Monkeys." *Int. J. Primatol.* 8:575–91.

Strier, K. B. 1993. "Growing Up in a Patrifocal Society: Sex Differences in the Spatial Relations of Immature Muriquis." Pp. 138–47 in *Juvenile Primates: Life History, Development and Behavior,* edited by M. E. Pereira and L. A. Fairbanks. New York: Oxford University Press.

Strum, S. C. 1982. "Agonistic Dominance in Male Baboons: An Alternative View." *Int. J. Primatol.* 3:175–202.

Strum, S. C. 2001. *Almost Human: A Journey into the World of Baboons.* Chicago: University of Chicago Press.

Sussman, R. W., O. Andrianasolondraibe, T. Soma, and S. Ichino. 2003. "Social Behavior and Aggression among Ringtailed Lemurs." *Folia Primatol.* 74: 168–72.

Sussman, R. W. and W. G. Kinzey. 1984. "The Ecological Role of the Callitrichidae: A Review." *Am. J. Phys. Anthropol.* 64:419–44.

Symington, M. M. 1988. "Demography, Ranging Patterns and Activity Budgets of Black Spider Monkeys (*Ateles paniscus chamek*) in the Manu National Park, Peru." *Am. J. Primatol.* 15:45–67.

Tattersall, I. 1977. "Ecology and Behavior of *Lemur fulvus mayottensis* (Primates, Lemuriformes)." *Anthropol. Pap. Am. Mus. Nat Hist.* 52:195–216.

Taylor, S. E., L. Cousino, B. Klein, T. L. Gruenewals, R. A. R. Gurung, and J. A. Updegraff. 2000. "Biobehavioral Responses to Stress in Females: Tend-and-Befriend, or Fight-or-Flight." *Psych. Rev.* 107:411–29.

Teleki, G. 1981. "The Omnivorous Diet and Eclectic Feeding Habits of Chimpanzees in Gombe National Park, Tanzania." Pp. 303–43 in *Omnivorous Primates: Gathering and Hunting in Human Evolution,* edited by R. S. O. Harding and G. Teleki. New York: Columbia University Press.

Treves, A. 1998. "The Influence of Group Size and Near Neighbor on Vigilance in Two Species of Arboreal Primates." *Behaviour* 135:453–82.

Treves, A. 2000. "Theory and Method in Studies of Vigilance and Aggregation." *Animal Behaviour* 60:711–22.

van Hooff, J. A. R. A. M. and C. P. van Schaik. 1994. "Male Bonds: Affiliative Relationships among Nonhuman Male Primates." *Behaviour* 130:309–37.

van Schaik, C. P. 1983. "Why Are Diurnal Primates Living in Groups?" *Behaviour* 87:120–22.

van Schaik, C. P. 1989. "The Ecology of Social Relationships amongst Female Primates." Pp. 195–218 in *Comparative Socioecology: The Behavioural Ecology of Humans and Other Animals,* edited by V. Standon and R. A. Foley. Oxford: Blackwell.

van Schaik, C. P. and J. A. R. A. M. van Hooff. 1983. "On the Ultimate Causes of Primate Social Systems." *Behaviour* 85:91–117.

Vasey, N. 1997. Community Ecology and Behavior of *Varecia variegata* and *Lemur fulvus albifrons* on the Masoala Peninsula, Madagascar. Ph.D. thesis, Washington University, St. Louis, MO.

Watts, D. 1988. "Environmental Influences on Mountain Gorilla Time Budgets." *Am. J. Primatol.* 15:195–211.

West, S. A., I. Pen, and A. S. Griffin. 2002. "Conflict and Cooperation: Cooperation and Competition between Relatives." *Science* 296:72–75.

Wilson, D. S. 1975. "A Theory of Group Selection." *Proc. Natl. Acad. Sci.* 72:143–46.

Wrangham, R. W. 1980. "An Ecological Model of Female-Bonded Primate Groups." *Behaviour* 75:262–300.

Wrangham, R. W. 1999. "Evolution of Coalitionary Killing." *Yrbk. Phys. Anthropol.* 42:1–30.

Wrangham, R. W. and D. Peterson. 1996. *Demonic Males: Apes and the Origins of Human Violence*. Boston: Houghton Mifflin.

Wright, R. 1999. "The Biology of Violence." Pp. 286–93 in *The Biological Basis of Human Behavior* (2nd edition), edited by R. W. Sussman. New Brunswick, NJ: Prentice Hall.

Zhang, S. Y. 1995. "Activity and Ranging Patterns in Relation to Fruit Utilization by Brown Capuchins (*Cebus apella*) in French Guiana." *Int. J. Primatol.* 16:489–508.

9

Sociality among Kin and Nonkin in Nonhuman Primate Groups

Karen B. Strier

Kinship has long been regarded as one of the fundamental elements of both human and nonhuman primate sociality. Social anthropologists (e.g., Morgan 1871) recognized its importance nearly a century before the late W. D. Hamilton (1964) invoked kin selection to explain the evolution of altruism based on the indirect fitness benefits that individuals gain by helping others with whom they share some proportion of their genes. Pioneering field primatologists studying howler monkeys (Carpenter 1934), Japanese macaques (Kawai 1958; Kawamura 1958), and rhesus macaques (Sade 1965) also saw extended kin bonds as central in primate societies before the evolutionary paradigm of kin selection had emerged.

Kin selection and its associated assumptions about how genetic relatedness affects behavior have played an influential role in the development of contemporary primatological theory and practice (Chapais and Berman, in press). Despite compelling evidence of limits on the abilities of primates to recognize distantly related kin or discriminate unfamiliar kin from nonkin, genetic relatedness has become one of the foundations for interpreting patterns of primate social interactions. Yet, as Robin Fox (1975) pointed out nearly three decades ago, human kinship systems are much more than the biological relationships that primatologists employ to describe nonhuman primate kinship. Efforts to compare human and nonhuman primates have appropriately sought "species neutral dimensions" (Rodseth, Wrangham, Harrigan, and Smuts 1991) to describe the continuities and discontinuities in human and nonhuman social relationships (Rodseth and Wrangham, in press), but subtle differences in the mechanisms by which biological kinship impacts human and nonhuman primate sociality still remain.

In his original comparative analysis of human sociality, Fox (1975) identified three basic components: (1) descent groups, which refer exclusively to extended lineal kin groups; (2) alliances, which refer to marriage, or long-term mate bonds; and (3) patterns of exogamy, which dictate the exchange of marriage partners (or mates) between descent groups. Other primates also maintain extended kin groups, long-term mate bonds, and dispersal patterns that lead to outbreeding, but only humans combine the first two components to produce the third. According to Fox (1967, 1991), human exogamy is not merely outmating, but instead, the systematic allocation of mates, which occurs within the larger network of differentiated relationships that humans maintain.

At the time of Fox's initial analysis, much less was known about the diverse configurations that we now know nonhuman primate sociality can take. In hamadryas baboons, for example, males establish one-male units with unrelated females (e.g., exogamy), while also maintaining alliances with related males when their units unite into clans and aggregate as larger bands. Yet, in contrast to humans, hamadryas baboons do not combine descent group-based exogamy to produce the systematic alliances that distinguish the social networks of humans (Fox 1991; Rodseth et al. 1991).

The ability to maintain ongoing relationships with a large number of both kin and nonkin in the absence of spatial proximity, or physical presence, has been attributed to the unique language capacities of humans. In contrast to humans, proximity appears to be a requisite for kin or mate-biased behavior in other primates. Language contributed to the "release from proximity" that characterizes human relationships (Fox 1979, 1980; Rodseth et al. 1991). It also provides a mechanism for naming categories of both biological and nonbiological kin, thereby reinforcing kin-group identities and the transmission of cultural rules that mandate kin to marry out. Thus, although the defining features of human sociality evolved from existing primate social patterns, language was the mechanism by which human social patterns could change.

Several conditions have been proposed to account for the transformation of nonhuman social patterns into their uniquely human form, including the ecological challenges of savanna-dwelling lifestyles (Fox 1975, 1980), the role of male provisioning and its effects on female life history and reproductive parameters (Alexander and Noonan 1979), and the benefits of expanded political alliances in intergroup competition (Rodseth and Wrangham, in press). Nonetheless, evidence is now accumulating that other primates also adjust their behavior in response to ecological and demographic fluctuations. Further investigations into the dynamic properties of nonhuman primate sociality can provide new insights into the conditions under which the unique features of human sociality emerged.

In this chapter, I examine the constraints on paternal and patrilineal kin recognition in nonhuman primates. First, I consider the implications of distinguishing between kin groups and descent groups. This distinction is rarely made explicit in comparative analyses of nonhuman primate kinship, yet it is crucial for evaluating the cognitive abilities that are not only necessary for humans to keep track of biological kin and individuals related through marriage, but may also correspondingly prohibit other primates from keeping track of the analogous relationships in their societies. The primary distinction between nonhumans and humans may not be so much the greater number of different relationships that humans can track without proximity, but rather, the *types of relationships* that humans recognize.

Next, I consider the conditions under which social groups composed of nonkin arise. Human mate bonds are embedded within the ongoing kinship networks to which both men and women belong, and are independent of residence. In nonhuman primates, by contrast, long-term mate bonds appear to compensate for the absence of extended kin in species in which both females and males disperse. Indeed, the association of long-term heterosexual mate bonds with one-male primate groups, versus those of long-term kin bonds with multimale groups (Fox 1975, 1980; Rodseth et al. 1991), may be confounded by the effects that the number of males in different kinds of groups have on mating and reproductive patterns (Strier 2000a). As genetic studies from a variety of wild primates have now demonstrated, the degree to which individual males can monopolize fertilizations may play a more significant role in the development of mate bonds than the number of males in groups per se.

In the final section, I examine how demographic constraints on dispersal can affect the composition of kin within groups and the exchange of mates between groups. Patrilocal societies, or those in which males remain and reproduce in their fathers' natal groups, provide the most appropriate points of departure for comparisons with humans, in part because they are found in our closest phylogenetic relatives, chimpanzees and bonobos (Wrangham 1987), and in part because they are the only kinds of societies in which the systematic exchange of mates between patrilineal descent groups, or groups of individuals related through a common male ancestor, could arise. Comparative data illustrate how extended kin groups can emerge in saturated habitats, where dispersal opportunities are limited, and how the nonrandom exchange of females between groups of patrilocal males can emerge in small, isolated populations. These examples of systematic "alliances" between "descent groups" provide insights into the demographic circumstances associated with human kinship and exogamy. They also imply that the distinctly human connection between kinship and exogamy can become an established pattern without the cognitive or linguistic skills associated with the naming of kin or the manipulation of alliances within and among groups.

Other models of human social evolution and contemporary behavior emphasize the importance of pair bonds (e.g., Alexander and Noonan 1979) and the trade-offs between male parenting effort versus mating effort (Marlowe 1999). Nonetheless, the role of paternity recognition in linking descent, alliances, and exogamy in nonhuman primate societies has not yet been fully explored. Understanding the conditions by which nonhuman primates recognize, or behave as if they recognize, paternal and patrilineal kin can therefore provide insights into the role of demography in shaping the evolution of both nonhuman and human primate sociality.

KIN GROUPS AND DESCENT GROUPS

Primates have long lifespans compared to most other animals in their size class. As a result, related individuals survive contemporaneously in overlapping generations. Even though individuals disperse, long-term associations among same-sexed kin can be maintained beyond the natal group when biological relatives transfer into the same groups as one another. These kin bonds are more likely to be paternal siblings that transfer together as age cohorts, than maternal sibs that, with the exception of twins, by definition will differ from one another in age. Male reproductive monopolies affect the distribution of paternal sibships, while female life histories affect the age distribution among maternal siblings and matrilineal kin. Yet, dispersal, mating, and life history patterns also impose constraints on the ability of primates to recognize paternity and the patrilineal kin bonds on which alliances between descent groups in humans are based (Strier, in press).

When females or males remain in their natal groups, the result is coresidence among extended matrilineal or patrilineal kin, respectively, and therefore greater opportunities for interactions among them than among members of the dispersing sex. Coresidence also promotes familiarity, which is a prerequisite for kin recognition and its manifestations as nepotism in primates. In the absence of prior familiarity, primates do not behave differently toward kin and nonkin (Fredrickson and Sackett 1984; Sackett and Fredrickson 1987; Erhart, Coelho, and Bramblett 1997).

Sex-Biased Dispersal and Philopatry in Primates

Primate kinship patterns have typically been inferred on the basis of sex-biased dispersal and residency, but we now know that there is much more variation in these patterns than was previously appreciated. Dispersal regimes, which ultimately determine the potential for the maintenance of descent groups, appear to be more phylogenetically conservative than behavioral responses to ecological conditions that affect group size or grouping patterns (Di Fiore and Rendall 1994; Strier 1999a). Nonetheless, dispersal patterns are not as categorically impermeable as most traditional classifications of primate kinship systems have implied (Moore 1984). Dispersal may be sex-biased, as when one sex disperses at higher frequencies or with greater regularity than the other. However, this does not preclude individuals of the dispersing sex from remaining in their natal groups or individuals of the so-called philopatric sex from dispersing. For example, among patrilocal chimpanzees at Gombe National Park, Tanzania, roughly half of the females born in the Kasakela community have remained and reproduced in it instead of transferring to another community to reproduce (Constable, Ashley, Goodall, and Pusey 2001). Conversely, although male bonobos, like male chimpanzees, also typically remain in their natal groups, recent field observations indicate that male bonobos may nonetheless transfer into other groups where adult sex ratios are more favorable to reproductive opportunities (Hohmann 2001). Thus, although in both species, males tend to be philopatric and females tend to disperse, there is still a great deal of behavioral flexibility at an individual and community level (Moore 1984).

There are many other examples of such facultative adjustments in primate dispersal patterns in response to local ecological and demographic conditions. In matrilocal societies, males routinely transfer groups throughout their lifetimes in response to more favorable sex ratios or greater reproductive opportunities elsewhere (Glander 1992; Moore 1992; Sussman 1992; Alberts and Altmann 1995). Nonetheless, habitat saturation and fragmentation have been associated with shifts toward bisexual dispersal in species in which dispersal is ordinarily sex-biased (e.g., Sterck 1998; Jones 1999; Sugiyama 1999).

The spatial distribution of groups also influences dispersal patterns. Physical proximity between adjacent groups promotes interactions, and therefore familiarity, which facilitates intergroup transfers by minimizing the risks of long-distance dispersal into unfamiliar groups (Perry 1996; Strier 2000b; Jack 2001). The genetic and social consequences of such "nonrandom" dispersal patterns have been well documented for a variety of species (e.g., vervet monkeys: Cheney and Seyfarth 1983; rhesus macaques: Melnick and Hoelzer 1992; long-tailed macaques: de Ruiter, van Hooff, and Scheffrahn 1994; de Ruiter and Geffen 1998; red howler monkeys: Pope 1992, 1998). Facultative adjustments in dispersal patterns in response to local ecological or demographic conditions can have similarly pronounced effects.

Both matrilocality and patrilocality have been associated with the respective benefits that females and males gain by having reliable allies on hand in competition against other groups over access to key resources. According to the tenants of kin selection, close kin should make more reliable allies than distant kin or nonkin because of their mutual genetic interests. Behavioral observations provide some support for predictions that females in matrilocal societies treat their close relatives better than distant relatives or nonkin. Such "nepotism" may be mutualistic when the benefits of discriminating in favor of kin are shared (Chapais 2001), and is evident among related females in the matrilocal societies of baboons, macaques, and vervet monkeys, where close relatives preferentially associate and interact affiliatively with one another, aid one another during conflicts with third parties, and join forces against outsiders to defend common food or other resources.

Like females in matrilocal societies, males in patrilocal societies also cooperate in aggressive confrontations against other groups of males (Morin et al. 1994). Cooperation among related males in intercommunity contests may be mutualistic (Wilson, Hauser, and Wrangham 2001), but whether nepotism is also involved would require evidence that closely related males cooperate preferentially with one another over nonkin. In addition, it would require distinguishing between direct and indirect fitness benefits to male kin that cooperate, a particularly challenging task in viscous populations where opportunities for dispersal are limited (Griffin and West 2002).

Kin-biased affinities among patrilocal males have been more difficult to evaluate than those among females in matrilocal societies because patrilineal relatedness cannot be inferred without genetic data. Efforts to evaluate the effects of biological relatedness on social relationships among patrilocal male chimpanzees have focused on maternal relatedness, but there is no evidence that males that share maternal haplotypes behave nepotistically toward one another (Goldberg and Wrangham 1997; Mitani, Merriwether, and Zhang 2000; Mitani, Watts, Pepper, and Merriwether 2002). Moreover, whether kin-biased behavior will emerge with analyses of patrilineal relatedness is not yet known.

There are other ways in which social bonds among matrilineal kin in matrilocal societies differ from those among patrilineal kin in patrilocal societies. For example, whereas males in matrilocal societies typically disperse facultatively throughout most of their lives, secondary dispersal by females in patrilocal societies appears to be rare. This difference can be attributed to the greater reproductive costs of dispersal on females than males (see below), but it results in different sets of opportunities for the maintenance of extended bilineal kin bonds. For example, matrilocal females will remain with their mothers, other matrilineal female kin, and paternal half-sisters, which are likely to be limited to females that are close to them in age. Males in matrilocal societies must repeatedly establish new sets of relationships each time they disperse (Strum 1982). Although primary dispersal often involves familiar age-mates, natal cohorts usually dissolve when males secondarily disperse (Cheney and Seyfarth 1983).

Patrilocal males, by contrast, remain in their natal groups with both their fathers and other paternally related and patrilineal male kin *and* their mothers and maternal brothers (Strier, in press). Mating and social patterns within and between such patrilocal societies may represent the basic elements from which human descent group-based exogamy evolved (Rodseth et al. 1991), but the transformation would have required an ability to differentiate among these categories of matrilineal and patrilineal kin.

Matrilocal Societies

Females in matrilocal primate societies remain with their matrilineal kin, in extended matrilineages, for the duration of their lives. When large matrilocal groups fission, they do so along matrilines and matrilineages (Sade 1991; Sauther, Sussman, and Gould 1999). Living in the same groups, matrilineal kin are familiar to one another and have ample opportunities to associate, affiliate, and aid one another preferentially. Nonetheless, there appear to be limits to their nepotistic behavior, which Chapais, Savard, and Gauthier (2001) identified as the "relatedness threshold for altruism," or RTA, based on the strength of affiliative interactions and social alliances.

Among female Japanese macaques, the RTA was lower for lineal kin ($r = 0.125$, corresponding to the degree of relatedness between a great grandmother and her great granddaughter) than for collateral kin ($r = 0.25$, corresponding to the degree of relatedness between a female and her mother's maternal sister). These behavioral differences could reflect the limitations of kin recognition beyond certain degrees of biological relatedness, as well as differences in the profitability of behaving nepotistically toward different categories and ages of kin, but both require familiarity (Chapais 2001; Chapais et al. 2001).

By definition, matrilineal kin are separated from one another by age, with the minimum being the age at which females reproduce. Collateral kin, by contrast, can be much closer to one another in age. In rhesus macaques, age proximity coincides with the strength of affiliative preferences among familiar paternal siblings (Widdig et al. 2001, 2002), and age proximity may also affect the profitability of nepotism (Chapais 2001). Like Japanese macaques, female rhesus macaques maintain the strongest affiliations with close maternally related kin, but they also exhibit significantly stronger preferences toward paternal half-sibs in their age cohorts than either paternal half-sibs that differ in age or nonkin (Widdig et al. 2001, 2002).

Shared paternity also confounds the assumption that matrilineal kinship predicts kin-biased behavior among females in matrilocal societies. Genetic analyses show that degrees of relatedness among long-tailed macaques (de Ruiter and Geffen 1998) and yellow baboons (Altmann et al. 1996) are higher than those based on matrilineal kinship alone whenever males succeed in fertilizing multiple females. The degree and duration of male fertilization monopolies will impact whether matrilineal kin are also paternally related, and therefore whether patrilineal kinship will affect social bonds in matrilocal societies (Strier, in press). For example, the similarly strong affiliative relationships that adult female yellow baboons maintain with both maternal and paternal half siblings may reflect a scarcity of kin due to birth intervals and mortality in wild groups (Smith et al. 2003).

Even when particular males succeed in monopolizing fertilizations with multiple females, the ability of primates to discriminate among familiar paternal siblings or father-offspring relationships is limited (Bernstein 1999). Paternal sibling recognition in opposite-sexed yellow baboons (Alberts 1999) and same-sexed rhesus macaques (Widdig et al. 2001) is stronger among age cohorts than across distant ages. Nonetheless, although there is some evidence that male Hanuman langurs protect the infants of females with which they have copulated (Borries et al. 1999), father-daughter matings are common among matrilocal Barbary macaques (Chapais 2001), where male investment in infants appears to be related to future mating opportunities instead of reproductive investment (Ménard et al. 2001). Recognition of paternal siblings appears to involve both familiarity and phenotypic matching, both of which may be enhanced by age proximity (Widdig et al. 2002). By contrast, recognition of paternity appears to be short-term, even when coresidence promotes familiarity and long-term opportunities for interactions (but see Buchan et al. 2003).

It would not be so surprising if mechanisms of paternity recognition were weak or nonexistent in most matrilocal primate societies. After all, in these societies, secondary dispersal by males ordinarily precludes fathers from maintaining long-term associations with offspring of either sex. In addition, the instability of male hierarchies means that even when high-ranking males can monopolize all or most fertilizations, the duration of their monopolies, and thus opportunities to recognize paternity, are limited in time. Nonetheless, there is no evidence that mechanisms of long-term paternity recognition are any stronger in patrilocal societies despite the greater opportunities that patrilocality should provide for lifelong interactions among paternally and patrilineally related males.

Patrilocal Societies

Affiliative interactions and coalitions among males in patrilocal societies differ from those among females in matrilocal societies in two important respects. First, there is little indication that maternally related males or maternal half-brothers preferentially associate or form coalitions with one another (Goldberg and Wrangham 1997; Mitani et al. 2000). The age differences between them may make maternal brothers less valuable as allies than males that are closer in age and social rank (Mitani et al. 2002; Strier, in press). Long interbirth intervals among dispersing females (compared to matrilocal females) contribute to the age differentials among maternal siblings in patrilocal societies (Strier, in prep.). Thus, although maternal brothers in patrilocal societies should be as familiar to one another as maternal sisters in matrilocal societies, the greater age differentials may explain why they fail to exhibit comparable levels of kin-biased behavior.

The absence of bonds among maternal brothers in patrilocal societies is paradoxical, considering the strength and longevity of their mother-son bonds. In bonobos, maternal support directly impacts the ability of sons to achieve high rank (Furuichi and Ihobe 1994; Furuichi 1997). In bonobos, as well as in patrilocal chimpanzees and muriquis, mother-son copulations are sufficiently rare to infer that they recognize one another as familiar kin to be avoided as mates (reviewed in Strier, in press). Indeed, mechanisms for mother-son inbreeding avoidance have been implicated in experiments showing that chimpanzees are able to match the phenotypes of unfamiliar mother-son, but not father-son or father-daughter, pairs (Parr and de Waal 1999).

The strength of mother-son kin recognition in patrilocal societies, as indicated by maternal support, inbreeding avoidance, and the ability to match unfamiliar phenotypes, differs dramatically from father-daughter kin recognition in matrilocal societies, which is minimal if it occurs at all. One reason for this difference may be that the dispersal of females into patrilocal groups of males tends to be for decades, if not life. In contrast to the short-term opportunities for father-daughter interactions in most matrilocal societies, the long-term coresidence of mothers and their sons in patrilocal societies creates opportunities for them to remain lifelong allies. Nonetheless, the strength of these mother-son bonds makes the lack of close affiliations among maternal brothers all the more curious.

The second way in which male relationships in patrilocal societies differ from those of females in matrilocal societies reflects the limitations of paternal and patrilineal kin recognition (Gouzoules 1984). For bonds among patrilineal male kin to resemble those among matrilineal female kin, patrilocal males would need to be able to discriminate among familiar paternally related kin in the same way that matrilocal females appear to discriminate among familiar matrilineal kin, at least within the constraints of their RTAs. That patrilocal males fail to do this is implied by the absence of mechanisms of patrilineal rank inheritance comparable to those of matrilineal rank inheritance found in the matrilocal societies of macaques and baboons. Instead, males that compete for rank in patrilocal societies do so with the aid of their mothers (Furuichi 1997) or of male allies that are close to them in age and rank, and therefore possibly paternal brothers but not close patrilineal kin (Mitani et al. 2002). Kin-biased relationships among patrilineal male kin in patrilocal societies require paternity recognition, not only of one's own biological offspring, but also of the offspring of the other males. Both kinds of recognition are similarly precluded by the mating patterns of females in patrilocal and matrilocal societies, as well as by the secondary dispersal of males in matrilocal societies.

MATE BONDS OR KIN BONDS

Although opportunities to interact with extended lineal kin are severed at the time of dispersal, primates that disperse into extant groups may reunite with older, same-sexed kin that previously joined and remained in these groups (Strier, in press). By contrast, when dispersing individuals must establish new groups, they are more likely to end up living in groups without access to close lineal kin. Mate bonds appear to compensate for the loss of close lineal kin bonds in new groups established by dispersing males and females, and result in the convergence of reproductive interests among mates.

Mate bonds are not restricted to pair-bonded primates, but the degree to which a male maintains exclusive access to females has undisputable direct fitness consequences for both. Males can guard females or resources that attract females and thereby increase their paternity probabilities. Males can also carry, provision, and protect offspring, thereby reducing the time and energetic costs of reproduction on their mates. These costs are higher for females that disperse from their natal groups than for matrilocal females, in part because dispersing females postpone reproduction beyond the age that philopatric females reproduce (Strier, in prep.).

Compensating for the Costs of Female Dispersal

Dispersal is costly for both males and females, who are similarly vulnerable to predators and to the difficulties of finding or defending food when they leave their natal groups (Pusey and Packer 1987; Alberts and Altmann 1995; Alberts 1999). Females also face additional costs associated with reproduction. Delaying the onset of puberty will reduce the risks of females dispersing into unfamiliar areas while bearing the energetic costs of gestation and lactation (Bronson and Rissman 1986). Delayed onset of puberty is consistent with evidence that dispersing females tend to be older at the time of their first reproductions than philopatric females in the same populations (red howler monkeys: Crockett 1984; Crockett and Pope 1993; Pope 2000; muriquis: Strier and Ziegler 2000; chimpanzees: Pusey, Williams, and Goodall 1997).

Establishing strong mate bonds may be one way of compensating for the reproductive costs that dispersal imposes on females. Male mates can defend food resources that females need, defend females from other males, and contribute to infant care. The most extreme examples of mate bonds are found in societies in which both sexes disperse from their natal groups, but even in these societies, male contributions to infant care increase with the degree to which they can affect female reproductive rates (Strier 1996). For example, male callitrichids and owl monkeys assume most of the nonlactational burden of caring for infants within days after parturition (Wright 1990). Indeed, the unique postpartum ovulation and simultaneous gestation and lactation that callitrichid mothers experience may reflect the relaxation of energetic costs that males and other subordinate caretakers provide (Garber 1997).

Strong mate bonds can also reduce birth intervals without compromising infant survivorship. The corresponding increase in female reproductive rates compensates for the reproductive costs of delayed puberty in dispersing females. Empirical tests of the compensatory relationship between mate bonds and kin bonds are now feasible with comparative data on the strength of mate bonds and individual female reproductive histories. For example, in mountain gorillas, infant survivorship is higher in multimale groups than in single-male groups because single-male groups are much less stable (Robbins 1995). One-male gorilla groups disintegrate when the silverback male dies, and they are more prone to disruption from challenges by extragroup males than are multimale groups. Multimale troops of red howler monkeys are also more stable than single-male troops, especially when the males that join forces are related to one another (Pope 1990). Red howler monkey females that are permitted to remain and reproduce in their natal troops also maintain more stable relationships with troop males and have correspondingly higher reproductive success than females that are expelled from their natal groups and must establish new groups with other males to reproduce (Pope 2000).

The availability of matrilineal kin as potential allies in matrilocal societies does not prevent females from soliciting resources from alpha males (e.g., brown capuchin monkeys: Janson 1984) or enlisting other males to assist in babysitting or providing protection against unrelated females and other males (e.g., Barbary macaques: Small 1990, 1992; olive baboons: Smuts 1985). Usually, the level of support these males provide corresponds with their future reproductive access to females. In Barbary macaques, male investment in infants is related to increased mating opportunities, but not to paternity (Ménard et al. 2001). In mountain baboons, males restrict their support to estrous or lactating females, which are the ones responsible for maintaining proximity with, and grooming, their male allies (Palombit 1999). Paternity studies indicate that multimale troops of red howler monkeys are genetically single-male troops because the alpha male can monopolize 100 percent of the fertilizations for the duration of his tenure (Pope 1990). Behavioral observations suggest similar reproductive monopolies by alpha males in multimale groups of golden lion tamarins (Baker et al. 1993).

Mating with Multiple Partners

An alternative to establishing strong mate bonds with particular males in multimale groups is to distribute matings across multiple partners. Confusing paternity does not promote male investment in infants, but it has been attributed to female strategies to reduce the risks of aggression directed by unrelated males toward their infants (Hrdy 1981; van Schaik et al. 1999; van Schaik et al. 2000). Yet, the need to reduce the risks of male aggression toward infants should not apply when females disperse into groups of patrilocal males because the biological relatedness among males should inhibit them from harming one another's offspring. The full implications of this distinction, which seems to represent an underlying contradiction to the benefits of confusing paternity in patrilocal societies, have not been fully considered in comparative models of primate reproductive strategies.

The degree to which patrilocal males compete overtly with one another for access to fertile females varies from total tolerance, as in the egalitarian societies of muriquis (Strier 1992; Strier et al. 2002), to direct contests, as in the hierarchical societies of chimpanzees (Watts 1998). Yet, females that disperse into patrilocal societies are as active as females in matrilocal societies in soliciting copulations with multiple partners. Either mating with multiple partners involves factors other than strategies for confusing paternity, or recognizing patrilineal kinship is not feasible or profitable among patrilocal males.

Mating with multiple partners could increase the probability of conception and thereby minimize birth intervals by reducing female cycling-to-conception delays in species in which the costs of female dispersal are not offset by male investment in mate bonds (Guimarães and Strier 2001; Strier et al. 2002). Increasing conception probabilities has also been proposed to account for why female gibbons solicit extrapair copulations (Palombit 1994, 1995) and may be a more parsimonious explanation than confusing paternity, considering that male gibbons are not known to direct aggression toward infants. Reductions in cycling-to-conception delays can help to reduce the reproductive costs that dispersal imposes on females, and mating with multiple partners may be an alternative means of increasing conception probabilities when ecological conditions preclude the maintenance of strong mate bonds that promote male investment, or when male investment in infants has little or any impact on female reproductive rates (Strier 1996). Increasing conception probabilities may also account for why matrilocal female ringtailed lemurs, which are individually and collectively dominant over males, nonetheless mate with multiple partners during their brief annual breeding season (Sauther et al. 1999).

Limited access to multiple mating partners may contribute to the much longer interbirth intervals of orangutans compared to those of chimpanzees and bonobos (nine years versus four to five years; Knott 2001). While negative energy balances might delay the resumption of ovarian cycling in female orangutans (Knott 1998), they should have little, if any, effect on cycling-to-conception delays. Comparative data on cycling-to-conception delays among orangutans (or other species) living at high and low densities with corresponding differences in access to multiple partners could provide insights into how variation in mating opportunities affects the variance in female reproductive rates.

Mating with multiple partners could only reflect counterstrategies against male aggression toward infants in patrilocal societies if patrilineally related males do not recognize one another as kin. Cognitive constraints on long-term paternity recognition would preclude patrilineal kin recognition, except perhaps among paternal half-brothers that are familiar to one another, as well as close in age. Indeed, the coalitions that male chimpanzees form to monopolize access to mates may prove to be an example of "mutualistic nepotism" (sensu Chapais 2001) instead of mutualism (Watts 1998) if coalition partners are more likely to be paternal half-brothers than nonkin. Nonetheless, although the genetic data necessary to evaluate cooperation among more extended patrilineal kin are still lacking, there is no reason to expect nepotism of any type among patrilineal kin whose relatedness to one another falls below established RTAs.

Mechanisms that might increase paternity recognition in patrilocal societies include both mate bonds and reproductive monopolies maintained by high-ranking males. For example, the reductions in cycling-to-conception delays that mate bonds could promote are consistent with ideas that have linked concealed ovulation and continuous receptivity with pair-bonding in humans (Alexander and Noonan 1979). Nonetheless, nepotistic behaviors associated with paternity recognition in nonhuman primates appear to be lacking among patrilineally related males in patrilocal societies.

Paternity studies indicate that high-ranking males sire a greater number of offspring than low-ranking males in the patrilocal societies of chimpanzees (Constable et al. 2001; Vigilant, Hofreiter, Siedel, and Boesch 2001) and bonobos (Gerloff et al. 1999). Yet, there is no evidence in these patrilocal societies that males invest differentially in infants they may have sired, or that males old enough to be fathers help males young enough to be their sons to rise in the adult male hierarchy (Gouzoules 1984). Indeed, even the meat obtained in successful hunts by male chimpanzees is not allocated preferentially to female mates or to the offspring of females with whom they have copulated (Mitani and Watts 2001).

Males in patrilocal societies do not appear to make the same kinds of trade-offs between parenting effort and mating effort as males in matrilocal societies (Gouzoules 1984; Ménard et al. 2001). This difference is consistent with the idea that mating with multiple partners may occur for different reasons in different kinds of societies, and that mechanisms of paternity recognition might be expected to evolve under demographic conditions in which male parenting effort *and* mating effort complement instead of conflict, with one another.

DEMOGRAPHY AND SOCIALITY

Increasing human pressures on many of the world's primates have led to severe habitat loss and disturbances and have altered primate population dynamics to an unprecedented degree. The effects of these anthropogenic pressures on primates range from isolating once-continuous populations from one another, to compressing surviving populations into small fragments at high densities, to pushing them into marginal habitats at the extremes of their species' historic distributions. Nonetheless, consistent patterns in how primates respond to such changes have begun to emerge.

Central among these responses is the tendency for dispersal patterns to shift. For example, both female callitrichids and male gibbons living at high densities are more likely to remain in their natal groups, along with one or both of their parents, than risk dispersing into saturated habitats. In callitrichids, this can result in extended matrilines (Ferrari and Digby 1996) similar to those that evolve in red howler monkeys in saturated habitats (Crockett 1996). Among so-called pair-bonded gibbons living at high densities, peaceful interactions and nonrandom exchanges of mates have been observed among adjacent groups, which may also contain two or more unrelated adults and subadults of the same sex (Reichard and Sommer 1997; Brockelman et al. 1998; Fuentes 2000; Oka and Takenaka 2001). Habitat fragmentation, by contrast, has been associated with increased tendencies toward bisexual dispersal in species in which dispersal is more commonly sex-biased and opportunities for joining extant groups or establishing new groups exist (Jones 1999).

Nonrandom Dispersal between Patrilocal Groups

Limited dispersal opportunities in small, fragmented populations can result in the systematic exchange of mates between groups. Such systematic exchanges have been well documented in populations of patrilocal chimpanzees at Gombe National Park, Tanzania (Pusey 2001), and muriquis at the Estação Biológica de Caratinga/RPPN-FMA, Brazil (Strier 2000b). Both populations inhabit forest fragments and are isolated from other populations of their kinds. Each of the two populations includes about 150–200 individuals and has been distributed into 2–3 communities or groups since the onset of field studies on the Gombe chimpanzees in 1960 (Goodall 1986) and on the Caratinga muriquis in 1982 (Strier 1999b). Recent genetic studies indicate that at least four of nine female chimpanzees born in the Kasakela community have remained and reproduced in their natal groups (Constable et al. 2001), while two of the twenty-eight female muriquis born in the Matão study group since 1982 have remained there and reproduced (Martins and Strier, in prep.). In both populations, the females that do disperse appear to remain and reproduce in their new groups for decades, if not life.

Female chimpanzees in the much larger populations at Mahale National Park, Tanzania (Nishida and Hosaka 1996), Kibale National Park, Uganda (Goldberg and Wrangham 1997; Mitani et al. 2000), and Tai National Park, Ivory Coast (Boesch and Boesch-Achermann 2000), do not appear to be as limited in their dispersal options as the Gombe chimpanzees. Indeed, the high diversity of maternal haplotypes found in the two Kibale chimpanzee communities that have been under intensive study (Goldberg and Wrangham 1997; Mitani et al. 2000) is similar to that found in a recent genetic study of bonobos (Hashimoto et al. 1996; Gerloff et al. 1999), where the females have all so far dispersed.

The limited dispersal opportunities available to Caratinga muriquis result in the dispersal of daughters into their mothers' natal groups, where both of their maternal grandparents, and their maternal and paternal uncles, still reside. Lack of familiarity should preclude them from recognizing their grandparents and uncles, even though these relatives fall within the RTAs among familiar kin. Moreover, females disperse into the same groups as both their maternal and paternal sisters. Age cohorts often disperse together or within weeks of one another and maintain associations after joining their new group, but whether they do so preferentially with their paternal half-sisters is not yet known. Matings between recent immigrants and males from their natal groups occur at lower frequencies than those between longer-term female residents and extragroup males, and mother-son matings are exceptionally rare (Strier 1997). It is clear that the social and mating patterns of these muriquis are embedded within complicated networks of both familiar and unfamiliar biological kin, but whether muriquis recognize these networks is not known.

Decoupling Dispersal from Patrilineal Descent

The demographic conditions that have led to the systematic exchange of females between groups of patrilocal male muriquis and chimpanzees provide clues into how similar patterns of exogamy between groups of patrilocal male hominids could have arisen. What is unique among humans is not that females move between groups of males in predictable, nonrandom ways, but rather that both human males and females remember where they and others came from.

In nonhuman primates, interactions among familiar matrilineal kin in matrilocal societies are consistent with the predictions based on kin selection, at least within established RTAs that reflect limitations on the ability of primates to either recognize distant matrilineal kin or profit by helping them. By contrast, neither maternally related males nor patrilineal kin in patrilocal societies behave as if they recognize one another as such. While differences in the profitability of nepotism could account for this divergence, it is also likely that mechanisms for recognizing paternal and patrilineal kin are weaker in patrilocal societies than they are for recognizing matrilineal kin and paternal half-siblings in matrilocal societies.

The absence of mechanisms for recognizing paternity or extended patrilineal kinship would preclude the development of alliances from the systematic exchanges of females among patrilineal descent groups. Conditions such as extreme and long-term reproductive monopolies maintained by particular males could facilitate paternity recognition, but females that transfer into groups of patrilocal males actively seek copulations with multiple males, with the effect of diluting paternity recognition. Demographic changes, such as sustained female-biased sex ratios resulting from stochastic processes, or behavioral innovations that lead to large and enduring asymmetries in male resource holding potentials, would limit female access to multiple partners and, thus, establish opportunities for mechanisms of paternity recognition, and ultimately patrilineal kin recognition, to evolve in patrilocal societies.

It may still be necessary to invoke uniquely human attributes, such as our capacity for language, to explain the size of human social networks and our ability to keep track of these networks without proximity. Nonetheless, understanding the limits on paternity and patrilineal kin recognition, and the demographic and behavioral mechanisms that extend these limits in other primates, provides insights into the processes by which human sociality could have been transformed. As in wild yellow baboon groups, where the small size of matrilines may underlie a relaxation in the age-sensitivity of paternal sibling affiliations (Smith et al. 2003), it is necessary to consider the demographic constraints on both males and females in patrilocal societies in order to identify the conditions that would strengthen selection pressures on the ability to recognize (or pretend to recognize) paternity and patrilineal kinship.

CONCLUSIONS

Dispersal and resident patterns directly affect opportunities for interactions among nonkin as well as different categories of biological kin. Familiarity, acquired and maintained through coresidence, appears to be critical for distinguishing close kin from distant kin or nonrelatives in matrilocal societies, but is insufficient for paternal and patrilineal kin recognition in patrilocal societies. This difference may be due to greater uncertainties about paternity (in contrast to maternity), as well as to fewer advantages of discriminating in favor of patrilineal kin.

We have seen how limited dispersal opportunities can lead to the systematic exchange of females between neighboring groups of patrilocal muriquis and chimpanzees. Nonetheless, females in these species neither maintain kin bonds with members of their natal groups after they emigrate, nor establish lasting mate bonds with males in their new groups, as humans do. Instead, the simultaneous maintenance of mate bonds and kin bonds within such patrilocal societies appears to require a distinct set of demographic conditions that facilitate the convergence of male and female reproductive interests in mutual offspring, and thus enhance opportunities for paternal kin to interact while also making it profitable for them to do so.

One of the key distinctions between the patrilocal societies of human and nonhuman primates lies in the evolution of mechanisms for paternal and patrilineal kin recognition in humans. These mechanisms could include increased familiarity between fathers and their biological offspring, perhaps as a consequence of closer or longer-term associations with female mates. However, familiarity can also lead to social mechanisms associated with treating a mate's offspring *as if* they were kin. The economic benefits of identifying familiar individuals as heirs to inheritable resources (Shennan 2003) could have positive impacts on fitness, just as the political benefits of relying on familiar individuals as allies are thought to have (Rodseth and Wrangham, in press). Under such circumstances, the advantages of recognizing familiar individuals as social kin could override the less reliable mechanisms for recognizing biological paternal and patrilineal kin. Familiar biological kin may be preferred over familiar nonkin, but only when and if they can be recognized as such.

ACKNOWLEDGMENTS

Many of the ideas developed in this chapter were developed for presentations at the 2000 and 2002 annual meetings of the American Association for the Advancement of Science (AAAS) and for the 2001 conference on "Origins and Nature of Sociality among Nonhuman and Human Primates" organized by Robert Sussman and Audrey Chapman. I thank them and the AAAS program for Dialogue on Science, Ethics, and Religion (DoSER) for their support. Conversations with John Moore, Robert Sussman, Paul Wason, and members of the AAAS panels and conference were influential. I am grateful to Robert Sussman and Adrienne Zihlman for helpful suggestions on an earlier version of this manuscript and Linda Sussman for careful editing.

REFERENCES

Alberts, S. C. 1999. "Paternal Kin Discrimination in Wild Baboons." *Proc. Royal Soc. London B* 266:1501–6.

Alberts, S. C. and J. Altmann. 1995. "Balancing Costs and Opportunities: Dispersal in Male Baboons." *Am. Nat.* 145:279–306.

Alexander, R. D. and K. M. Noonan. 1979. "Concealment of Ovulation, Parental Care, and Human Social Evolution." Pp. 436–53 in *Evolutionary Biology and Human Social Behavior,* edited by N. A. Chagnon and W. Irons. North Scituate, MA: Duxbury.

Altmann, J., S. C. Alberts, S. A. Haines, J. Dubach, P. Muruthi, T. Coote, E. Geffen, D. J. Cheesman, R. S. Mututua, S. N. Saiyalele, R. K. Wayne, R. C. Lacy, and M. W. Bruford. 1996. "Behavior Predicts Genetic Structure in a Wild Primate Group." *Proc. Natl. Acad. Sci. USA* 93:5797–5801.

Baker, A. J., J. M. Dietz, and D. G. Kleiman. 1993. "Behavioural Evidence for Monopolization of Paternity in Multi-Male Groups of Golden Lion Tamarins." *Anim. Behav.* 46:1091–1103.

Bernstein, I. S. 1999. "Kinship and the Behavior of Nonhuman Primates." Pp. 202–5 in *The Nonhuman Primates,* edited by P. Dolhinow and A. Fuentes. Mountain View, CA: Mayfield.

Boesch, C. and H. Boesch-Achermann. 2000. *The Chimpanzees of the Taï Forest: Behavioural Ecology and Evolution.* New York: Oxford University Press.

Borries, C., K. Launhardt, C. Epplen, J. T. Epplen, and P. Winkler. 1999. "Males as Infant Protectors in Hanuman Langurs (*Presbytis entellus*) Living in Multi-Male Groups—Defence Pattern, Paternity, and Sexual Behaviour." *Behav. Ecol Sociobiol.* 46:350–56.

Brockelman, W. Y., U. Reichard, U. Treesucon, and J. J. Raemakers. 1998. "Dispersal, Pair Formation and Social Structure in Gibbons (*Hylobates lar*)." *Behav. Ecol. Sociobiol.* 42:329–39.

Bronson, F. H. and E. F. Rissman. 1986. "The Biology of Puberty." *Biological Reviews* 61:157–95.

Buchan, J. C., S. C. Alberts, J. B. Silk, and J. Altmann. 2003. "True Paternal Care in a Multi-male Primate Society." *Nature* 425:179-181.

Carpenter, R. C. 1934. "A Field Study of the Behavior and Social Relations of Howling Monkeys." *Comp. Psychol. Monog.* 48:1–168.

Chapais, B. 2001. "Primate Nepotism: What Is the Explanatory Value of Kin Selection?" *Int. J. Primat.* 22:203–29.

Chapais, B. and C. Berman (Eds.). In press. *Kinship and Behavior in Primates.* New York: Oxford University Press.

Chapais, B., L. Savard, and C. Gauthier. 2001. "Kin Selection and the Distribution of Altruism in Relation to Degree of Kinship in Japanese Macaques (*Macaca fuscata*)." *Behav. Ecol. Sociobiol.* 49:493–502.

Cheney, D. L. and R. M. Seyfarth. 1983. "Non-Random Dispersal in Free-Ranging Vervet Monkeys: Social and Genetic Consequences." *Am. Nat.* 122:392–412.

Constable, J. L., M. V. Ashley, J. Goodall, and A. E. Pusey. 2001. "Noninvasive Paternity Assignment in Gombe Chimpanzees." *Molec. Ecol.* 10:1279–1300.

Crockett, C. M. 1984. "Emigration by Female Red Howler Monkeys and the Case for Female Competition." Pp. 159–73 in *Female Primates: Studies by Women Primatologists,* edited by M. E. Small. New York: Alan R. Liss.

Crockett, C. M. 1996. "The Relation between Red Howler Monkey (*Alouatta seniculus*) Troop Size and Population Growth in Two Habitats." Pp. 489–510 in *Adaptive Radiations of Neotropical Primates,* edited by M. A. Norconk, A. L. Rosenberger, and P. A. Garber. New York: Plenum.

Crockett, C. M. and T. Pope. 1993. "Consequences of Sex Differences in Dispersal for Juvenile Red Howler Monkeys." Pp. 104–18 in *Juvenile Primates: Life History, Development, and Behavior,* edited by M. E. Pereira and L. A. Fairbanks. New York: Oxford University Press.

de Ruiter, J. R. and E. Geffen. 1998. "Relatedness of Matrilines, Dispersing Males and Social Groups in Long-Tailed Macaques (*Macaca fascicularis*)." *Proc. Royal Soc. Lond. B* 265:79–87.

de Ruiter, J. R., J. A. R. A. M. van Hooff, and W. Scheffrahn. 1994. "Social and Genetic Aspects of Paternity in Wild Long-Tailed Macaques." *Behaviour* 129:203–24.

Di Fiore, A. and D. Rendall. 1994. "Evolution of Social Organization: A Reappraisal for Primates by Using Phylogenetic Methods." *Proc. Natl. Acad. Sci. USA* 91: 9941–45.

Erhart, E. M., A. M. J. Coelho, and C. A. Bramblett. 1997. "Kin Recognition by Paternal Half-Siblings in Captive *Papio cynocephalus*." *Am. J. Primatol.* 43:147–57.

Ferrari, S. F. and L. J. Digby. 1996. "Wild *Callithrix* Groups: Stable Extended Families?" *Am. J. Primatol.* 38:19–27.

Fox, R. 1967. *Kinship and Marriage.* Baltimore, MD: Penguin.

Fox, R. 1975. "Primate Kin and Human Kinship." Pp. 9–35 in *Biosocial Anthropology,* edited by R. Fox. New York: Halsted.

Fox, R. 1979. "Kinship Categories as Natural Categories." Pp. 132–44 in *Evolutionary Biology and Human Social Behavior,* edited by N. A. Chagnon and W. Irons. North Scituate, MA: Duxbury.

Fox, R. 1980. *The Red Lamp of Incest.* New York: E. P. Dutton.

Fox, R. 1991. "Commentary." *Curr. Anthropol.* 32:242–43.

Fredrickson, W. T. and G. Sackett. 1984. "Kin Preferences in Primates, *Macaca nemestrina*: Relatedness or Familiarity?" *J. Comp. Psych.* 98:29–34.

Fuentes, A. 2000. "Hylobatid Communities: Changing Views on Pair Bonding and Social Organization in Hominoids." *Yrbk. Phys. Anthropol.* 43:33–60.

Furuichi, T. 1997. "Agonistic Interactions and Matrifocal Dominance Rank of Wild Bonobos (*Pan paniscus*) at Wamba." *Int. J. Primatol.* 18:855–75.

Furuichi, T. and H. Ihobe. 1994. "Variation in Male Relationships in Bonobos and Chimpanzees." *Behaviour* 130:212–28.

Garber, P. A. 1997. "One for All and Breeding for One: Cooperation and Competition as a Tamarin Reproductive Strategy." *Evol. Anthropol.* 5:187–99.

Gerloff, U., B. Hartung, B. Fruth, G. Hohmann, and D. Tautz. 1999. "Intracommunity Relationships, Dispersal Pattern and Paternity Success in a Wild Living Community of Bonobos (*Pan paniscus*) Determined from DNA Analyses of Faecal Samples." *Proc. Royal Soc. Lond. B* 266:1189–95.

Glander, K. E. 1992. "Dispersal Patterns in Costa Rican Mantled Howling Monkeys." *Int. J. Primatol.* 13:415–36.

Goldberg, T. L. and R. W. Wrangham. 1997. "Genetic Correlates of Social Behaviour in Wild Chimpanzees: Evidence from Mitochondrial DNA." *Anim. Behav.* 54:559–70.

Goodall, J. 1986. *The Chimpanzees of Gombe: Patterns of Behavior*. Cambridge, MA: Harvard University Press.

Gouzoules, S. 1984. "Primate Mating Systems, Kin Associations, and Cooperative Behavior: Evidence for Kin Recognition?" *Yrbk. Phys. Anthropol.* 27:99–134.

Griffin, A. S. and S. A. West. 2002. "Kin Selection: Fact and Fiction." *Trends Ecol. Evol.* 17:15–21.

Guimarães, V. O. and K. B. Strier. 2001. "Adult Male-Infant Interactions in Wild Muriquis (*Brachyteles arachnoides hypoxanthus*)." *Primates* 42(4):389–93.

Hamilton, W. D. 1964. "The Genetical Evolution of Social Behaviour." I and II. *J. Theoret. Biol.* 7:1–52.

Hashimoto, C., T. Furuichi, and O. Takenaka. 1996. "Matrilineal Kin Relationship and Social Behavior of Wild Bonobos (*Pan paniscus*): Sequencing the D-Loop Region of Mitochondrial DNA." *Primates* 37:305–18.

Hohmann, G. 2001. "Association and Social Interactions between Strangers and Residents in Bonobos (*Pan paniscus*)." *Primates* 42(1): 91–99.

Hrdy, S. B. 1981. *The Woman That Never Evolved*. Cambridge, MA: Harvard University Press.

Jack, K. 2001. "Effect of Male Emigration on the Vigilance Behavior of Coresident Males in White-Faced Capuchins (*Cebus capucinus*)." *Int. J. Primatol.* 22:715–32.

Janson, C. H. 1984. "Female Choice and Mating System of the Brown Capuchin Monkey *Cebus apella* (Primates: Cebidae)." *Zeitschrift Tierpsychologie* 65: 177–200.

Jones, C. 1999. "Why Both Sexes Leave: Effects of Habitat Fragmentation on Dispersal Behavior." *Endangered Species UPDATE* 15:70–73.

Kawai, M. 1958. "On the System of Social Ranks in a Natural Group of Japanese Monkeys." *Primates* 1:11–48.

Kawamura, S. 1958. "Matriarchal Social Ranks in the Minoo-B Troop: A Study of the Rank System of Japanese Monkeys." Pp. 105–12 in *Social Communication among Primates*, edited by S. A. Altmann. Chicago: University of Chicago Press.

Knott, C. 1998. "Changes in Orangutan Caloric Intake, Energy Balance, and Ketones in Response to Fluctuating Fruit Availability." *Int. J. Primatol.* 19:1029–43.

Knott, C. 2001. "Female Reproductive Ecology of the Apes." Pp. 429–63 in *Reproductive Ecology and Human Evolution,* edited by P. T. Ellison. Hawthorne, NY: Aldine de Gruyter.

Marlowe, F. 1999. "Male Care and Mating Effort among Hadza Foragers." *Behav. Ecol. Sociobiol.* 46:57–64.

Martins, W. P. and K. B. Strier. In press. "Age at First Reproduction in Philopatric Female Muriquis (*Brachyteles arachnoides hypoxanthus*)." *Primates* 45(1).

Melnick, D. J. and G. A. Hoelzer. 1992. "Differences in Male and Female Macaque Dispersal Lead to Contrasting Distributions of Nuclear and Mitochondrial DNA Variation." *Int. J. Primatol.* 13:379–93.

Ménard, N., F. von Segesser, W. Scheffrahn, J. Pastorini, D. Vallet, B. Gaci, R. D. Martin, and A. Gautier-Hion. 2001. "Is Male-Infant Caretaking Related to Paternity and/or Mating Activities in Wild Barbary Macaques (*Macaca sylvanus*)?" *Life Sciences* 324:601–10.

Mitani, J. C., D. A. Merriwether, and C. Zhang. 2000. "Male Affiliation, Cooperation and Kinship in Wild Chimpanzees." *Anim. Behav.* 59:885–93.

Mitani, J. C. and D. P. Watts. 2001. "Why Do Chimpanzees Hunt and Share Meat?" *Animal Behaviour* 61:915–24.

Mitani, J. C., D. P. Watts, J. W. Pepper, and D. A. Merriwether. 2002. "Demographic and Social Constraints on Male Chimpanzee Behaviour." *Anim. Behav.* 64:727–37.

Moore, J. 1984. "Female Transfer in Primates." *Int. J. Primatol.* 5:537–89.

Moore, J. 1992. "Dispersal, Nepotism, and Primate Social Behavior." *Int. J. Primatol.* 13:361–78.

Morgan, L. H. 1871. *Systems of Consanguinity and Affinity of the Human Family.* Washington, DC: Smithsonian Institution Press.

Morin, P. A., J. J. Moore, R. Chakraborty, L. Jin, J. Goodall, and D. S. Woodruff. 1994. "Kin Selection, Social Structure, Gene Flow, and the Evolution of Chimpanzees." *Science* 265:1193–1201.

Nishida, T. and K. Hosaka. 1996. "Coalition Strategies among Adult Male Chimpanzees of the Mahale Mountains, Tanzania." Pp. 114–34 in *Great Ape Societies,* edited by W. C. McGrew, L. F. Marchant, and T. Nishida. New York: Cambridge University Press.

Oka, T. and O. Takenaka. 2001. "Wild Gibbons' Parentage Tested by Non-Invasive DNA Sampling and PCR-Amplified Polymorphic Microsatellites." *Primates* 42:67–73.

Palombit, R. A. 1994. "Extra-Pair Copulations in a Monogamous Ape." *Anim. Behav.* 47:721–23.

Palombit, R. A. 1995. "Longitudinal Patterns of Reproduction in Wild Female Siamang (*Hylobates syndactylus*) and White-Handed Gibbons (*Hylobates lar*)." *Int. J. Primatol.* 16:739–60.

Palombit, R. A. 1999. "Infanticide and the Evolution of Pair Bonds in Nonhuman Primates." *Evol. Anthropol.* 7:117–29.

Parr, L. A. and F. B. M. de Waal. 1999. "Visual Kin Recognition in Chimpanzees." *Nature* 399:647–48.

Perry, S. 1996. "Intergroup Encounters in Wild White-Faced Capuchins (*Cebus capucinus*)." *Int. J. Primatol.* 17:309–30.

Pope, T. R. 1990. "The Reproductive Consequences of Male Cooperation in the Red Howler Monkey: Paternity Exclusion in Multi-Male and Single-Male Troops Using Genetic Markers." *Behav. Ecol. Sociobiol.* 27:439–46.

Pope, T. R. 1992. "The Influence of Dispersal Patterns and Mating System on Genetic Differentiation within and between Populations of the Red Howler Monkey (*Alouatta seniculus*)." *Evolution* 46:1112–28.

Pope, T. R. 1998. "Effects of Demographic Change on Group Kin Structure and Gene Dynamics of Populations of Red Howling Monkeys." *J. Mammal.* 79:692–712.

Pope, T. R. 2000. "Reproductive Success Increases with Degree of Kinship in Cooperative Coalitions of Female Red Howler Monkeys (*Alouatta seniculus*)." *Behav. Ecol. Sociobiol.* 48:253–67.

Pusey, A. E. 2001. "Of Genes and Apes: Chimpanzee Social Organization and Reproduction." Pp. 9–37 in *Tree of Origin*, edited by F. B. M. de Waal. Cambridge, MA: Harvard University Press.

Pusey, A. E. and C. Packer. 1987. "Dispersal and Philopatry." Pp. 250–66 in *Primate Societies*, edited by B. B. Smuts, D. L. Cheney, R. M. Seyfarth, T. T. Struhsaker, and R. W. Wrangham. Chicago: University of Chicago Press.

Pusey, A. E., J. Williams, and J. Goodall. 1997. "The Influence of Dominance Rank on the Reproductive Success of Female Chimpanzees." *Science* 277:828–31.

Reichard, U. and V. Sommer. 1997. "Group Encounters in Wild Gibbons (*Hylobates lar*): Agonism, Affiliation, and the Concept of Infanticide." *Behaviour* 134:1135–74.

Robbins, M. M. 1995. "A Demographic Analysis of Male Life History and Social Structure of Mountain Gorillas." *Behaviour* 132:21–47.

Rodseth, L. and R. W. Wrangham. In press. "Human Kinship: A Continuation of Politics by Other Means?" In *Kinship and Behavior in Primates*, edited by B. Chapais and C. Berman. New York: Oxford University Press.

Rodseth, L., R. W. Wrangham, A. M. Harrigan, and B. B. Smuts. 1991. "The Human Community as a Primate Society." *Curr. Anthropol.* 32:221–54.

Sackett, G. P. and W. T. Fredrickson. 1987. "Social Preferences by Pigtailed Macaques: Familiarity versus Degree and Type of Kinship." *Anim. Behav.* 35:603–7.

Sade, D. S. 1965. "Some Aspects of Parent-Offspring and Sibling Relations in a Group of Rhesus Monkeys, with a Discussion of Grooming." *Am. J. Phys. Anthropol.* 23:1–17.

Sade, D. S. 1991. "Kinship." Pp. 229–41 in *Understanding Behavior: What Primate Studies Tell Us About Human Behavior*, edited by J. D. Loy and C. V. Peters. New York: Oxford University Press.

Sauther, M. L., R. W. Sussman, and L. Gould. 1999. "The Socioecology of the Ring-tailed Lemur: Thirty-Five Years of Research." *Evolutionary Anthropology* 8(4):120–32.

Shennan, S. 2003. *Genes, Memes, and Human History: Darwinian Archaeology and Cultural Evolution*. London: Thames and Hudson.

Small, M. F. 1990. "Alloparental Behaviour in Barbary Macaques, *Macaca sylvanus*." *Anim. Behav.* 39:297–306.

Small, M. F. 1992. "Female Choice in Mating: The Evolutionary Significance of Female Choice Depends on Why the Female Chooses Her Reproductive Partner." *Am. Sci.* 80:142–51.

Smith, K., S. C. Alberts, and J. Altmann. 2003. "Wild Female Baboons Bias Their Social Behaviour towards Paternal Half-Sisters." *Proc. Royal Soc. London B* 270:503–10.

Smuts, B. B. 1985. *Sex and Friendship in Baboons.* Hawthorne, NY: Aldine de Gruyter.

Sterck, E. H. M. 1998. "Female Dispersal, Social Organization and Infanticide in Langurs: Are They Linked to Human Disturbance?" *Am. J. Primatol.* 44:235–54.

Strier, K. B. 1992. "Causes and Consequences of Nonaggression in Woolly Spider Monkeys." Pp. 100–16 in *Aggression and Peacefulness in Humans and Other Primates,* edited by J. Silverberg and J. P. Gray. New York: Oxford University Press.

Strier, K. B. 1996. "Male Reproductive Strategies in New World Primates." *Human Nature* 7:105–23.

Strier, K. B. 1997. "Mate Preferences of Wild Muriqui Monkeys (*Brachyteles arachnoides*): Reproductive and Social Correlates." *Folia Primatol.* 68:120–33.

Strier, K. B. 1999a. "Why Is Female Kin Bonding So Rare: Comparative Sociality of New World Primates." Pp. 300–19 in *Primate Socioecology,* edited by P. C. Lee. Cambridge: Cambridge University Press.

Strier, K. B. 1999b. *Faces in the Forest: The Endangered Muriqui Monkeys of Brazil.* Cambridge, MA: Harvard University Press.

Strier, K. B. 2000a. "From Binding Brotherhoods to Short-Term Sovereignty: The Dilemma of Male Cebidae." Pp. 72–83 in *Primate Males: Causes and Consequences of Variation in Group Composition,* edited by P. M. Kappeler. Cambridge: Cambridge University Press.

Strier, K. B. 2000b. "Population Viabilities and Conservation Implications for Muriquis (*Brachyteles arachnoides*) in Brazil's Atlantic Forest." *Biotropica* 32(4b):903–13.

Strier, K. B. In press. "The Impact of Patrilineal Kinship on Behavior." In *Kinship and Behavior in Primates,* edited by B. Chapais and C. Berman. New York: Oxford University Press.

Strier, K. B., L. T. Dib, and J. E. C. Figueira. 2002. "Social Dynamics of Male Muriquis (*Brachyteles arachnoides hypoxanthus*)." *Behaviour* 139:315–42.

Strier, K. B. and T. E. Ziegler. 2000. "Lack of Pubertal Influences on Female Dispersal in Muriqui Monkeys, *Brachyteles arachnoides*." *Anim. Behav.* 59:849–60.

Strum, S. C. 1982. "Agonistic Dominance in Male Baboons: An Alternative View." *Int. J. Primatol.* 3:175–202.

Sugiyama, Y. 1999. "Socioecological Factors of Male Chimpanzee Migration at Bossou, Guinea." *Primates* 40:61–68.

Sussman, R. W. 1992. "Male Life History and Intergroup Mobility among Ring-tailed Lemurs (*Lemur catta*)." *Int. J. Primatol.* 13:395–413.

van Schaik, C. P., J. K. Hodges, and C. L. Nunn. 2000. "Paternity Confusion and the Ovarian Cycles of Female Primates." Pp. 361–87 in *Infanticide by Males,* edited by C. P. van Schaik and C. H. Janson. Cambridge: Cambridge University Press.

van Schaik, C. P., M. A. van Noordwijk, and C. L. Nunn. 1999. "Sex and Social Evolution in Primates." Pp. 204–40 in *Primate Socioecology,* edited by P. C. Lee. Cambridge: Cambridge University Press.

Vigilant, L., M. Hofreiter, H. Siedel, and C. Boesch. 2001. "Paternity and Relatedness in Wild Chimpanzee Communities." *Proc. Natl. Acad. Sci. USA*, 98:12890–95.

Watts, D. P. 1998. "Coalitionary Mate Guarding by Male Chimpanzees at Ngogo, Kibale National Park, Uganda." *Behav. Ecol. Sociobiol.* 44:43–55.

Widdig, A., P. Nürnberg, M. Krawczak, W. J. Streich, and F. M. Bercovitch. 2001. "Paternal Relatedness and Age Proximity Regulate Social Relationships among Adult Female Rhesus Macaques." *Proc. Natl. Acad. Sci. USA* 98: 13769–73.

Widdig, A., P. Nürnberg, M. Krawczak, W. J. Streich, and F. M. Bercovitch. 2002. "Affiliation and Aggression among Adult Female Rhesus Macaques: A Genetic Analysis of Paternal Cohorts." *Behaviour* 139:371–91.

Wilson, M. L., M. D. Hauser, and R. W. Wrangham. 2001. "Does Participation in Intergroup Conflict Depend on Numerical Assessment, Range Location, or Rank for Wild Chimpanzees?" *Anim. Behav.* 61:1203–16.

Wrangham, R. W. 1987. "The Significance of African Apes for Reconstructing Human Social Evolution." Pp. 51–71 in *The Evolution of Human Behavior: Primate Models*, edited by W. G. Kinzey. Albany: SUNY Press.

Wright, P. C. 1990. "Patterns of Paternal Care in Primates." *Int. J. Primatol.* 11:89–102.

10

Revisiting Conflict Resolution

Is There a Role for Emphasizing Negotiation and Cooperation Instead of Conflict and Reconciliation?

Agustin Fuentes

> What is the value in all this fighting? In nature, fighting is such an ever-present process, its behavior mechanisms and weapons are so highly developed and have so obviously arisen under the selection pressure of a species-preserving function, that it is our duty to ask this Darwinian question.
> —Konrad Lorenz, *On Aggression* (1966:23)

Much in our current set of assumptions and theories regarding aggression and conflict can be seen as rooted in Konrad Lorenz's notions about the intrinsic and ever-present nature of aggression. Aggression and conflict have been and continue to be of great interest to anthropologists, psychologists, biologists, philosophers, and others who study the human condition. Although most studies of aggression have relied on nonhuman organisms as their study species, their results have frequently been applied to human society. The late 1960s and early 1970s saw the results of earlier nonhuman aggression studies used to make statements about human aggression and society. Overcrowding in the inner cities, gang warfare, and sports teams were seen as expressions of an animal continuum of violence and aggression. E. O. Wilson's 1975 *Sociobiology* spurred investigation into the evolutionary nature of human aggression. Explanations for homicide rates, spousal and sexual abuse, and other aggressive elements of human society became the foci of intensive interest and discussion.

Behavioral scientists have frequently used the nonhuman primates as potential models for the discussion of human evolution. As the role of field research in primatology expanded throughout the last three decades, it became readily apparent that aggression in nonhuman primates was more complex than the simplistic assumptions of alpha and beta males battling for control of females or other resources (Strier 1994; Strum and Fedigan 1999). The concept of aggression in the nonhuman primates can be complex, with multiple components relevant to evolutionary assessments. There are diverse types of aggression that can achieve similar ends. Aggression is neither simply defined nor easily quantified. For example, primate aggression can range from simple threats to contact fighting to vocal contests between opponents who cannot see each other. These aggression types can have similar outcomes, depending on the interactive history of the individuals and their species (and even on the specific environment in which the behaviors are exhibited). There is evidence of a wide continuum of conflict across all age/sex classes, and this conflict has a complex relationship to the daily lives of highly social group-living organisms such as primates (Mason and Mendoza 1993; Aureli and de Waal 2000; Silk 2002).

THE "RESPONSE TO CONFLICT" PARADIGM IN PRIMATE POSTCONFLICT BEHAVIOR STUDIES

> Evolutionary approaches to animal social behavior have been dominated by a false dichotomy between aggression and sociality. Struggle-for-life language was directly transferred to the social domain, resulting in an overemphasis on clashing individual interests. Theorists insisted on cost-benefit analysis, whereas in reality benefit-benefit arrangements seemed quite common. The possibility of shared interests was so far from the minds of evolutionary biologists (except with regard to kin) that when it came to the accounting for the rarity of lethal violence, rather than assuming a need for cooperation and stable group life, explanations focused exclusively on the physical risks of combat.
>
> — de Waal 2000a:24–25

The history of conflict-related studies in primatology began with the notions of intrinsic aggression, focusing investigations on the many facets of aggression as a driving force in animal behavior. By the 1970s and 1980s there was a focus on conflict itself. Increasing complexity in field studies revealed insight into the types and characteristics of conflict in primate societies, moving the focus from aggressive drives and generalized patterns to a specific examination of the conflict and its potential costs to the participants' fitness. During the 1980s and 1990s the work by Frans de Waal and his students and colleagues initiated a series of investigations into the immediate postconflict period. The relatively high frequency of friendly

- Aggression as antisocial tendency model:
  - **Dispersal:** aggression reduces interaction/association/proximity
  - **Motivational continuity:** postconflict contact is residual antagonism
  - **Bonding excludes aggression:** aggression is rare among socially close and cooperative individuals
- Relational model
  - **Social repair:** increased friendly contact postaggression, especially between opponents, serves a relationship repair function
  - **Motivational shift:** postaggression contact can be unusually intensive
  - **Conflict as negotiation:** aggression and postconflict contact common among socially close and cooperative individuals
  - **Calming effect:** postconflict contact reduces anxiety and restores tolerance

This "relational model envisions" a tripartite set of responses to conflict

Figure 10.1 A summary of de Waal's (2000a) models.

postconflict reunions between conflict participants (termed "reconciliations") in a large group of chimpanzees at the Arnhem Burger's Zoo, coupled with changing theoretical concepts regarding the function of grooming and affiliation (McKenna 1978, 1983), suggested to de Waal that there might be mechanisms at play specifically geared toward the amelioration of the costs and damage imposed by social conflict (de Waal and van Roosmalen 1979; de Waal and Yoshihara 1983). A derivation of this assumption is that cooperative and affiliative behaviors arise (in large part) as repair mechanisms to fix the damage caused by conflict. This resulted in a heavy emphasis on reconciliation as a specific set of behavior in primates, especially chimpanzees (due to de Waal's publications, and chimpanzees' heralded role as humans' closest cousins). Reconciliation is defined as a specific increase in affiliation between opponents shortly after conflict. The current definition of reconciliation focuses on attracted, dispersed, and neutral interactions between dyads (de Waal and Yoshihara 1983). de Waal and Yoshihara suggested that following conflict, attraction rather than dispersal is likely to occur and former opponents are likely to interact affiliatively through close proximity and body contact. Veenema, Das, and Aureli (1994) refined reconciliation data collection methodology by comparing the postconflict behavior with a matched control period to establish baseline parameters. They added a correction factor that removed baseline behavior from postconflict interaction data (see also Veenema 2000). This resulted in a more robust measure of reconciliation called the *corrected conciliatory tendency* (CCT). Here reconciliation is assumed to serve a relationship repair, or cost-reducing function, and is recognizable as a characteristic set of behaviors.

de Waal's and others' work with multiple primate species (primarily macaques) led de Waal to propose his "relational model" to counter what he refers to as a Lorenzian aggression-as-antisocial force model (see Figure 10.1). de Waal has suggested that it is through the examination of conflict and the evolution of behaviors to ameliorate it that we can better understand our own evolutionary "peacemaking" histories (de Waal and Aureli 1996; de Waal 2000a).

UNCERTAINTY REDUCTION

> Although we have studied peaceful post-conflict behavior in dozens of species, we do not yet have enough information about its form, pattern, and function to draw firm conclusions about the selective forces that have shaped its evolution.
>
> —Silk 2000:181

Counter to the relationship-repair function championed by de Waal and colleagues, Joan Silk has argued, alternatively, that the postconflict friendly reunions called "reconciliations" by many researchers can actually be seen as simply indicating a cessation of aggression (Silk 1996, 1998, 2000, 2002). Silk proposes that natural selection, rather than favoring a suite of specific reconciliatory behavior, has resulted in the use of simple signals of benign intent that can act to reduce the uncertainty about intentions of former opponents and thereby relieve the stress that is associated with not knowing whether hostilities will continue (Silk 2002). Here the argument is that conflict can have a negative immediate impact, but rather than acting in a relationship restoring function, the postconflict friendly reunions (reconciliations) merely facilitate a cessation of aggression and potential acquisition of short-term objectives (such as access to a desired resource or a preferred partner) (Silk 2002).

INTERINDIVIDUAL RELATIONSHIPS, HISTORIES, AND RELATIONSHIP VALUES

To date, a number of postconflict behavior studies have been undertaken across primate taxa (de Waal 1989, 1993; Kappeler and van Schaik 1992; Aureli and de Waal 2000; Silk 2002). In general the results have demonstrated a range in postconflict behavior, particularly in rates and styles of reconciliation [see Kappeler and van Schaik (1992), de Waal (1993), and Aureli and de Waal (2000) for overviews of the datasets]. Other postconflict behavior patterns investigated include consolation (affiliative interaction with third party after a conflict) and redirection (redirecting aggression toward other group members) (see Aureli and de Waal 2000). One of the major findings emerging from the growing number of postconflict behavior studies is that of variability in the types and patterns of behavior related to conflict negotiation and resolution (Castles, Aureli, and de Waal 1996; Cords and Killen 1998; Schino, Rosati, and Aureli 1998; Aureli and de Waal 2000; Thierry 2000; de Waal and Aureli 2000; Silk 2002).

Emerging from this recognition of complexity and variety in behavior after conflict, some researchers have recently begun to examine "relationship qualities" as major factors influencing postconflict behavior. Cords and Aureli (2000) summarize their recent efforts and those of others by proposing a tripartite relationship quality-measuring tool including relationship "value," "security," and "compatibility" (Table 10.1). In this system, the authors attempt to assess the relative value of a dyadic relationship in the sense of a cost-benefit paradigm implying an optimality driver influencing the assessment of each individual's relative contribution to increasing success versus incurring reproductive cost via conflict and/or loss of the coalitionary relationship (van Schaik and Aureli 2000). However, a recent review of the existing data on relationship quality concluded that the current dataset offers mixed support for the role of the three proposed relationship qualities in explaining postconflict interactions (Silk 2002).

THE PRIMARY FOCI

Today the initial emphasis on reconciliation has grown to include the complexities involved in conflict negotiation and has begun to concede that the uncertainty reduction hypothesis (that friendly postconflict behaviors only indicate a cessation of aggression, not more) may be a strong potential explanation for the observed postconflict behavior (de Waal and Aureli 2000; Silk 2000, 2002). However, although both of the orientations (de Waal's reconciliation and Silk's uncertainty reduction) have been published and are part of the discourse, the de Waalian focus remains the most popular and prevalent in the literature. In general, the predominant thesis

Table 10.1   The Three Relationship Qualities (Cords and Aureli 2000)

Value	level of overall benefits derived from the association and the accessibility of the partner in the relationship
Security	perceived probability of a partner's consistency in behavioral patterns
Compatibility	the "tenor" of social interactions between the dyad, resulting from both "temperament" and history of social exchanges

of postconflict research in nonhuman primates is that postconflict behavior, especially reconciliation, is functional and has evolved as a response to curtail the damage to interindividual relationships caused by conflict.

THE BOTTOM LINE

> . . . without denying the human heritage of aggression and violence, this research demonstrates an equally old heritage of countermeasures that protect cooperative arrangements against the undermining effects of competition.
>
> —(de Waal 2000b:590)

Over the last decade and a half we have been seeing a growing integration of the concept of aggression and conflict into the larger context of social interactions (Mason and Mendoza 1993). Investigations are starting to shift from a primary focus on conflict to an examination of the full range of cooperative and affiliative behaviors characteristic of the primates (this volume, for example). However, despite this change in emphasis, the current impetus for most postconflict research remains rooted in the notion that postconflict behavior has evolved as a response to curtail the damage caused by conflict. That is, under current hypotheses the majority of cooperative and affiliative behaviors surrounding conflict are seen as adaptations to prevent, ameliorate, and/or otherwise deal with the evolutionary "costs" imposed by aggression (de Waal and Aureli 2000; Silk 2002). The main foci of this current discourse surrounding conflict rest soundly upon the Dawkinsian "selfish gene"/individualistic focus concept: cooperation as driven by selfish propagation (Wilson 1975; Dawkins 1976; van Schaik and Aureli 2000). This orientation regards specific conflicts and the potential costs of aggression, but not a lifetime of cooperation, as the most important focus when assessing overall costs and their related evolutionary strategies relative to conflict negotiation. In short, most theoretical explanations rest on the assumption that cooperation and affiliation have evolved to deal with damage "costs" caused by conflict (Aureli and de Waal 2000; Silk 2002).

Results from some current research on nonhuman primates suggest that the focus on conflict as selecting for a set of behaviors that repair the damage it causes may not provide the best set of explanations for observed postconflict behavior patterns. I suggest that there has been an overemphasis on a few short-term, but visibly costly events (such as conflicts) as driving the natural selection for behavioral traits. In particular, it is possible that the hypothesis that the potential costs of conflict have led to the selection of a functional suite of behavior termed "reconciliation" is incorrect or incomplete (Silk 2002).

ARE THERE ALTERNATIVES TO THE "RESPONSE TO CONFLICT" SCENARIO?

Behavior being observed in postconflict situations might not be the direct result of selection. That is, these behaviors may be nonfunctional and/or neutral relative to lifetime reproductive success, similar to Joan Silk's arguments (Silk 2002). It is also possible that there may not be uniform sets of postconflict behavior. Although some researchers report consistent sets of postconflict behavior, others do not. If "reconciliation" is not characterized by uniform or at least consistent sets of behavior, then we may not be able to characterize it as a specific adaptation resulting from the costs imposed by conflict. Finally, it is possible that we cannot effectively uncouple postconflict behavior from ongoing behavioral patterns such that isolating it presents a false, or at least incomplete, picture of the actual energetic investment patterns (behavior) of any individual.

Rather than a focus on reconciliation, I suggest that by examining the longer-term behavioral patterns and relationship histories as a package we can better understand which postconflict behaviors are, and which are not, the result of natural selection.

COOPERATION AND FLEXIBILITY AS SELECTIVELY IMPORTANT?

"Who are the fittest: those that are continually at war with each other, or those who support one another?" ... Those animals which acquire habits of mutual aid are undoubtedly the fittest.
                                        —Kropotkin (1902) 1955:6

The evolutionary biologist Lee Dugatkin (1999) recently suggested that Prince Petr Kropotkin's notions of a primacy of cooperation in natural selection were idealistic and misdirected, but is a central role for cooperation and affiliation possible in a functional paradigm? Multilevel selection might be an overlooked factor in the examination of postconflict behavior. That is, intergroup selection may favor types of cooperation that result in what appears to be conflict amelioration but in fact is a set of behavior favored under specific population-level selective conditions. This is what Sober and Wilson (1998) have termed "unselfish" behavior. In other words, behavior currently considered to appear in order to ameliorate the damage caused by intragroup conflict could actually be part of a larger set of behavior favored in response to intergroup competition. Rather than seeing a few short-term, but visibly high-cost events (intragroup conflicts) as driving selection for behavioral traits, what if we examine the longer term behavioral patterns as a package, related to lifetime reproductive success, that selection can act upon? That is, currently we see all/most cooperative, affiliative behaviors as adaptations to prevent, ameliorate, and/or otherwise deal with aggressive conflict. What if we shift to a holistic focus on multiple facets of long-term relationships? Conflicts might then be seen as perturbations in an overall system, not necessarily the drivers of such a system. While intraindividual conflict is an important component of social life, it might not be the prime driving force behind the evolution of affiliation.

Nonfunctional behaviors occur, and it is possible that some of the behavior we see in postconflict situations is not functional. Could postconflict behavior consist of patterns that include elements that are invisible to selection (for whatever reasons)? Selection might favor a certain behavioral flexibility that is capable of dealing with a multiplicity of social situations fairly well rather than specific patterns or specific behaviors related to repairing conflict extremely well. For example, Richard Potts's (1998, chapter 12 in this volume) proposed variability selection hypothesis for evolution in the genus *Homo* hypothesizes that selection favored variable behavior patterns and a range of flexibility rather than specific localized adaptations in early human behavior. What if this notion is extended into the context of exhibited circum-conflict behavior? Could humans, and other highly social primates, have evolved a pattern in which they have a certain flexibility in behavioral response not tied directly to achieving optimal (in an evolutionary sense) effects? Along similar lines, Virginia Vitzthum (2001) has proposed a flexible response model for human reproduction. In this model, the evolution of flexibility in function and response, rather than specific optimality, is invoked to explain the observed variability in human female reproductive patterns. Could it be that natural selection is favoring a pattern of social organization in primates that is relatively flexible in demographic and behavioral variables in order to rapidly, behaviorally, adapt to changing social and ecological conditions (Fuentes 1999)?

TWO EXAMPLES

There is currently a fairly robust and growing set of data regarding primate postconflict behavior. However, a focus on reconciliation as a prime characteristic of this behavior may mask some of the underlying complexities. Here I present two examples that demonstrate variability in nonhuman primate postconflict behavior as an illustration of the possibility that the focus on the conflict-ameliorating functional response (reconciliation) may be misdirecting the broader inquiry into cooperative relationships and social complexity in primates. While these two examples are not meant to represent the entire diversity of postconflict studies, they do accentuate the role of variation in dyadic interactions and variability in postconflict behavior in two primate species.

Chimpanzees as a Case Study in Variable Postconflict Behavior

Chimpanzees are central to the current investigation into postconflict behavior, as they were the focus of de Waal's early investigations and many of his subsequent books (de Waal and van Roosmalen 1979; de Waal 1982, 1987, 1989; de Waal and Aureli 1996). Chimpanzees' well-documented social complexity has made them appropriate targets for the examination of reconciliation. Until recently, however, only a few captive studies conducted by de Waal and colleagues made up the available database for chimpanzee postconflict behavior. Recently three new studies, conducted with similar, controlled methodologies, have emerged showing a somewhat different picture of chimpanzee postconflict behavior (Arnold and Whiten 2001; Fuentes et al. 2002; Preuschoft, Wang, Aureli, and de Waal 2002). Table 10.2 shows the published studies of *Pan troglodytes* and *Pan paniscus* that have examined postconflict behavior and reported reconciliation. Three of the studies (Arnold and Whiten 2001; Fuentes et al. 2002; Preuschoft et al. 2002) utilized the more robust CCT methodology (Veenema et al. 1994), one contained sufficient information to calculate a CCT, and two others did not report controls for baseline interindividual interaction patterns. Most of the literature on chimpanzee reconciliation relied solely on the studies at the Arnhem Burger's Zoo and at the San Diego Zoo (de Waal and van Roosmalen 1979; de Waal 1982, 1987, 1989; de Waal and Aureli 1996). From these initial studies chimpanzees became heralded as frequent reconcilers and the model organism for the importance and function of reconciliation as a relationship repair mechanism. However, the studies with appropriate data to calculate CCTs show marked variability in rates of mean group reconciliation (14.4–41.2 percent; Table 10.2). Surprisingly, three of the four studies, including the only study of free-ranging chimpanzee postconflict behavior (Arnold and Whiten 2001), report mean group corrected recon-

Table 10.2   Description of Reconciliation and Corrected Conciliatory Tendencies in Captive Chimpanzee Studies (Fuentes et al. 2002)

Species	Location/Setting	Group Size	Opponent Pairs	Reconciliation (%, uncorrected)	Corrected Conciliatory Tendency (%, corrected reconciliation)
Pan troglodytes	**Arnhem**[a]				
	Indoor, 1979	20	150	34.7	—
	Outdoor, 1979	20	200	29.5	—
	Outdoor, 1981	20	95	26.6	—
	Detroit[b]				
	Indoor/outdoor, 1994	11	43	49.7	14.4
	CHCI[c]				
	Indoor/outdoor 1997–2000	5	262	25.1	17.25
	Budongo[d]				
	Free-ranging, Uganda	51	120		19.2
	Yerkes[e]				
	Outdoor	16	401	44.8	41.2
Pan paniscus	**San Diego Zoo**[f]				
	Indoor/outdoor, 1987	6	333	43.8	—
	Outdoor, 1997	4	179	55.9	—

[a]de Waal and van Roosmalen (1979); Griede (1981).
[b]Baker and Smuts (1994).
[c]Fuentes, Malone, Sanz, Matheson, and Vaughn (2002).
[d]Arnold and Whitten (2001).
[e]Preuschoft, Wang, Aureli, de Waal (2002).
[f]de Waal (1987).

ciliation values of less than 20 percent. This suggests that reconciliation may not be as common in chimpanzees as previously thought. Also, in all chimpanzee studies where the information is available, there was no correlation between conflict intensity and rates of reconciliation.

The study at the Central Washington University Chimpanzee and Human Communication Institute (CWU-CHCI) (Fuentes et al. 2002), recent work published by Preuschoft et al. (2002) at Yerkes, and work at Budongo by Arnold and Whiten (2001) also suggest that there is enormous variation in postconflict attraction (CCT) between dyads within groups. For example, over a four-year study at CWU-CHCI (Fuentes et al. 2002), some individuals reconciled many conflicts (such as the females Moja and Tatu), others only a few (such as the males Dar and Loulis) (Table 10.3).

Table 10.3   CWU-CHCI Interindividual Data: Number of Conflicts per Dyad and Corrected Conciliatory Tendency (CCT) between Individual Conflict Pairs (Attracted-Dispersed/Total Conflicts for that Pair) (Fuentes et al. 2002)

	Dar	*Loulis*	*Moja*	*Tatu*	*Washoe*
Individual CCT (%)	*8.1*	*6.4*	*32.0*	*17.6*	*5.8*
Dyadic CCT					
Dar	—	90 (4.4%)	5 (20.0%)	13 (30.8%)	3 (−33.0%)[a]
Loulis	—	—	9 (33.0%)	50 (10.0%)	78 (3.8%)
Moja	—	—	—	8 (37.5%)	3 (33.0%)
Tatu	—	—	—	—	3 (33.0%)
Washoe	—	—	—	—	—

[a]A negative CCT is reported for this dyad due to a higher frequency of dispersal following the observed conflicts.

Different dyads exhibited very different patterns. Comparisons by sex of adult dyadic reconciliation at CWU-CHCI and Yerkes (Table 10.4) show broad variation in these patterns. Unlike the captive studies, the Budongo study shows very infrequent female-female reconciliation and no significant differences in reconciliation rates between female-male and male-male dyads (Arnold and Whiten 2001). Both the Yerkes and Budongo studies also demonstrated variation across age groups in dyadic patterns, with dyads involving immature individuals showing higher reconciliation rates than adult-only dyads (Arnold and Whiten 2001; Preuschoft et al. 2002). These data suggest that while some individuals do reconcile, many do so very infrequently, and there is a great amount of variation in dyadic reconciliatory patterns by age, sex, and individual.

The chimpanzees at CWU-CHCI displayed redirection of aggression and the maintenance of visual contact more frequently after conflicts than during baseline behavior (Table 10.5). Redirection may be a by-product of the conflict; however, it is likely that it is not a specifically selected response (Malone, Vaughn, and Fuentes 2000). The maintenance of visual contact may reflect a monitoring factor that is related to the conflict and might best be seen as part of the behaviors of the conflict. Also, in contrast to earlier statements by de Waal and colleagues (de Waal 1989), consolation behaviors (affiliative behavior with a third after conflict) were not significantly more common after conflicts in either the CWU-CHCI or the Budongo studies (Arnold and Whiten 2001; Fuentes et al. 2002).

In addition to some reconciliation, visual monitoring, and redirection, all five members of the group at CWU-CHCI exhibited a range of behaviors that increased significantly after conflict relative to matched controls (Table 10.6). While this may be an artifact of the small sample size, it is also

Table 10.4   Corrected Conciliatory Tendency (%) by Age/Sex
            Class Dyad at CWU-CHCI (Fuentes et al. 2002) Compared
            with the Yerkes Study (Preuschoft et al. 2002) Values for
            Same Age/Sex Classes

Age/sex class	CWU-CHCI	Yerkes
Adult male-adult male	4.4	NA
Adult male-adult female	16.3	45.7
Adult female-adult female	34.5	57.9

possible that each individual is behaving somewhat differently after conflicts across years as relationships and circumstances change and as chance
events surrounding conflicts come and go. Interestingly, the group totals
in Table 10.6 are obviously artificial, as some individuals impact the numbers much more than others. This suggests that comparisons of lumped
behaviors at the group level can be misleading when attempting to find
patterns and trends in dyadic behavior exhibited after conflicts. While the
Yerkes study reports some specific postconflict "reconciliatory" behaviors
such as kissing, hand-holding, and embracing, neither the Budongo nor
the CWU-CHCI studies found these behaviors, or any specific set of "reconciliatory" behavior, occurring at higher than matched control or baseline levels. Given the current set of published data available for
chimpanzees, there is currently little quantified evidence for a specific set
of reconciliation behavior in chimpanzees. The wide array of nonreconcil-
iatory postconflict behavior suggests that there is a range of behavior
exhibited postconflict and that reconciliation is a relatively minor element
for the chimpanzees, at least in three of five groups studied.

The CWU-CHCI study also produced a preconflict dataset (Mulcahy
2001). Video data collection demonstrated the possibility that behavior
prior to a conflict impacts the behavior afterwards, or, that the conflicts
themselves impact behavior pre- and postconflict. Mulcahy (2001) reports
a higher frequency of conflict when chimpanzee group members were
closer together spatially and when visitors (humans outside the chimpanzee enclosure) were present. Mulcahy also reports a dampening of certain behaviors and substantial overall variation between individuals in
preconflict behavior. Again, this suggests that, for at least the CWU-CHCI
chimpanzees, there is a wide array of individual response before a conflict
that may, or may not, impact the conflict and the behaviors after it.

What Might This Mean for Chimps?

Chimps do reconcile after some conflicts, but three of the four published controlled studies show fairly low rates of reconciliation. Individu-

Table 10.5   CWU-CHCI Group Totals for Redirection, Consolation, Overall Proximity, and Maintenance of Visual Contact for Both Postconflict (PC) and Matched-Control (Baseline)(MC) Periods Expressed as a Percentage of Total Conflicts (Fuentes et al. 2002)

	PC	MC	PC vs. MC (chi square test)
Consolation	40.0	35.9	NS
Redirection	30.5	8.0	$p < .01$
Proximity (closer sooner)	67.2	31.7	$p < .05$
Maintenance of visual contact	79.1	64.6	$p < .01$

als vary in their behavior after conflicts, and their response patterns are not necessarily consistent across years. Individual patterns also varied across studies. It is not clear that reconciliation is a major, functional postconflict pattern in chimpanzees. For chimpanzees, patterns of association, dyadic histories, individual variation in behavior, and use of space might predict behavior surrounding conflict better than a focus on the conflict itself or on specific postconflict behavior sets.

Hanuman Langurs: Sex Differences and Lack of Functional Postconflict Behavior

A study by Phyllis Dolhinow, Elsworth Ray, and myself (Fuentes, Ray, and Dolhinow 1996; Ray, Fuentes, and Dolhinow 1996) compared an all-male group and a mixed-sex group of Hanuman langurs (*Semnopithecus entellus*) during June-August 1996 at the University of California, Berkeley, behavioral field research station. A total of 386 conflicts, and the behaviors surrounding those conflicts, were recorded during the study involving eleven adult and two subadult subjects. The mixed-sex group had six subjects (four female and two male) for a total of fourteen dyads, and the all-male group had seven subjects and twenty-one dyads.

The females did participate in postconflict social interactions, with only 32 percent of their conflicts *not* followed by social interaction. The females had an overall conciliatory response (CR-uncorrected reconciliation) of 40 percent, and when compared to baseline data, three out of the six female dyads had significantly higher percentages of affiliation after conflicts than during baseline observations. Overall, the females had a corrected conciliatory tendency (CCT) of 11.4 percent.

While individual females varied in their responses to conflict, all females exhibited a tendency to interact with another female in a postconflict situation. For the females it is highly likely that affiliative behavior immediately following agonism may function to lessen the tension caused by the agonistic episode [such as proposed by Silk (1996, 2000)]. It may be

Table 10.6   CWU-CHCI Individual and Group Patterns for Behaviors Exhibited Significantly More Often in the PC Than in the MC Periods ($p < .01$)[a] (Fuentes et al. 2002)

	Dar	Loulis	Moja	Tatu	Washoe	Group
Agonism						3
Affinitive social				1,4	2	
Bad observation	4			5		
Coprophagy			2,3,5			2,3,5
Display		1,2,3,5,6			2,3,5,6	1,2,3,5,6
Feeding	6			1,6	3	6
Groom		3	4		4	
Play			4	1		1,4
Object manipulation	3		4	2	3,4,6	3,4
Other	5			2		2
Reassurance				3		3
Self-groom	3,5,6	3,4,5	1	3		3,5,6
Threat	5	1,2,4,5,6		3	1,2,5,6	1,2,4,5,6
Travel	2	1,4		3		1,2,4

[a]Data collection periods: 1997, 1; 1998, 2; 1999a, 3; 1999b, 4; 2000a, 5; 2000b, 6.

that in the females we are seeing a form of restorative or diffusing behavior. Interaction with both participants and other animals in postconflict periods may decrease overall levels of tension, or a simple lack of agonism might restore the preconflict state of affairs within the group. That is, these females might not need to increase levels of affiliative interactions (rates) or even interact with the individual with whom they have engaged in a conflict. It may be that simply participating in affiliative interactions restores preconflict levels of social tension among these female langurs.

The all-male group displayed very low rates of postconflict affiliation (CR = 8 percent). None of the male-male dyads had significantly higher rates of affiliation after conflict when compared with baseline, and, in fact, the males in this group generally had no affiliative response at all to agonistic events. Sixty-nine percent of all conflict interactions in the all-male group were followed by no social interaction by the individuals involved in the conflict with either the conflict participant or a third party. For six of the seven males no social interaction was the most common reaction after a conflict. Two males redirected agonism frequently when they did respond; another ate and avoided other group members as a main response. Only the oldest of the males in the all-male group (a male more than sixteen years of age) had a tendency to interact affiliatively with his participant in a postconflict situation (he had a CR of 27 percent). This situation arose because he was consistently approached by opponents after agonistic episodes (100 percent opponent-initiated contact).

Interestingly, the two males in the mixed-sex group, with a CR of 5 percent and no social response to conflict in 73 percent of events, behaved more like the males in the all-male group than they did like the female members of their own group.

The overall conciliatory response measure suggests that these langur monkeys generally did not engage in postconflict ameliorative behavior at a level significantly distinct from baseline interactions. Reconciliation, as a substantial phenomenon, was not found in this study. However, overall numbers distort the actual patterns (or lack of patterns) in response to conflict in this case. Although specific reconciliation was very low between females, there was consistent affiliative behavior after conflict, although this behavior did not differ dramatically from their baseline patterns. Alternatively, the males tended not to interact after conflicts. This study presents evidence of divergent patterns of behavior exhibited by the male and female langurs in postconflict contexts. At least in this mixed group, general patterns of behavior for the female-female relationships were the best indicator of behavior around conflict, and conflict in the all-male group was relatively common and not reconciled. Because of matrilineal kin clusters in this species, aspects of female cooperative relationships may help explain the postconflict behaviors. For the all-male group it appeared that postconflict affiliation with the oldest male was important for group members, but that conflicts in general did not require a set of postconflict ameliorative behaviors.

RECONCILIATION IN PERSPECTIVE

Table 10.7 provides a brief summary of the existing postconflict reconciliation data sets from the appendix of Aureli and de Waal (2000). Highly variable results and the fact that different methodologies were used across the different studies make it difficult to compare these numbers. At least a few of the species' percentages represented in Table 10.7 come from only one study, and the other species display a wide range of reconciliation (controlled and uncontrolled) across the different study groups. The majority of these studies are of captive groups.

The results of the studies reviewed here (CWU-CHCI chimpanzees, chimpanzee overview, and the Hanuman langurs) plus the general results of many postconflict studies (Castles et al. 1996; Aureli and de Waal 2000; Thierry 2000; Silk 2002) reveal a wide degree of variability in the postconflict behavior, especially in behavior termed "reconciliation," both within and between genera/species. This high variation in reconciliatory tendencies and the variety of postconflict behavior patterns reported for primates raise an important functional question: Is anything being "selected" for in postconflict behavior?

Table 10.7   Reconciliation (%) in Other Primate Species (from Controlled and Non-
controlled Studies, Thus Not Actually
Comparable) (Aureli and de Waal 2000)

Species	Reconciliation
Macaques	7–53
Lemurs	14–21
Langurs	9–54
Baboons	10–45
Capuchins	21
Vervets	14

Given the current datasets available for reconciliation in primates, it appears unclear if there is a pattern of behavior that has resulted from selection that serves the function of relationship-repair/conflict amelioration (Silk 2002). Many primate species do exhibit reconciliation behaviors at variable frequencies, but these are not necessarily specific sets of behavior. In the species studied there is a high degree of individual variation in postconflict response, especially in flagship species such as *Pan troglodytes*. This variability in response coupled with the inherent dangers of presenting group-level patterns when examining individual dyadic datasets suggests that, at present, we cannot fully disentangle the cooperative relationships and relationship histories of individuals from the conflicts they engage in and how they behave before and after those conflicts. It is not clear that reconciliation is necessarily the (or a) primary response to conflict in most nonhuman primate species. This suggests that the need to repair the potential damage to interindividual relationships caused by conflict may not be as universal or imperative as previously thought. Conflicts can be damaging, but how damaging they are is both debatable and dependent on multiple factors. It appears, at least in chimpanzees, that different dyads/relationships will respond differently to conflicts. Individual variation in response to conflicts suggests that interindividual relationships are key to understanding circum-conflict behavior. Therefore understanding the history and context of relationships, which can be envisioned as long-term cooperative interactions, may tell us more about the individual's behavior and strategies than short-term responses after conflicts, especially if those responses are predicated on the long-term relationships. To better understand these differences we need to examine not just the type/situation/context of the conflict but also interindividual relationship patterns, dyadic cooperative and conflict histories, and other aspects of the dyadic and groupwide relationships (in a sense, an expansion on the "relationship value" measures proposed by Cords and Aureli 2000).

Can shifting to a holistic focus on multiple facets of long-term relationships be a beneficial path of investigating the behavior surrounding conflict? Is it possible that conflicts can be seen as perturbations in an overall system, not necessarily the drivers of such a system?

I propose that current data available on postconflict interactions for chimpanzees, and possibly other primates, suggest that the answer to these questions is a tentative yes. While conflict is an important component of social life, it might not be the prime driving force behind the evolution of affiliative postconflict behavior. Rather than having evolved as a set of specific behavioral responses to repair the damage caused by conflict, interindividual patterns of cooperation and affiliative relationships may be important causal factors behind observed postconflict behavior.

ACKNOWLEDGMENTS

I would like to thank Nicholas Malone and Dr. Megan Matheson for the discussions and collaborations that impacted the ideas and context of this chapter. I also thank Dr. Roger Fouts and Deborah Fouts for permission to conduct research at the Chimpanzee and Human Communication Institute at Central Washington University, and Dr. Phyllis Dolhinow for research access and support with the Hanuman langur colony at the University of California, Berkeley, Behavioral Research Station. I am especially thankful to Dr. Robert Sussman for the invitation to participate in this volume.

REFERENCES

Arnold, K. and A. Whiten. 2001. "Post-Conflict Behaviour of Wild Chimpanzees (*Pan troglodytes schweinfurthii*) in the Budongo Forest, Uganda." *Behaviour* 138:649–90.

Aureli, F. and F. B. M. de Waal (Eds.). 2000. *Natural Conflict Resolution*. Berkeley: University of California Press.

Baker, K. C. and B. B. Smuts. 1994. "Social Relationships of Female Chimpanzees: Diversity between Captive Social Groups." Pp. 227–42 in *Chimpanzee Cultures*, edited by R. W. Wrangham, W. W. McGrew, F. B. M. de Waal, and P. G. Heltne. Cambridge, MA: Harvard University Press.

Castles, D., F. Aureli, and F. B. M. de Waal. 1996. "Variation in Conciliatory Tendency and Relationship Quality across Groups of Pigtail Macaques." *Animal Behavior* 52:389–403.

Cords, M. and F. Aureli. 2000. "Reconciliation and Relationship Qualities." Pp. 177–98 in *Natural Conflict Resolution*, edited by F. Aureli and F. B. M. de Waal. Berkeley: University of California Press.

Cords, M. and M. Killen. 1998. "Conflict Resolution in Human and Nonhuman Primates." Pp. 193–219 in *Piaget, Evolution and Development*, edited by J. Langer and M. Killen. Hillsdale, NJ: Lawrence Erlbaum and Associates.

Dawkins, R. 1976. *The Selfish Gene*. New York: Oxford University Press.

de Waal, F. B. M. 1982. *Chimpanzee Politics*. New York: Harper and Row.

de Waal, F. B. M. 1987. "Tension Regulation and Nonreproductive Functions of Sex Among Captive Bonobos (*Pan paniscus*)." *National Geographic Research* 3:318–35.

de Waal, F. B. M. 1989. *Peacemaking Among Primates*. Cambridge, MA: Harvard University Press.

de Waal, F. B. M. 1993. "Reconciliation Among Primates: A Review of Empirical Evidence and Unresolved Issues." Pp. 111–44 in *Primate Social Conflict*, edited by W. Mason and S. Mendoza. Albany: State University of New York Press.

de Waal, F. B. M. 2000a. "The First Kiss: Foundations of Conflict Resolution Research in Animals." Pp. 15–33 in *Natural Conflict Resolution*, edited by F. Aureli and F. B. M. de Waal. Berkeley: University of California Press.

de Waal, F. B. M. 2000b. "Primates—A Natural Heritage of Conflict Resolution." *Science* 289:586–90.

de Waal, F. B. M. and F. Aureli. 1996. "Consolation, Reconciliation, and a Possible Cognitive Difference Between Macaques and Chimpanzees." Pp. 80–110 in *Reaching Into Thought: The Mind of the Great Apes*, edited by A. E. Russon, K. A. Bard, and S. T. Parker. Cambridge: Cambridge University Press.

de Waal, F. B. M. and F. Aureli. 2000. "Shared Principles and Unanswered Questions." Pp. 375–82 in *Natural Conflict Resolution*, edited by F. Aureli and F. B. M. de Waal. Berkeley: University of California Press.

de Waal, F. B. M. and A. Van Roosmalen. 1979. "Reconciliation and Consolation Among Chimpanzees." *Behav. Ecol. Sociobiol.* 5:55–66.

de Waal, F. B. M. and D. Yoshihara. 1983. "Reconciliation and Redirected Affection in Rhesus Monkeys." *Behaviour* 85:224–41.

Dugatkin, L. 1999. *Cheating Monkeys and Citizen Bees: The Nature of Cooperation in Animals and Humans*. New York: Free Press.

Fuentes, A. 1999. "Variable Social Organization in Primates: What Can Looking at Primate Groups Tell Us about the Evolution of Plasticity in Primate Societies?" Pp. 183–89 in *The Nonhuman Primates*, edited by P. Dolhinow and A. Fuentes. Mountain View, CA: Mayfield.

Fuentes, A., N. Malone, C. Sanz, M. Matheson, and L. Vaughn. 2002. "Conflict and Post-Conflict Behavior in a Small Group of Chimpanzees." *Primates* 43(3): 233–35.

Fuentes, A., E. Ray, and P. Dolhinow. 1996. "Post-Agonistic Interactions in the Hanuman Langur: 'Reconciliation' or Not?" *American Journal of Physical Anthropology* Supplement 22:107.

Griede, T. 1981. "Involed op verzoening bij chimpansees." Research Report of University of Utrecht, Utrecht, The Netherlands.

Kappeler, P. M. and C. P. van Schaik. 1992. "Methodological and Evolutionary Aspects of Reconciliation Among Primates." *Ethology* 92:51–69.

Kropotkin, P. [1902] 1955. *Mutual Aid: A Factor of Evolution*. Boston: Extending Horizon Books.

Lorenz, K. 1966. *On Aggression*. New York: Harcourt Brace.

Malone, N., L. Vaughan, and A. Fuentes. 2000. "The Role of Human Caregivers in the Post-Conflict Interactions of Captive Chimpanzees (*Pan troglodytes*)." *Laboratory Primate Newsletter* 39(1):1–3.

Mason, W. and S. Mendoza. 1993. *Primate Social Conflict*. Albany: State University of New York Press.

McKenna, J. J. 1978. "Biosocial Functions of Grooming Behavior among Common Indian Langur Monkeys (*Presbytis entellus*)." *American Journal of Physical Anthropology* 48:503–10.

McKenna, J. J. 1983. "Primate Aggression and Evolution: An Overview of Sociobiological and Anthropological Perspectives." *Bulletin of the American Academy of Psychiatry and the Law* 2(2):105–30.

Mulcahy, J. B. 2001. *Preconflict Behavior in a Small Group of Chimpanzees*. Unpublished master's thesis, Central Washington University.

Potts, R. 1998. "Variability Selection in Hominid Evolution." *Evolutionary Anthropology* 7(3):81–96.

Preuschoft, S., X. Wang, F. Aureli, and F. B. M. de Waal. 2002. "Reconciliation in Captive Chimpanzees: A Reevaluation with Controlled Methods." *International Journal of Primatology* 23(1):29–50.

Ray, E., A. Fuentes, and P. Dolhinow. 1996. "Patterns of Agonism in a Captive Colony of *Presbytis entellus* Langurs." *American Journal of Physical Anthropology* Supplement 22:195.

Schino, G., L. Rosati, and F. Aureli. 1998. "Intragroup Variation in Conciliatory Tendencies in Captive Japanese Macaques." *Behaviour* 135:897–912.

Silk, J. 1996. "Why Do Primates Reconcile?" *Evolutionary Anthropology* 5:39–42.

Silk, J. 1998. "Making Amends: Adaptive Perspectives on Conflict Remediation in Monkeys, Apes and Humans." *Human Nature* 9(4):341–68.

Silk, J. 2000. "The Function of Peaceful Post-Conflict Interactions: An Alternate View." Pp. 179–81 in *Natural Conflict Resolution*, edited by F. Aureli and F. B. M. de Waal. Berkeley: University of California Press.

Silk, J. 2002. "The Form and Function of Reconciliation in Primates." *Annual Reviews in Anthropology* 31:21–44.

Sober, E. and D. S. Wilson. 1998. *Unto Others: The Evolution and Psychology of Unselfish Behavior*. Cambridge, MA: Harvard University Press.

Strier, K. B. 1994. "Myth of the Typical Primate." *Yearbook of Physical Anthropology* 37:233–71.

Strum, S. C. and L. M. Fedigan. 1999. "Theory, Method, Gender, and Culture: What Changed our Views of Primate Society?" Pp. 67–105 in *The New Physical Anthropology: Science, Humanism, and Critical Reflection*, edited by Shirley C. Strum and Donald G. Lindburg. New Brunswick, NJ: Prentice Hall.

Thierry, B. 2000. "Covariation of Conflict Management Patterns across Macaque Species." Pp. 106–28 in *Natural Conflict Resolution*, edited by F. Aureli and F. B. M. de Waal. Berkeley: University of California Press.

van Schaik, C. P. and F. Aureli. 2000. "The Natural History of Valuable Relationships in Primates." Pp. 307–33 in *Natural Conflict Resolution*, edited by F. Aureli and F. B. M. de Waal. Berkeley: University of California Press.

Veenema, H. C. 2000. "Methodological Progress in Post-Conflict Research." Pp. 21–23 in *Natural Conflict Resolution*, edited by F. Aureli and F. B. M. de Waal. Berkeley: University of California Press.

Veenema, H. C., M. Das, and F. Aureli. 1994. "Methodological Improvements for the Study of Reconciliation." *Behav. Proc.* 31:29–38.

Vitzthum, V. J. 2001. "Why Not So Great Is Still Good Enough." Pp. 179–202 in
    Reproductive Ecology and Human Evolution, edited by P. T. Ellison. Hawthorne,
    NY: Aldine de Gruyter.
Wilson, E. O. 1975. *Sociobiology: The New Synthesis.* Cambridge, MA: Harvard Uni-
    versity Press.

V

Evolution of Sociality

11

Emergent Behaviors and Human Sociality

Ian Tattersall

Human beings are storytelling creatures, with a deeply ingrained penchant for reductionist explanations. This is seen particularly clearly in the recent popularity (particularly, it seems, with the press) of "evolutionary psychological" scenarios of human behavior. Yet I will argue that taking the reductionist route of the evolutionary psychologists in explaining the unfathomable *Homo sapiens* to itself involves ignoring the self-evident complexities of the evolutionary process by which we—like all other living organisms—came to be what we are. And if this is indeed the case, clearly the place to start in evaluating such explanations is with the evolutionary process itself. Mistake how evolution proceeds and you are forever condemned to misinterpret its results. It is particularly important to realize this crucial linkage in the context of evolutionary psychology, since notions of the role of "the genes" in behavioral evolution are heavily dependent on the associated concepts of process.

There are two basic themes that must be addressed by any account of evolution that pretends to completeness. One of these is indeed change, at both population and individual levels, in genes and their products. And the other factor, more commonly ignored, is the origination of new species (Eldredge 1979) and taxic diversity. These two processes are not only conceptually distinct, but they routinely require the operation of entirely different mechanisms. Yet, particularly in anthropology and primatology, this crucial matter of species origins is often considered to be no more than a passive sequel to gradual morphological change: accumulate enough tiny genetic differences over enough generations, under the beneficent aegis of natural selection, and a new (and improved) species will inevitably result. The signal of the fossil record, however, indicates that this is not the case. Not only do environments as a whole fluctuate more rapidly than natural selection as traditionally conceived of could be expected to track effectively,

237

but the fossil record shows clearly that, historically, species have tended to come and go rather abruptly, rather than displaying gradual changes from one to another over the eons. What is more, external events that are totally random with respect to adaptation also enter the picture, rendering natural selection as traditionally conceived only one more influence among many. In this more complex view, then, species as wholes, as well as their component units at various levels from the individual to the local population, are critical players in the evolutionary drama [see Tattersall (2002) for a more extended discussion].

In concocting evolutionary scenarios it is thus critically important to take into account mechanisms operating above the level of the individual organism; for macroevolution is emphatically not simply microevolution writ large. In the context of behavior it is also particularly important to note that, individually, organisms spend by far the greater proportion of their lives performing economic activities, those related to survival, rather than indulging in those related to reproduction. Further, in a perplexingly complex and mysterious organism such as *Homo sapiens*, it is inevitable that even reproductive activities should become entangled in politicoeconomic contexts.

Nonetheless, since the overwhelming triumph during the years following the end of World War II of the movement known as the *evolutionary synthesis* (see Mayr 1982), which portrayed the process of evolution as little if anything more than gradual shift in population gene frequencies under the guiding hand of natural selection, the tendency in anthropology in general, as well as among evolutionary psychologists, has been to portray human evolution as a linear struggle from primitiveness to perfection (Tattersall 1995). The flourishing of this tendency, even in the face of rapidly accumulating evidence to the contrary, reflects a durable and deeply ingrained anthropological mindset, which those who favor unwieldy names might dub the *transformationist/adaptationist paradigm*. Of course, if it were actually true that evolution simply sums out as a matter of adaptive gene frequency shifts in populations over long periods of time, then it would follow that we could understand all evolutionary phenomena strictly in terms of adaptation. But is this view in fact permissible, or is it ruled out when the complexities of the situation are accounted for?

THE SHIBBOLETH OF ADAPTATION

There is a very strong temptation, particularly among those interested in how organisms function, to regard behavioral and anatomical characters as somehow "optimized" by natural selection. In this view, the particular character state(s) under consideration are inevitably superior in some way to those that preceded them. But superior to what? In the evolutionary arena what is important is simply succeeding, rather than being (relatively) perfect; and what it takes to succeed (or even to get by) will vary wildly depending on external environmental circumstances, which have a tendency to change abruptly on rather short timescales. What's more, it is hard to imagine an organism that is optimized in all characters (relative to what?) or, given what we know of inheritance, a process that could possibly give rise to such a result. Of course, all organisms that have not gone extinct are by definition in some way "adapted" to their environments. However, when we speak of *adaptations* we almost invariably refer not to the adaptedness of organisms (or species) as functioning units, but to specific attributes of those organisms: in other words, to parts of organisms, and not the integrated wholes. Equally, the concept of adaptation itself is taken to refer to change in such specific structures (or behaviors) over time. This, however, is dangerous; and my feeling is that, while *adaptation* is a useful word when employed as an abstract noun, it should be eschewed in its concrete form, since this usage has permitted the emergence of a tendency to identify unitary attributes and then to study them as if they could be independently tracked through the ages, in isolation from the organisms of which they form part. This is particularly true, it turns out, of behavioral attributes, if only because we already have a useful vocabulary for singling them out. Hence we tend to speak easily of the "evolution of bipedality," of the "evolution of cooperation," of the "evolution of the behavior," and even of the "evolution of infidelity," while rarely bothering to consider the wider implications of such broad-brush characterizations.

Evolution seen this way becomes essentially a process of fine-tuning, over the ages, of huge numbers of separate characters. Yet the underlying reality is that the "adaptations" thus defined cannot even in concept be independent entities, each with its own evolutionary history that can be pursued at will (Tattersall 1998b). For every adaptation is embedded in an individual organism (or, at another level, is the property of a particular taxon), each of which incorporates hundreds or, more likely, thousands of other attributes that could also be regarded as adaptations. And natural selection in the strict Darwinian sense can only vote up or down on individuals (or taxa) as the sum of their parts; it cannot single out particular features to promote or disfavor. Individuals certainly vary in their reproductive success, to which one specific "adaptation" may even make a crucial contribution, positive or negative; but it is the whole organism that succeeds or fails in the reproductive stakes. For even if we can identify individual "adaptations" that plausibly contribute to reproductive success, the rest of the genome is inevitably carried along simultaneously in the same winnowing process, and all of the components of the same individual will share the same evolutionary fate. In short, if we ignore the phylogenetic histories of the "packages" into which all adaptations are incorporated, we expose ourselves to a severe danger of distorting the evolutionary scenarios we produce.

THE ROLE OF TAXA

To avoid this trap, we have to take into account the fact that evolutionary histories are in large part the story of the differential successes of taxa that succeeded or failed in the ecological arena as the overall sum of their parts. Of course, nobody would nowadays disagree that natural selection, acting at the level of the local population through the reproductive winnowing of individuals, plays a crucial role in establishing gene frequency changes in those populations over time. Indeed, random sampling aside, there is no other obvious mechanism for producing the genetic and phenotypic diversification of populations *within* species that provides the basic material for evolutionary change.

At the same time, however, we have to acknowledge that evolution is a complex affair in which various levels of action combine to produce the results observed. What is more, we are not speaking here of a hierarchy just of heredity. Certainly, genes, chromosomes, individuals, demes, species, and higher taxa all play their differing parts in a hierarchy of roles. But, in producing phylogenetic histories, all of these factors act within a conceptually and functionally separate economic hierarchy (individuals, local populations, species, ecological communities, environments). To reduce this complex interaction to little more than gene frequency shifts or to other forms of genetic innovation is to lose sight of the greater part of the story. Novelties arise within individuals and local populations, it is true; but in the longer term new structures and behaviors are triaged by selection not simply inside those populations, but are sorted on a broader level by competition (with related or unrelated species) and by larger environmental changes (including shifts that are random with respect to a particular population's adaptations). And in an unpredictably varying environment, today's adaptation may well be tomorrow's liability.

EVOLUTIONARY PSYCHOLOGY AND SELFISH GENES

Evolutionary psychologists like to claim that the ancestry of their field dates back to Darwin's 1859 assertion that "In the distant future . . . psychology will be based on a new foundation, that of the necessary acquirement of each mental power and capacity by gradation" ([1859] 1985:458). In reality, however, the search for intellectual antecedents for this "discipline" cannot functionally extend back beyond Hamilton's kin selection notions of the mid-1960s. These would probably have remained largely unremarked and ungeneralized had they not been publicized a decade later by Ed Wilson in his remarkable work *Sociobiology: The New Synthesis* (1975). But what really gave evolutionary psychology as practiced today its initial impetus was the association of Wilson's ideas on the biological bases of behavior with the mechanisms of the "selfish gene" espoused by Richard Dawkins (1976) in his book of that name.

In this work Dawkins claimed that the essential targets of evolution by natural selection are not individual organisms, as Darwin had it, and certainly not species as wholes; rather, the basic evolutionary unit is constituted by the genes themselves: the "immortal replicators." In Dawkins's view, individual organisms are simply transitory vehicles for these replicators, which vie among themselves for evolutionary success. Life histories are reduced to little if anything more than the struggle of hapless individuals to maximize the number of genes each one contributes to the next generation. Here, then, is the basis for the notion that individual behaviors are determined by identifiable "genes" that are sufficiently discrete to be in competition with those for alternative behavioral patterns. In the evolutionary psychological view, the search has become one for "mental adaptations" that have resulted from the action of natural selection during our species' past.

Heavily influenced by Darwin's view that the evolutionary process consists of little other than slow, gradual changes in gene frequencies over vast periods of time, many evolutionary psychologists have taken to emphasizing the significance of the "ancestral environment" in which current genotypes allegedly evolved (e.g., Wright 1994, and references therein; Tooby and Cosmides 1990; Irons 1998). Evolution, they believe, takes place too slowly to keep up with rapidly changing environments such as those in which *Homo sapiens* lives today. To find the "adaptive" causes of specific behaviors exhibited by modern humans, we need not look at the world around us, but should instead direct our attention toward the ancestral environment (effectively, that of hunter/gatherers, hardly a unitary category) in which our genes became established.

That this "environment" was as monolithic (in structure or in time) as the choice of term suggests is, of course, highly debatable; and in any event, our view is skewed by the fact, among other things, that the few hunting/gathering societies that survived to form the subject of anthropological study had already been peripheralized to a handful of marginal habitats. Significantly, Dawkins himself took a distinctly different tack when it came to the analysis of human behaviors. Noting that many behaviors are learned within cultures, Dawkins derived the parallel notion (to genes) of culturally transmitted (and fast-changing) behavioral units that he called "memes." And interestingly, this concept has not found universal favor in mainstream evolutionary psychology. The central problem with all of this is, of course, that exactly as in the case of adaptations, genes do not have independent existences. Just as adaptations are inextricably embedded in organisms that must succeed or fail as the sum of their parts, genes are embedded in genotypes of staggering complexity that must inevitably have greater or lesser reproductive success as functioning wholes. Significantly, these genotypes are passed along through the generations, if not intact, at least with strong associations among their parts. Further, most units that we might identify as "genes" cooperate with others (often many others) in determining particular phenotypes, just as many of the same units influence more than one developmental process in the individual to whom they belong. Geneticists have known this for years, in characterizing most genes as pleiotropic and most characters as polygenic. It is, indeed, significant that remarkably few geneticists have joined the sociobiological camp.

Thus, while it is fruitless to deny that the behaviors and capacities of organisms are in some way influenced by genetic heritage, it is usually equally unproductive to identify or even to infer genes "for" particular behaviors. Even in those cases where strenuous efforts have purported to locate genes for specific human behavioral proclivities (for example, homosexuality in males, e.g., Byne and Parsons 1993), such findings have been equally vigorously contested (see Marks 2002); and it is in any event usually impossible to know whether the genes involved also play other roles in the development, life, and behaviors of the individual. None of this is, of course, to deny the indirect influence of the genes, and of the phylogenies they embody, upon individual behaviors. The genotype is, after all, the blueprint upon which each new individual is built; and at some level it certainly determines the range of behaviors possible. The question is simply one of specificity of action; and where evolutionary psychologists are clearly correct is in their perception that species are the products of long evolutionary histories.

But to equate genes with particular behaviors in the case of an organism as complex cognitively as *Homo sapiens* is to ask a little much, no matter how closely controlled "fight-or-flight" responses may be among certain kinds of fish. What genes actually do is to guide developmental processes; and in the behavioral arena they specifically channel the development of the controlling organ, the brain (Tattersall 1998a). The incomparably complex human brain is the product of a long phylogenetic history, and has evolved through a drawn-out process of intermittent accretions of new structures and differential expansions of preexisting components. The resulting "layered" organization ensures that our behaviors may be affected by some very ancient structures indeed, as well as by more recent acquisitions; and if we are ever to understand the mysteries of human consciousness, we are going to have to do it in the context of understanding how our cognitive function is influenced by the phylogenetic history of the human brain. Whatever the outcome, the notion of "modular" intelligence or behavioral control (implying a kind of "lateral" division of brain functions, rather than the complex results of "layering") is very much at variance with the actual results of humankind's long phylogeny.

THE PATTERN OF HUMAN BEHAVIORAL EVOLUTION

A critical factor to consider in the context of human behavioral evolution is that our fossil and particularly archaeological records eloquently indicate that, unlike its predecessors, *Homo sapiens* is not simply an improvement upon what went before. As a cognitive and behavioral entity, our species is truly unprecedented in the entire history of life (Tattersall 1998a). And, if so, it is to this recently acquired quality of uniqueness, not to mythical "ancestral environments," that we must look in the effort to understand our often unfathomable behaviors.

To elaborate on this insight, it is worth pointing out that the achievement of fully human status was quite clearly not only an unprecedented event, but also a relatively recent one. The human fossil record is now documented back to seven to six million years (7–6 myr) ago (Brunet et al. 2002). In the ensuing period there has been a substantial amount of taxic diversification (see Tattersall and Schwartz 2000), but significant anatomical modifications have been rare, and by the time the first stone toolmaking started about 2.5 myr ago (see Klein 1999) body proportions (short stature, long arms and extremities, short legs, narrow shoulders, and so forth) seem still to have been much as they were in the earliest postcranially known hominids millions of years before. Modern body proportions appeared abruptly in the record after about 2 myr ago; and while there was certainly an average brain volume increase in the subsequent period (modern size having been reached by the Neanderthals at about 200 thousand years ago), we have at present no way of knowing how this increase was configured either taxically or over time (Tattersall 1998a), and certainly no reason to believe that there was a linear intralineage brain expansion. Throughout, the pattern has been one of taxic diversification, in the absence, for most of the time, of functionally significant anatomical innovation.

Similarly with the technological record, which shows the original Oldowan stoneworking tradition (which focused on the production of simple small, sharp, but unstandardized flakes) persisting with only minor refinement for a million years before being joined by the Acheulean, which introduced the first bifacially flaked stone tools produced to con-form to a "mental template." It was again over a million years before a sub-stantially new tool type was produced (the "prepared-core" tool, made by elaborately shaping a stone core until a single blow could detach a more or less finished implement); and it was several hundred thousand years, again, before this tradition was finally replaced by the blade-based toolkit possessed by, among others, the earliest modern people to occupy Europe.

Thus in technology and inferred behavior, as well as in hominid physi-cal structure, business was functionally much as usual for vast spans of time during hominid evolutionary history, most innovations being little more than minor improvements on long-running major themes.

Humans of modern bony structure appeared at some time between about 150 and 100 thousand years (150–100 kyr) ago (see discussion in Tat-tersall and Schwartz 2000). But early evidence for the appearance of mod-ern human symbolic behavior patterns is much more recent yet—and there is plenty of evidence for the former existence of anatomically modern pop-ulations that for many tens of millennia behaved, as far as can be told, pretty much like their predecessors. The most dramatic evidence for early modern human behaviors, of course, comes from the Aurignacian invasion of Europe by *Homo sapiens* in the period between about 40 and 30 kyr ago [see, among other sources, Tattersall (1998a) and Tattersall and Schwartz (2000)]. The abrupt appearance in this interval of cave and portable art and symbolic notation, followed rapidly by impressive evidence of elaborate burial customs with grave goods, of sumptuous personal ornamentation and adornment, and of technological innovations in such areas as the pro-duction of kiln-baked ceramic objects and couture using delicate eyed bone needles, bears witness to the fully fledged existence of the modern human capacity among the first *Homo sapiens* to occupy Europe.

Yet it is clear that the majority of these new behaviors were already pres-ent among these early moderns when they entered the subcontinent, and that the capacity underlying them had been acquired elsewhere. As recently emphasized by McBrearty and Brooks (2000), it is most likely that this new potential emerged somewhere in the huge continental mass of Africa. It is certainly in Africa that both the first direct hints of symbolic behaviors such as pattern-incision on ostrich eggshells and other materi-als, and more indirect indicators such as the long-distance transport of materials and flint-mining, are found (Tattersall 1998a; McBrearty and Brooks 2000). The most dramatic piece of evidence for early symbolic behavior in Africa comes from close to the continent's southern tip, at Blombos Cave, where ochre plaques deliberately engraved with geomet-ric designs occur in Middle Stone Age deposits over 70,000 years old (Hen-shilwood et al. 2002). Unfortunately, Africa is huge, and the evidence is so far sparse; but whether one accepts interpretations of the symbolic use of living space as long ago as about 120 kyr at sites such as Klasies River Mouth (e.g., Deacon and Deacon 1999), or insists upon having more direct symbolic indicators such as the Blombos plaques at a little more than half that age, it is Africa that consistently proffers evidence of the earliest emergence of the modern human sensibility.

It is important to note that, if the earliest stirrings of modern human symbolic consciousness resulted from the behavioral "discovery" of an exapted underlying biological potential (and there really is no plausible alternative to this), one would not expect this new capacity to be revealed immediately in all of its dimensions. Indeed, the history of symbolically mediated mankind has in an important sense been the history of discovery of what this new capacity can do—a process that is still continuing today. McBrearty and Brooks (2000) argue that the sporadic nature of early signs of behavioral "modernity" in Africa implies a long period of transition from more archaic to more modern human behaviors; but this does not mean that the emergence of the biological underpinning was not a relatively short-term event that took place in a particular location, as would be expected for any biological innovation. What's more, there has never been any reason to expect that the earliest human forebears with symbolic reasoning skills should have employed them to do everything that we use them for today, and we should not be surprised that the earliest evidence we have for symbolic behaviors appears to be restricted to one or only a few of the domains of human experience. What we can now be sure of, however, is that by 70 kyr ago, and in Africa, the cognitive equipment was in place to allow the beginning of the process of discovering what can be done with it.

But while we glimpse only dimly the very first expressions of the unique modern human consciousness, the general pattern over the long term is clear: human behavioral evolution was not a gradual, steady, and incremental process that chugged along steadily over millions of years. Rather—as indeed with our physical evolution—it was an episodic affair, with long periods of effective nonchange punctuated by rare bursts of innovation (see also discussion by Tattersall 1998a, 2002). And if the appearance of the modern human capacity was in evolutionary terms an abrupt event (even if its sequelae—and the distinction between capacity and sequelae is an important one—are still accumulating), it is vanishingly unlikely that any of our unique behaviors today have been programmed by natural selection and our genes to give us little volition about the ways in which we comport ourselves.

There can be little doubt that, in its power of symbolic reasoning and associated behaviors, behaviorally modern *Homo sapiens* is an entity of a kind totally unprecedented in nature. And the acquisition of this uniqueness was quite evidently an emergent event, resulting from a chance coincidence of innovations. Natural selection is not a creative force, conjuring innovations into existence because of their potential adaptive advantage. It cannot work that way; it can only promote the success or failure of novelties that have arisen spontaneously and are already there. Our vaunted capacities are the result of an exaptation, a novelty all of whose uses we did not discover until long after our ancestors acquired it.

In sum, we do not see a pattern of individual fine-tuning over the eons either of our behaviors or of our physical selves. In the behavioral realm, what we see instead is evidence for the short-term emergence of an unprecedented, generalized, and flexible capacity, at some time probably about 120–70 kyr ago, that underpins the huge variety evident in human behaviors today. This new capacity was added, of course, to a preexisting substrate that had been shaped by an enormously long vertebrate evolutionary story, one marked by the sporadic accretion, over a vast span of time, of new brain structures and behavioral innovations. Nothing happens in evolution that is independent of prior history; and human cognition, remarkable and unprecedented as it may be, is simply one more (albeit highly unusual) riff on the basic higher primate (diurnal, group-living, intensely social) quality that was already possessed by the common hominoid ancestor. Still, *Homo sapiens* displays an unprecedented voluntarity of behavior compared even to its closest relatives. Individual human beings, similar as their physiological requirements may be, are not condemned to behave in identical ways; and although their comportments are—on average—limited both by anatomy and by social conventions, it is fair to point out that both components of virtually any thinkable pair of antitheses can be employed to characterize the specific displayed behaviors of some human individuals. Nature may have written the menu—in our case, a menu of astonishing range of mix-and-match flexibility—but it is random genetic chance, in combination with our experience, that chooses the dishes. Further, it seems likely that some of the potential aspects of human behavior that are made possible by this extraordinary general capacity of ours have yet to be discovered. Clearly, our future as individuals or as a species is not preordained by the limiting canons of evolutionary psychology. Instead, it is up to us.

SUMMARY

If we are to seek biological bases for behavioral proclivities, it is essential that we have an accurate model of the evolutionary process. Nowhere is this more important than when we are trying to trace potential links between human behavioral patterns and our genetic heritage. Yet the notion of evolution on which gene-based "evolutionary psychological" scenarios of human behavioral evolution are based is skewed, or is at the very least dangerously incomplete. The tendency is to view evolution as a process of fine-tuning over time of individual anatomical and behavioral characters, in isolation from the taxa in which they are embedded. Yet adaptations do not have independent existences. All such attributes are intimately tied up with whole organisms, and with taxa. And natural selection can only vote up or down on the reproductive success of whole organisms, not of their individual constituent parts. Similarly, it is taxa, not attributes, that succeed or fail in the evolutionary stakes. Further, examination of the record of human behavioral evolution indicates that the acquisition of our unique human cognition was a recent, and fairly abrupt, event. Our consciousness is an emergent quality (albeit one that was an addition to a cognitive apparatus already shaped by a long evolutionary history), not the product of thousands of generations of gradual fine-tuning. It is to this recently acquired capacity, the ultimate result of a long and untidy accretionary history, and not to ill-defined "ancestral environments," that we should turn in the effort to understand the origin of our remarkable human behaviors.

ACKNOWLEDGMENTS

I thank Audrey Chapman, Bob Sussman, and Jim Miller for their invitation to participate in the AAAS seminar, *Primatology and Human Sociality,* on which this volume is based, and all its participants for lively discussion. Special thanks to Wentzel van Huyssteen, my discussant, to Bryan Ferguson, and to Bob Sussman for his comments on the manuscript.

REFERENCES

Brunet, M. et al. 2002. "A New Hominid from the Upper Miocene of Chad, Central Africa." *Nature* 418:145–51.

Byne, W. and B. Parson. 1993. "Human Sexual Orientation: The Biologic Theories Reappraised." *Archives of General Psychology* 50:228–38.

Darwin, Charles. [1859] 1985. *The Origin of Species by Means of Natural Selection.* London: Penguin.

Dawkins, Richard. 1976. *The Selfish Gene.* New York: Oxford University Press.

Deacon, H. and J. Deacon. 1999. *Human Beginnings in South Africa: Uncovering the Secrets of the Stone Age.* Cape Town: David Philip.

Eldredge, N. 1979. "Alternative Approaches to Evolutionary Theory." *Bulletin of the Carnegie Museum of Natural History* 13:7–19.

Henshilwood, C. S., F. d'Errico, R. Yates, Z. Jacobs, C. Tribolo, G. A. T. Duller, N. Mercier, J. C. Sealy, H. Valladas, I. Watts, and A. G. Wintle. 2002. "Emergence of Modern Human Behavior: Middle Stone Age Engravings from South Africa." *Science* 295:1278–80.

Irons, W. 1998. "Adaptively Relevant Environments versus the Environment of Evolutionary Adaptiveness." *Evolutionary Anthropology* 6:194–204.

Klein, R. 1999. *The Human Career: Human Biological and Cultural Origins*, 2d. ed. Chicago: University of Chicago Press.

Marks, J. 2002. *What It Means to Be 98% Chimpanzee: Apes, People, and Their Genes.* Berkeley: University of California Press.

Mayr, E. 1982. *The Growth of Biological Thought: Diversity, Evolution, Inheritance.* Cambridge, MA: Harvard University Press.

McBrearty, S. and A. Brooks. 2000 "The Revolution That Wasn't: A New Interpretation of the Origin of Modern Human Behavior." *Jour. Hum. Evol.* 39:453–563.

Tattersall, I. 1995. *The Fossil Trail: How We Know What We Think We Know About Human Evolution.* New York: Oxford University Press.

Tattersall, I. 1998a. *Becoming Human: Evolution and Human Uniqueness.* New York: Harcourt Brace.

Tattersall, I. 1998b. "The Abuse of Adaptation." *Evolutionary Anthropology* 7:115–16.

Tattersall, I. 2002. *The Monkey in the Mirror: Essays on the Science of What Makes Us Human.* New York: Harcourt.

Tattersall, I. and J. H. Schwartz. 2000. *Extinct Humans.* Boulder, CO: Westview.

Tooby, J. and L. Cosmides. 1990. "The Past Explains the Present: Emotional Adaptations and the Structure of Ancestral Environments." *Ethology and Sociobiology* 11:375–424.

Wilson, E. 1975. *Sociobiology: The New Synthesis.* Cambridge, MA: Harvard University Press.

Wright, R. 1994. *The Moral Animal: Evolutionary Psychology and Everyday Life.* New York: Vintage.

12

Sociality and the Concept of Culture in Human Origins

Richard Potts

The concept of culture used to be applied exclusively to human beings. Cracks in this crucible of human uniqueness became widely noticeable when chimpanzees were first reported to make tools (Goodall 1968), thus destroying the elegant simplicity of "man the toolmaker" (Oakley 1965), and when the learning capabilities of many species began to be ascribed to culture (Bonner 1980). Field research on great apes has now demonstrated geographic variation in the behavioral repertoires of chimpanzees (Whiten et al. 1999) and orangutans (van Schaik et al. 2003). This variation is described as "cultural" because these apes are able to invent new customs and pass them on independently in different social groups, while neither ecological nor genetic factors can account for the behavioral differences manifested between the groups. Accordingly, "culture" has a wider zoological background regardless of whether cultural transmission depends on imitation, teaching, or language (de Waal 1999). It is commonly assumed, consequently, that a common ancestry of culture exists among at least great apes and humans, including the oldest human ancestors. This assumption begs the question, What can we reconstruct about the evolution of cultural capacities in humans?

With this question in mind, I have written this chapter with four principal goals. The first is to define the common ground in the cultural behavior of modern humans and other great apes, notably chimpanzees and orangutans. The second is to summarize briefly the record of behavior in early hominins. (Hominini has begun to replace Hominidae as the name of the evolutionary group of bipedal apes to which humans belong.) As we will see, early hominin cultural behavior, expressed in the archaeological record, differed from "culture" in the modern human sense evident beginning in the late Pleistocene, and may also have differed from that apparent in living apes.

The third goal is to summarize the adaptive conditions under which modern human cultural behavior emerged. Paleoenvironmental records for the period of human evolution are numerous; the approach here is to illustrate the environmental dynamics in which distinctive forms of human behavior—home-based activity, long-distance exchange between social groups, and heightened cultural diversification, for example—first occurred. This discussion leads to the fourth goal, which is to outline how the cultural behavior characteristic of modern humans emerged from a *paleocultural* system of earlier humans.

This analysis shows that a distinctive suite of archaeologically detectable behaviors marked the emergence of modern humans, and these behaviors were first manifested during one of the most volatile eras in Earth's environmental history. Modern human behavior emerged during a long time span of ecological unpredictability that impacted the social fabric of hominin populations. In this context we see the earliest examples of complex symbolic expression of events, objects, and diverse phenomena (e.g., meanings, planning) that were not visible, i.e., temporally or spatially separated from the humans voicing or otherwise representing them. This expression of the nonvisible continues to impart a peculiar quality to the social behavior of *Homo sapiens*.

CULTURAL BEHAVIOR: THE COMMON GROUND

Across the biological and anthropological sciences, multiple concepts of culture exist that are inconsistent with one another. Cultural anthropologists, primatologists, and archaeologists tend to construe culture in different ways, a practice that devalues the utility of the term. If the idea of culture will ever again prove useful in understanding human sociality and behavioral evolution, it will be essential for scholars in many fields to grasp these diverse uses of the culture concept.

For ethnographers, cultural studies are generally confined to social phenomena called institutions. In the general sense, institutions are complex systems of economic exchange, marriage rules, ideology, religion, laws, political hierarchy, and so on, that are underpinned and maintained by symbolic behavior. In a narrower sense, institutions are particular expressions (e.g., governments, legal and ethical systems, trade networks) of a set of complexly organized social relationships and values that people create through accepted bodies of thought, creed, and activity. In either the general or narrow sense, institutions represent the highest order of human social complexity, and they channel people's search for meaningful social action and personal belief. They also take on a life of their own, bolstered by rituals, art, sayings, flags, songs, dress, colors, and other symbolic forms, which enable institutions to persist or change independently of the life and death of individuals (Potts 1996).

To primatologists, cultural behavior need not involve such complex phenomena. Relatively simple differences between populations in grooming and foraging activity qualify as culture. Such behavioral differences are sometimes considered to define a crucial baseline for the general concept of culture (McGrew 1992). In a recent comparative study of chimpanzee populations (Whiten et al. 1999), culture is represented by distinctive behaviors manifested in geographically separate populations. A common example is termite feeding. At Gombe, Tanzania, chimpanzees practice termite fishing, in which slender twigs are prepared and inserted into termite nests. By contrast, chimps in Equatorial Guinea practice termite digging, where stout digging sticks are used to break open the nests. The two customs are specific to the groups living in the different locales. Demonstration of ape culture requires that such different customs exist in the face of ecological similarities between the regions or other kinds of evidence indicating that physical and biotic settings play no role in explaining the behavioral variations.

Finally, a different culture concept is employed by Paleolithic archeologists. When they say "culture," they typically mean *stone tools*. Different artifacts and manufacturing techniques are equated with distinct cultures and are considered to signal cultural change. Oldowan culture, Acheulean culture, Mousterian culture, and similar terms are still commonly used by archaeologists. In applying these terms, archaeologists assume that change in toolmaking methods and in the proportion of distinct artifact types over time and space reflects the development of new bodies of cultural information, which evolving hominins became capable of inventing and sustaining.

Despite these differences in the culture concept, the quest for common ground is scientifically sound and potentially defines a way of studying the continuities between human and nonhuman primate behavior. In anthropological and primatological studies, there are at least four elements of agreement concerning what constitutes cultural behavior:

1. Culture is a system of *nongenetic information transfer,* which occurs across generations and among individuals of the same generation.

2. Cultural behavior is manifested as *discrete forms*—specific types of activity, implement, or systems of belief. These discrete forms are readily identifiable behaviors, artifacts, mental output, or social manifestations that are both separable from other such forms and distinctive of particular populations. In chimpanzees, besides specific methods of termite foraging,

certain forms of grooming behavior—e.g., use of the above-the-head, hand-clasp position—are considered to be cultural in part because they represent distinct forms of behavior unique to particular social groups.

3. *Geographic differentiation* of behavior—i.e., differences between separated populations—is another element in the common ground of cultural behavior. This characteristic has been well documented in chimpanzees, orangutans, and humans.

4. The potential for *change across generations* is a final hallmark of primate cultural behavior. In modern humans, cultural variations are cumulative; the inventions of many generations are stockpiled, creating vast repositories of social information. In nonhuman primates, there is less evidence for such accumulation. Nonetheless, new behavioral variations arise frequently, well within the decadal time frame of long-term field studies. Separation or semi-isolation of populations allows such novel variations to become adopted, sustained, and manifested by different social groups.

THE ARCHAEOLOGICAL RECORD

How does this "common ground" fare when we look at the behavioral deposits, or artifacts, left behind by early humans? The oldest material record of behavior is about 2.6 million years old (Ma), and it consists of assemblages (not just isolated or rare, selected pieces) of precisely and repetitively chipped and battered rocks that define Oldowan toolmaking. It used to be thought that Oldowan cores (the sharp-edged rocks that bear multiple flake scars) represented purposeful tool designs, i.e., target forms that were in the minds of *Homo habilis* toolmakers. Detailed studies by Toth (1985) and others (Wynn 1989; Potts 1991) have shown, however, that these idealized forms were mainly "stopping points" in a continuous process of knocking sharp flakes from rocks of varying original shape. There is a continuum between so-called choppers, discoids, scrapers, and other forms—a continuum that can be followed through the products of Oldowan flaking (Figure 12.1). The continuum among these chipped products is evident statistically and is manifested as a gradual increase in the number of flake scars and edge length, and a concomitant reduction in the mass of these pieces (Potts 1991).

Over a span of about 800 thousand years (ca. 2.5 to 1.7 Ma), stretching from the Afar rift of Ethiopia (e.g., Kimbel et al., 1996; Semaw et al., 1997) to northwest Africa (Sahnouni et al. 2002) and southward to the Transvaal (Kuman 1994), the Oldowan largely, if not entirely, consists of continuously varying artifact forms that resulted from the repetitive process of flaking stone, making sharp edges, and using rounded stones or the cores

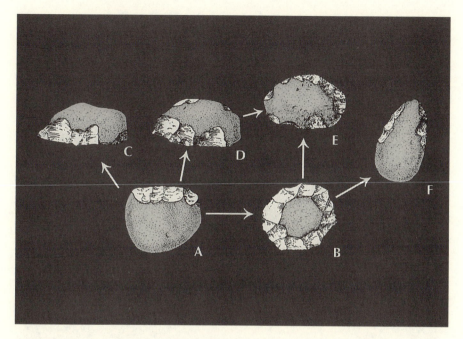

Figure 12.1 Continuity of Oldowan flaked pieces (typically referred to as cores or tools). (A) chopper, (B) discoid, (C) heavy-duty scraper, (D) polyhedron, (E) subspheroid, (F) protohandaxe. Illustration drawn by J. Clark (from Potts 1991).

themselves to hammer and crush other objects. Stone tool assemblages found at different sites do vary considerably. For example, unmodified manuports (hand-transported rocks) and battered subspheroids are common in Olduvai Beds I and II, respectively, whereas both types of objects are rare at East Turkana, Kenya (Leakey 1971; Isaac and Harris 1997). Stone flaking at Lokalalei, West Turkana, was more intensive, producing more complexly flaked stone cores than at other Oldowan sites (Roche et al. 1999), and the diversity of lithic materials used by Oldowan toolmakers at Kanjera South, Kenya, was considerably higher than elsewhere (Plummer, Ferraro, Ditchfield, Bishop, and Potts 2001), especially compared to near-contemporaneous sites in Omo Member F, where the single material quartzite was shattered into sharp flakes (Merrick and Merrick 1976).

In each of these cases, however, careful exploration of the variation among Oldowan sites seems to be explicable less by cultural idiosyncrasy than by variation in the available stone material, e.g., different treatment of different raw materials (Potts 1988; Kimura 1999) or variation in the

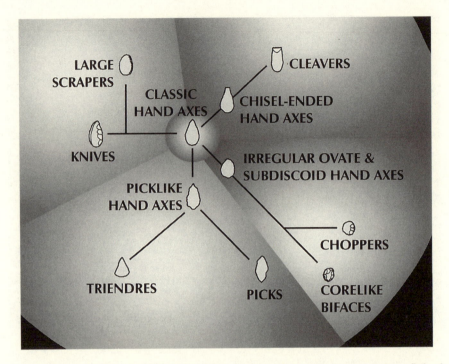

Figure 12.2 Continuity of Acheulean large cutting tools and other flaked
   pieces (from Isaac 1977). Although handaxes represent a morphological
   mode, Isaac proposed that Acheulean artifact types are actually part of a
   continuous gradient of form.

distance between sites and source outcrops (Isaac and Harris 1997).
Although raw materials may explain much of the diversity in Oldowan
tool assemblages, it is possible that raw material selection by the hominin
toolmakers was subject to cultural variation. Nonetheless, the entire
process of acquiring rocks and making tools did not result in the consistent
manufacture of discrete forms typical of the later archaeological record.

   Cultural factors may begin to play a larger role in the time period of the
oldest Acheulean. Between 1.8 and 1.6 Ma, Karari-type scrapers and bifacial flaking of large cores and flakes that anticipate Acheulean handaxes
may reflect cultural variants in the Turkana Basin relative to contemporaneous sites elsewhere (Isaac and Harris 1997). Interestingly, examples of
late Oldowan flaking behavior appear to be more complex [e.g., Peninj ST
site (de la Torre et al. 2003)] than exhibited at some contemporaneous
Acheulean sites [e.g., Olduvai Bed II, EF-HR site (Leakey 1971)].

The earliest Acheulean, by about 1.7 Ma, reflected an apparent break-through in the process of stone flaking. Hominin toolmakers were, for the first time on a regular basis, able to detach very large flakes—greater than 10 cm in maximum dimension. When this was done and the piece was then flaked around the perimeter, the first handaxes, or large cutting tools (LCTs), were made. LCTs characterize human stone technology for nearly 1.4 million years. By no means do they occur everywhere, but they are the most prominent element in the record of stone toolmaking of Africa, Europe, and much of Asia between 1.7 and 0.3 Ma. Over most of that time span, handaxes made later in time were not necessarily more refined than ones made earlier, a point documented by Isaac's work in East Africa (Isaac 1972, 1977).

There is very little evidence of innovation or distinct trends over vast periods of time. Instead, the tool products of this period, especially between 1.7 and 0.5 Ma, exhibit wide, continuous variation. Around the modal form of the handaxe, there is a graded continuum with other tool forms, as Isaac (1977) pointed out (Figure 12.2). His interpretation was that, rather than directional change in cultural behavior, handaxe technology exhibited "random walk" variation over time. In his model there is no evidence of divergence of behavior over time, merely random change in LCT morphology (Figure 12.3).

A recent study by Noll (2000), which (like Isaac's work) focused on the handaxe site of Olorgesailie, Kenya, shows that variation among stone tool assemblages reflected the intensity of stone flaking and the mechanical properties of diverse stone raw materials; both the flaking intensity and the raw material percentages varied idiosyncratically from site to site. Thus random variation in LCT shape, size, and other morphological characteristics over time does not appear to be explainable as a cultural signal, i.e., the observed variation was not determined primarily by variation in cultural norms.

During the period of Acheulean stone flaking, handaxes and other LCTs (called cleavers, picks, and knives) predominate in many stone tool assemblages, while other excavation sites within the same region exhibit very few or no LCTs. Could this variation reflect a cultural difference? That is, did certain social groups prefer to make LCTs, while others preferred not to? If this were the case, it would represent a significant geographic differentiation in toolmaking behavior. This turns out, however, not to be the case. Landscape-scale analysis at Olorgesailie is the first to systematically test this possibility (Potts, Behrensmeyer, and Ditchfield 1999). Following specific soil and other stratigraphic layers, we have documented the distribution of stone tools and ancient habitats across the Olorgesailie region in four separate time periods between 1.0 and 0.78 Ma. Our research shows that handaxe-rich and handaxe-poor assemblages were part of the same

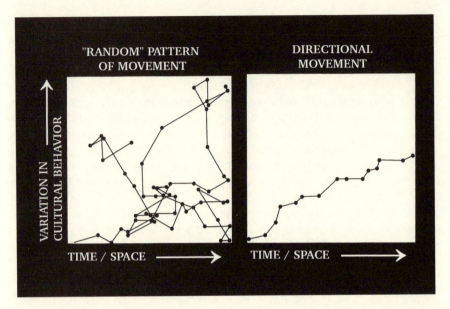

Figure 12.3 Isaac's (1969) portrayal of random walk variation in Acheulean
     handaxe morphology (left) versus directional change, typical of more
     recent human technological change (right).

behavioral system (Potts et al. 1999). In Member 1 of the Olorgesailie Formation (ca. 990,000 years old), for example, the one site known to be very
rich in handaxes is part of a continuum of stone-discarding behavior
across the entire paleolandscape. Handaxes were carried around, and
stone refuse from flaking them was strewn across the region; yet the handaxes themselves were curated and almost always deposited in one small
area close to where the lake basin and volcanic highlands met. Thus the
handaxe-rich and handaxe-poor assemblages were spatial variations of a
single system of stone flaking behavior.

Perhaps the most famed geographic distinction in early human behavior is that known as the Movius Line (Movius 1969). This line marks a separation between handaxe and non-handaxe technologies across the Old
World. It has long been supposed that handaxe-making prevailed only
west of the line (Africa, Europe, and western, central, and southern Asia),
while simpler forms of stone tools persisted in the east (eastern and southeastern Asia), even beyond the end of the Acheulean handaxe tradition in
the west.

Recent finds, however, violate the central idea of the Movius Line, particularly its assumption that cognitive and technological competence

related to LCT toolmaking was largely absent from East Asia. In southern China, sites of early Pleistocene age (dated 803 ± 3 thousand years ago), contemporaneous with Olorgesailie, preserve abundant LCTs of equal sophistication to those in the west (Hou et al. 2000). These LCTs need not imply the presence of Acheulean toolmaking populations in China, but they do point to similar cognitive and technological capabilities on both sides of the line. The distance between Africa and East Asia did not prohibit or hinder the making of tools very similar to one another. Even over this large geographic scale, and long stretches of time, cultural differentiation of populations is not as apparent as once believed.

A PALEOCULTURAL BEHAVIOR PATTERN

When we compare the rich body of archaeological evidence to the common ground of modern ape and human culture, we see several intriguing differences. Over the first 2 million years of the archaeological record, early humans manifested a behavior system that differed from great ape or human culture. I refer to it as a "paleocultural system," which has the characteristics listed in Table 12.1. It was indeed characterized by the transmission of nongenetic information; yet until roughly 500 thousand years ago, the artifacts and inferred behaviors of early humans exhibited continuous variation. Discrete tool types—e.g., so-called awls and "burins" in Bed II Olduvai (Leakey 1971) and Karari scrapers in the Okote tuff complex at Koobi Fora (Isaac and Harris 1997)—did sometimes occur but were rare. Tool assemblages vary from one site to another in the intensity of tool modification or in the proportion of handaxes versus scrapers, for example. Yet it remains difficult to discount ecological factors, including the availability of different stone raw materials, to explain these variations. Consistent geographic differentiation of behavior is either not present, weakly represented, or undocumented for much of the earliest 2 million years of the archaeological record. Paleolithic toolmaking was also marked by long periods of stasis—hundreds of thousands of years over which stone flaking behavior did not change in any appreciable or systematic manner. In other words, separation in space and time did not necessarily yield distinct behavioral packages—or cultures as the term is used by anthropologists and primatologists. This comparison suggests therefore that, on three out of four points, this paleocultural system of behavior contrasts with certain defining features of culture common to living humans, chimpanzees, and orangutans.

Detection of this paleocultural system throughout much of the span of hominin evolution has interesting implications. Figure 12.4 helps to visualize one of them. Similarities between living chimps and humans are

Table 12.1   Early Human Paleocultural Behavior
(ca. 2.6 to 0.5 Million Years Ago)

1. Transmission of nongenetic information
2. Typified by continuously varying forms
3. Little evidence of geographic differentiation
4. Long periods of stasis

generally thought to arise from the last common ancestor of the two. The occurrence of toolmaking and other cultural behaviors in both living species would imply that the last common ancestor also possessed such behaviors. Evidence of geographic traditions in orangutans might further indicate that culture is a shared feature of all great ape species and humans (van Schaik et al. 2003). In fact, the existence of geographically distinctive social conventions in capuchin monkeys (Perry et al. 2003) implies that behaviors sometimes described as "cultural" are by no means unique to apes and humans. The addition of archaic hominins to the comparison, however, helps establish the evolutionary timing of certain aspects of human cultural behavior. If key elements in the common ground of chimp-human cultural behavior were absent in early human toolmakers, it means that while nongenetic transmission of habits exists in all great apes, other commonalities of extant ape and human culture are not necessarily homologous (Figure 12.4B). Although bonobos and gorillas exhibit complex social learning, tool-assisted behavior in the wild and consistent geographic variations in social behavior have yet to be demonstrated in these species. Standardization of discrete tool forms, novelty, and geographic differentiation of behavior all become increasingly manifested in the human archeological record after 500,000 years ago. It is possible, then, that a greater capacity for innovation and geographic differentiation of discrete behavioral variants emerged independently in chimpanzees and orangutans during the Pleistocene, parallel to the pattern in the human archaeological record.

EVOLUTION OF MODERN HUMAN CULTURAL BEHAVIOR: THE ADAPTIVE CONTEXT

When was it, then, that modern human cultural behavior emerged? And under what conditions did this process of emergence take place? Past records of many types—e.g., oxygen stable isotopes (deep-sea $\delta^{18}O$), continental aeolian dust, fossil pollen, Mediterranean sapropels, and basin sedimentary sequences—show that vast environmental fluctuations have characterized the past several million years and provided the critical context in which humans evolved. The past 700,000 years, in particular,

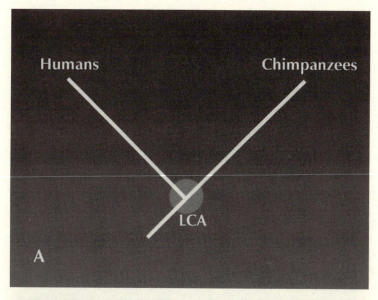

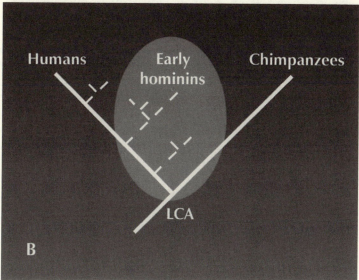

Figure 12.4 (A) Cladogram showing the relationship between chimpanzees and humans. This cladogram implies that features found in both living species were also represented in the last common ancestor (LCA). (B) Cladogram that adds ancient hominin lineages, in which certain aspects of the time-space patterning in cultural behavior do not easily fit the common ground of culture in modern chimpanzees and humans. This cladogram suggests that a distinctive paleocultural system of behavior characterized early human ancestors and related lineages.

have been one of the most turbulent periods of environmental instability in Earth's history (Potts 1996, 1998).

Virtually all organisms were affected by these environmental dynamics. As a consequence, the Pleistocene was a period of high species extinction, and lineages that did not become extinct had two options. First, mobility and wide dispersal allowed organisms to keep up with geographic shifts in their preferred climatic zone or food resources. Second, organisms in some cases appear to have evolved a greater degree of versatility or adaptability to a wider range of environmental conditions.

I have previously proposed that versatility is the astonishing hallmark of modern humanity (Potts 1996, 1998, 2002). Never before has a single species of such ecological adaptability evolved, at least among vertebrates. *Homo sapiens* (the extant human species) emerged as a result of its ancestral lineage having persisted and changed in the face of dramatic environmental variability. Extinction was prevalent in our evolutionary tree, apparently as common as in other groups of large mammals (Potts 1996). Persistence of any population required the means to accommodate to instability in the surroundings. While variability occurred on seasonal and decade scales, the largest environmental disruptions, including reconfiguration of resource landscapes, took place over many thousands of years. The process of evolution in the lineage leading to modern humans was one that essentially decoupled the human organism from any single ancestral environment (Potts 1996, 1998).

In this dynamic prehistoric context, a suite of anatomical and behavioral shifts occurred. The fastest rate of increase in brain size relative to body size took place over the past 700 thousand years (ka) (Ruff, Trinkaus, and Holliday 1997). Around 500 ka, the stone tools made by early hominins began to become more diversified with increasingly standardized forms. The continuum of handaxes and other tools became divided into more discrete types. Regional variation in stone toolkits increased during the African Middle Stone Age (MSA), beginning around 250–300 ka. Innovations in foraging behavior, indicated by specially designed throwing spears (Thieme 1997) and bone points designed for fishing (Yellen et al. 1995), became evident between 400 and 70 ka, whereas no older instances of these specific technologies are known.

Social interactions also intensified. Although it has been suggested that home bases existed earlier in time, the primary archaeological signals of modern human home-base behavior—namely, hearths and shelters—do not become apparent in the prehistoric record prior to about 400–300 ka. The sites of Vérteszöllös (Hungary), Terra Amata (France), and Zhoukoudian (China) are arguably the first sites from a narrow slice of time that demonstrate the occurrence of such behaviors. If so, there was an intensification of social activity centered at distinctly humanlike home bases starting around 400 ka.

Concerning symbolic behavior, a few artifacts between 250 and 70 ka indicate that symbolic behavior began to be reflected in the things hominins made. The presence of pigment back to at least 230 ka in central Africa is also suggestive of decorative capabilities in some early human populations (McBrearty and Brooks 2000; Barham 2002). A tremendous expansion of symbolic behavior occurred at 50–30 ka, manifested in body ornamentation (e.g., beadwork), cave paintings, sculpture, and bone flutes—symbolic activities of many different types. In this same period, we see a further increase in discrete artifact forms and in the geographic differentiation of behavior. Innovation started to soar, exemplified by the routine use of new materials such as antler and bone, in addition to stone, to create novel types of implements (e.g., spear-throwers, sewing needles) and aesthetic objects.

Figure 12.5 presents a simple model of the timing of these novel developments in human cultural behavior. Tool forms became more diversified and geographic differentiation more embellished prior to the expansion of symbolic activity. Yet the latter had a powerful positive feedback effect on these two other signals of modern cultural behavior. The developments in this feedback model occurred in the context of intensified social interaction.

They also arose during a lengthy period of environmental instability. The qualities of social life distinctive to our species thus appear to relate to the capacity of humans to buffer survival risks and resource uncertainty. Four key examples are offered here. First, intricate social networks in the form of economic pair bonds (marriage), gift exchange, and trade enable humans to spread the risks of hard times and the abundance of good times, largely by creating bonds of social reciprocity.

Second, language—one of the most elaborate forms of symbolic coding imaginable—has many functions, one of which may have made the difference between extinction and survival in dynamic Pleistocene environments, namely, the ability to refer to distant sources of water, food, and other nonvisible aspects of one's natural and social environment. Symbolic language enables human beings to create complex mental maps, to think in terms of contingencies, and to imagine "what if. . . ," i.e., to plan and create strategies for events that have not yet happened. A recent analysis of the language faculty argues that *recursion*—the ability to turn a finite set of elements (phonemes, words, syntactical rules, mental activities) into a potentially infinite array of discrete expressions and internal representations—lies at the heart of uniquely human language (Hauser, Chomsky, and Fitch 2002). It is difficult to see how this infinitely inventive aspect of language could have evolved via habitat-specific, directional selection (ibid.). Recursion is necessary, however, for humans to imagine and communicate an infinite array of objects and possible outcomes, to plan for contingencies, and to think socially about past and future events. Recursion thus enables us to deal creatively with matters that are possible

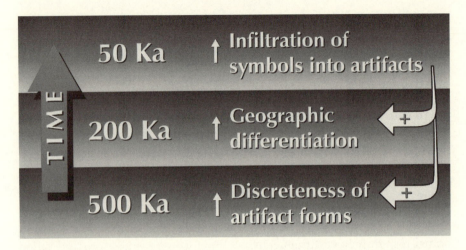

Figure 12.5   A simple feedback model of the transition from paleocultural behavior to modern cultural behavior in humans. Discrete artifact forms and geographic differentiation became much more strongly manifested starting 500,000 to 200,000 years ago. The infiltration of symbolic behavior into artifact manufacture magnified the degree of systematic cultural variation in time and space. The positive feedback was especially notable after 50,000 years ago; at this point relatively homogeneous cultural behavior gave way to cultural diversification—cultures, the plural.

but otherwise cannot be seen, and thus it does make sense as an evolved response to the type of inconsistent, uncertain, and disruptive adaptive settings in which *Homo sapiens* emerged (Potts 1996).

Third, technology—not merely stone flaking but rather the social and mental means of manipulating one's habitat in ever more intricate ways—represents the capacity for creatively altering the environment, and thus moderating the risks and uncertainty of dynamic surroundings. That technology may change very rapidly, over periods of years or centuries, rather than over hundreds of thousands of years, indicates the enormously greater adaptability of *Homo sapiens* relative to its surroundings compared with much earlier hominins.

Fourth, cultural institutions represent the most vigorous means by which individuals buffer environmental risks and personal debility, i.e., via complex social activity powered by accepted symbols and shared meanings. Modern human cultures are an amazing blend of economic, political, religious, ethical, and legal rules and beliefs, which organize human social relationships in extremely complex, symbolic ways. These relationships (reflected even in simple words like mother, father, president, and professor) are often abstractions; they do not appear and disappear as individuals are born and die. Rather, they transcend the time and space boundaries of individuals' lives, creating mental and social stability in a dynamic world.

The argument here, in short, is that the origin of cultural capacities distinctive to living humans embellished the chances of adapting to environmental instability, and this enhancement decoupled the human organism from any single ancestral milieu.

SYMBOLIC REFERENCE TO THE NONVISIBLE

In discussing the evolution of human cultural capacities, the overarching influence of symbolic ability (the means by which humans create meaning) is inescapable. Human cultural behavior involves not only the transmission of nongenetic information but also the coding of thoughts, sensations, and things, times, and places that are not visible. All the odd elaborations of human life, socially and individually, including the heights of imagination, the depths of depravity, moral abstraction, and a sense of God, depend on this *symbolic coding of the nonvisible*.

The most exaggerated manifestations of human sociality may involve the gathering of many hundreds to tens of thousands of people to witness events and participate in rituals of celebration, competition, and mourning. Such exaggerations are only possible because of the ability of many individuals to conceptualize an immediate future they have not yet witnessed but wish to be a part of, and to hold this conceptualization in their brains and plan to act accordingly. Morality, a sense of life's purpose, and acts of mass annihilation are all conveyed symbolically, made possible by mental abstraction and imagination. They depend on coding the nonvisible, requiring humans to believe in abstract terms: "if *A* happens, then *B* will be the larger event that will occur, or the larger goal I/we will obtain." The concept of God itself follows from the ability to abstract and to conceive of "person" and possibilities neither present nor visible, and to conceive of exceptional times of turmoil and blessing that have not yet occurred.

None of these aspects of human sociality makes sense as a response to a world of certainty and stability. They all make sense as psychological and societal responses to a world of contingency and attempts to make sense and stability in a world prone to perturbation. The need to create meaning (religious, ethical, philosophical, artistic, etc.) is part of the toolkit *Homo sapiens* has evolved to deal with disturbance and uncertainty in the external and social domains. The more we investigate the environmental context of human origins, the more we see that the elaboration of human sociality, the ability to accommodate to contingencies and disruption, and coding of the nonvisible are all intimately related, having evolved together.

CONCLUSION

There are several noteworthy implications of the "paleocultural system" of early human behavior I have sketched here, particularly regarding the emergence of modern human sociality from this behavioral background.

First, evidence of gradational cultural forms (e.g., continuity among tool types), weak geographic differentiation, and long periods of stasis in the archaeological record of human origins raises the question of whether some aspects of cultural behavior in humans, chimpanzees, and orangutans were independently evolved. Evidence that crows (*Corvus moneduloides*) make distinct tool designs that have diversified and cumulatively changed over time, a pattern analogous to cumulative technological evolution in humans (Hunt and Gray 2003), suggests that parallel evolution of cultural abilities is plausible and likely to have occurred more widely than currently accepted.

Second, the cognitive and social underpinnings of modern human cultural behavior apparently emerged in the context of highly unstable adaptive settings. The elaboration of culture involved a heightened degree of social interdependence and exchange of resources in response to uncertain, unstable landscapes. The contraction and expansion of African forests and woodlands were also part of the evolutionary background of chimpanzees. However, like other apes, chimpanzees remained tied to certain habitat characteristics, such as the availability of ripe fruit, whereas human evolutionary history appears to reflect a series of narrow escapes from habitat specialization (Potts, in press).

Third, a redefined concept of culture is profoundly needed, one that promotes evolutionary analysis. The diverse concepts currently employed by different disciplines do not offer particularly useful analytical opportunities. When asked what they know about the evolution of cultural abilities, some students of primatology tend to say that culture is a feature common to all great apes; students of Pliocene archaeology believe that culture began with the first stone tools; students of later Pleistocene archaeology increasingly think that the evolutionary transitions crucial to human culture took place over the past few hundred thousand years; and students versed in cultural anthropology look for evidence of the complex symbolic and organizational behaviors apparent mainly over the past 40,000 years. None of these answers, by itself, is complete. Although applying the label of "culture" to chimpanzees and orangutans will help open avenues for studying the evolution of cultural behavior, we sorely

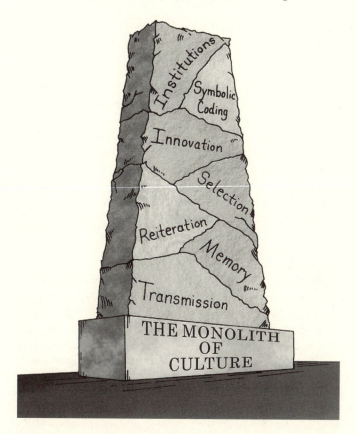

Figure 12.6   A conceptual framework for the evolutionary analysis of cultural behavior (Potts 1993, 1996). The monolithic concept of culture can be dissected into components that reflect the processes of nongenetic *transmission* of information; its retention (*memory*) and reproduction by individuals (*reiteration*); the capacity to generate novel information (*innovation*) and to retain or reject novelty (*selection*); and, finally, the capacity to code information symbolically (*symbolic coding*) and the social systems that result (*institutions*) when complex symbolic codes (e.g., language, cultural symbolism of all forms) are developed and create social order underlain by creed and socially channeled values. Human and nonhuman primates share the lower components of the culture monolith; symbolic coding and institutions represent the distinctively human dimension of cultural behavior. Illustration drawn by J. Clark.

need better strategies of studying the cognitive and social components of cultural behavior—that is, the mechanisms that underlie the transmission and use of nongenetic information.

One way of formulating the problem is sketched in Figure 12.6, where the monolithic term "culture" is actually divided into building blocks: the transmission, retention, and reiteration of nongenetic information; the production and selection of novel information; the symbolic coding of information and its translation into complex social activity (Potts 1993, 1996). Humans share with chimpanzees, orangutans, and other animals the fundamental building blocks near the base of the cultural monolith. However, in applying the term "culture" to both chimps and humans, we would be foolish to miss the intriguing emergence of distinctive cultural properties during the course of human evolution. And researchers study-ing the upper echelon of the monolith would do well to weigh both the evolutionary continuities and disjunctions in the cultural behavior of *Homo sapiens*, earlier humans, and other primates.

Finally, evolution of the uniquely human ability to refer symbolically to nonvisible events and possibilities has greatly magnified the complexity of human life and society. Sharing of complex mental images of the world enables human beings to transcend the immediate moment. We can imag-ine ourselves and others across time and space, which we express to one another as stories of past and future. Language engages the interactive minds of the social group and enables the social world beyond an indi-vidual's own lifetime to be defined symbolically. This experience, unique to humans and obvious archaeologically in the burials of late Pleistocene modern humans, suggests the origin of a spiritual sense—the ability to engender and instill in others a belief in powers beyond one's immediate sight or control. A sense of the ineffable, the sacred, the spiritual—what-ever one calls it—is part and parcel of how human beings mend the tears in their personal and social universe. It is the imaginative means by which forces that lie outside direct human experience and society are made real and brought into focus. The evolutionary origin of *Homo sapiens* in a con-text of unpredictable environmental dynamics and ecological disruption, with critical implications in the social domain, makes considerable sense of this astonishing dimension of human cultural behavior.

ACKNOWLEDGMENTS

I am grateful to the participants of the conference on Primatology and Human Nature: Sociality, organized by the AAAS Program of Dialogue on Science, Ethics, and Religion, and to Robert Sussman and Audrey Chapman for their role in devel-oping the conference and this volume. This chapter is partly based on a presenta-tion at the 2000 AAAS Annual Meeting, Washington, D.C. I thank Jennifer Clark for her expertise in preparing the figures.

REFERENCES

Barham, L. S. 2002. "Systematic Pigment Use in the Middle Pleistocene of South-Central Africa." *Current Anthropology* 43:181–90.

Bonner, J. T. 1980. *The Evolution of Culture in Animals.* Princeton, NJ: Princeton University Press.

de la Torre, I., R. Mora, M. Dominguez-Rodrigo, L. de Luque, and L. Alcalá. 2003. "The Oldowan Industry of Peninj and Its Bearing on the Reconstruction of the Technological Skills of Lower Pleistocene Hominids." *Journal of Human Evolution* 44:203–24.

de Waal, F. B. M. 1999. "Cultural Primatology Comes of Age." *Nature* 399:635–36.

Goodall, J. L. 1968. "The Behaviour of Free-Living Chimpanzees in the Gombe Stream Reserve." *Animal Behaviour Monographs* 1:161–311.

Hauser, M. D., N. Chomsky, and T. W. Fitch. 2002. "The Faculty of Language: What Is It, Who Has It, and How Did It Evolve?" *Science* 298:1569–79.

Hou, Y., R. Potts, B. Yuan, Z. Guo, A. Deino, W. Wang, J. Clark, G. Xie, and W. Huang. 2000. "Mid-Pleistocene Acheulean-like Stone Technology of the Bose Basin, South China." *Science* 287:1622–26.

Hunt, G. R. and R. D. Gray. 2003. "Diversification and Cumulative Evolution in New Caledonian Crow Tool Manufacture." *Proc. R. Soc. Lond. B* (online) 02PB0739.1–8.

Isaac, G. Ll. 1969. "Studies of Early Culture in East Africa." *World Archaeology* 1:1–27.

Isaac, G. Ll. 1972. "Chronology and Tempo of Cultural Change during the Pleistocene." Pp. 381–430 in *Calibration of Hominid Evolution,* edited by W. W. Bishop and J. A. Miller. Edinburgh: Scottish Academic Press.

Isaac, G. Ll. 1977. *Olorgesailie.* Chicago: University of Chicago Press.

Isaac, G. Ll. and J. W. K. Harris. 1997. "The Stone Artefact Assemblages: A Comparative Study." Pp. 262–362 in *Koobi Fora Research Project,* Volume 5, *Plio-Pleistocene Archaeology,* edited by G. Ll. Isaac. Oxford: Clarendon.

Kimbel, W. H., R. C. Walter, D. C. Johanson, K. E. Reed, J. L. Aronson, Z. Assefa, C. W. Marean, G. G. Eck, R. Bobe-Quinteros, E. Hovers, Y. Rak, C. Bondra, T. Yemane, D. York, Y. Chen, N. M. Evenson, and P. E. Smith. 1996. "Late Pliocene *Homo* and Oldowan Tools from the Hadar Formation (Kada Hadar Member, Ethiopia)." *Journal of Human Evolution* 31:549–61.

Kimura, Y. 1999. "Tool-Using Strategies by Early Hominids at Bed II, Olduvai Gorge, Tanzania." *Journal of Human Evolution* 37:807–31.

Kuman, K. 1994. "The Archaeology of Sterkfontein—Past and Present." *Journal of Human Evolution* 27:471–95.

Leakey, M. D. 1971. *Olduvai Gorge,* Volume 3. Cambridge: Cambridge University Press.

McBrearty, S. and A. S. Brooks. 2000. "The Revolution That Wasn't: A New Interpretation of the Origin of Modern Human Behavior." *Journal of Human Evolution* 39:453–563.

McGrew, W. C. 1992. *Chimpanzee Material Culture.* Cambridge: Cambridge University Press.

Merrick, H. V. and J. P. S. Merrick. 1976. "Archeological Occurrences of Earlier Pleistocene Age from the Shungua Formation." Pp. 574–84 in *Earliest Man and*

Environments in the Lake Rudolf Basin, edited by Y. Coppens, F. C. Howell, G. Ll. Isaac, and R. E. F. Leakey. Chicago: Chicago University Press.

Movius, H. L., Jr. 1969. "Lower Paleolithic Archaeology in Southern Asia and the Far East." Pp. 17–77 in *Early Man in the Far East*. Studies in Physical Anthropology, Volume 1, edited by W. W. Howells. New York: Humanities Press.

Noll, M. P. 2000. *Components of Acheulean Lithic Assemblage Variability at Olorgesailie, Kenya*. Ph.D. dissertation. University of Illinois, Urbana.

Oakley, K. P. 1965. *Man the Tool-Maker*. London: British Museum (Natural History).

Perry, S., M. Baker, L. Fedigan, J. Gros-Louis, K. Jack, K. C. MacKinnon, J. H. Manson, M. Panger, K. Pyle, and L. Rose. 2003. "Social Conventions in Wild White-Faced Capuchin Monkeys." *Current Anthropology* 44:241–68.

Plummer, T., J. Ferraro, P. Ditchfield, L. Bishop, and R. Potts. 2001. "Late Pliocene Oldowan Excavations at Kanjera South, Kenya." *Antiquity* 75:809–10.

Potts, R. 1988. *Early Hominid Activities at Olduvai*. Hawthorne, NY: Aldine de Gruyter.

Potts, R. 1991. "Why the Oldowan? Plio-Pleistocene Toolmaking and the Transport of Resources." *Journal of Anthropological Research* 47:153–76.

Potts, R. 1993. "Archeological Interpretations of Early Hominid Behavior and Ecology." Pp. 49–74 in *The Origin and Evolution of Humans*, edited by D. T. Rasmussen. Boston: Jones and Bartlett.

Potts, R. 1996. *Humanity's Descent: The Consequences of Ecological Instability*. New York: Avon.

Potts, R. 1998. "Environmental Hypotheses of Hominin Evolution." *Yrbk. Phys. Anth.* 41:93–136.

Potts, R. 2002. "Complexity and Adaptability in Human Evolution." Pp. 33–57 in *Probing Human Origins*, edited by M. Goodman and A. S. Moffat. Cambridge, MA: American Academy of Arts and Sciences.

Potts, R. In press. "Paleoenvironments and the Evolution of Adaptability in Great Apes." In *The Evolution of Great Ape Intelligence*, edited by A. E. Russon and D. R. Begun. Cambridge: Cambridge University Press.

Potts, R., A. K. Behrensmeyer, and P. Ditchfield. 1999. "Paleolandscape Variation and Early Pleistocene Hominid Activities: Members 1 and 7, Olorgesailie Formation, Kenya." *Journal of Human Evolution* 37:747–88.

Roche, H., A. Delagnes, J.-Ph. Brugal, C. Feibel, M. Kibunjia, V. Mourre, and P.-J. Texier. 1999. "Early Hominid Stone Tool Production and Technical Skill 2.34 Myr Ago in West Turkana, Kenya." *Nature* 399:57–60.

Ruff, C. B., E. Trinkaus, and T. W. Holliday. 1997. "Body Mass and Encephalization in Pleistocene *Homo*." *Nature* 387:173–76.

Sahnouni, M., D. Hadjouis, J. van der Made, A.-e-K. Derradji, A. Canals, M. Medig, and H. Belahrech. 2002. "Further Research at the Oldowan Site of Ain Hanech, North-Eastern Algeria." *Journal of Human Evolution* 43:925–37.

Semaw, S., P. Renne, J. W. K. Harris, C. S. Feibel, R. L. Bernor, N. Fesseha, and K. Mowbray. 1997. "2.5-Million-Year-Old Stone Tools from Gona, Ethiopia." *Nature* 385:333–36.

Thieme, H. 1997. "Lower Palaeolithic Hunting Spears from Germany." *Nature* 385:807–10.

Toth, N. 1985. "The Oldowan Reassessed: A Close Look at Early Stone Artifacts." *Journal of Archaeological Science* 12:101–20.

van Schaik, C. P., M. Ancrenaz, G. Borgen, B. Galdikas, C. D. Knott, I. Singleton, A. Suzuki, S. S. Utami, and M. Merrill. 2003. "Orangutan Cultures and the Evolution of Material Culture." *Science* 299:102–5.

Whiten, A., J. Goodall, W. C. McGrew, T. Nishida, V. Reynolds, Y. Sugiyama, C. E. G. Tutin, R. W. Wrangham, and C. Boesch. 1999. "Cultures in Chimpanzees." *Nature* 399:682–85.

Wynn, T. 1989. *The Evolution of Spatial Competence.* Urbana: University of Illinois Press.

Yellen, J. E., A. S. Brooks, E. Cornelissen, M. H. Mehlman, and K. Stewart. 1995. "A Middle Stone Age Worked Bone Industry from Katanda, Upper Semliki Valley, Zaire." *Science* 268:553–56.

13

Large-Game Hunting and the Evolution of Human Sociality

Christopher Boehm

METHODOLOGICAL INTRODUCTION

We humans are a species naturally given to curiosity, and the question of moral origins has stirred such curiosity all over the world. This is no recent phenomenon, for tens of thousands of years ago people surely possessed oral traditions that dealt with the origin of our uniquely "right and wrong" way of doing business socially. But methodologically we are just now in a position such that scientists may enter this behavioral arena and hypothesize about the dawn of social control, doing so on a basis that is substantive and plausible if not ultimately "testable."

For some time, archaeology has afforded a reasonably good view of our *physical* predecessors, that is, of human cranial and postcranial evolution and prehistoric subsistence patterns (e.g., Tattersall 1998). However, it has been only with the advent of a conservative cladistic methodology (Wrangham 1987) that we have been able to reconstruct the evolutionary beginnings of social behavior in the human line, and do so without undue speculation. Wrangham's rule of thumb was that if humans, bonobos, gorillas, and chimpanzees unanimously shared a behavior, and if there were no good evidence to the contrary (see Wrangham and Peterson 1996), their Common Ancestor (CA) should have had that behavior as well.

Elsewhere, by building on Wrangham's original social reconstruction I have fashioned an ancestral political portrait that is quite detailed (see Boehm 1999b, 2000). As of 7 million years ago (MYA), we have a group-living species with female dispersion and political tensions between groups. Within groups, we have social dominance hierarchies, formation of coalitions that tended to reduce the power of alpha individuals, and conflict interventions by the powerful that served to pacify fights of subordinates.

More recently, at 5 MYA we have a shared ancestor of humans, chimpanzees, and bonobos, which I have designated the *Pan*-Human Ancestor (see Boehm 2000). All the behavioral features of the earlier CA were retained in this successor species, but when vegetarian, harem-dwelling gorillas are removed from the picture, we may reconstruct additional features that are of interest in thinking about the evolution of morals, especially "hunting" and the sharing of meat within the group. (In addition the *Pan*-Human Ancestor had fission-fusion groups, with defense of resources that might be called "territorial.")

If such a comprehensive political picture is remarkable given its antiquity, the two social portraits are far from being "complete." For instance, it is impossible to say anything about the role of pair-bonding in the family structure of either ancestor. This is because even though gorillas and humans have long-term social associations of parents who produce offspring together, chimpanzees and bonobos breed promiscuously and fail to form such bonds between adults. So with respect to the nuclear family we must ascribe to both ancestors a major question mark. On the other hand, a protracted mother-infant bond that lasts for several years (until weaning) is found in all four extant species. It is extremely unlikely that such a basic social feature could have been absent ancestrally and then evolved independently in all four species on a convergent basis, so we may affirm that the ancestor in question did have protracted mother-offspring bonding.

That is how our methodology works. It is true that purely by chance behavioral convergences, such as the winged flight of moths, birds, and bats, do occur in nature. But such accidents of natural selection are relatively rare. On the other hand, if the species involved are known to be close molecularly, as with all four African apes, then we may make definite assumptions about such behavioral patterns being homologous as opposed to their having resulted from convergent evolution.

This methodology provides some major clues with respect to the dawn of social control, and to the extent that it is used rigorously the clues are quite solid. There are, in fact, two specific rules that we will be following here in reconstructing behavior patterns for both of these ancestors, and they are simple as well as conservative. We have just dealt with the unanimity principle, but this requires some clarification. If the extant descendants exhibit a major behavioral trait, and exhibit it *unanimously* either in the wild or in captivity (Boehm 1999b), that same behavioral potential is assumed to have been present ancestrally. This does not mean, necessarily, that the behavior was actually being expressed in ancestral environments, which surely varied from those found today. What it does mean is that long ago the same potential existed as a building block for blind natural selection processes to work with.

A second rule is that when we are looking for behavioral similarities among four (or three) African great apes to reconstruct the Common Ancestor or the *Pan*-Human Ancestor, we must conservatively go with the least common denominator (see Boehm 2000). For instance, consider the fact that chimpanzees and humans engage in perimeter defense using patrols and kill their enemies on sight, whereas bonobos merely show hostility on the parts of males when neighboring communities meet, seldom engage in intergroup fighting, and are not reported so far to engage in lethal aggression (see Stanford and Allen 1991; Wrangham and Peterson 1996). The most we can say about the *Pan*-Human Ancestor is that it would have had moderate and probably nonlethal territorial tendencies—limited to the bonobo level.

There is also a second methodology I shall employ here, which goes against some earlier archaeological thinking (e.g., Foley 1988). In working back from today's humans to a much more recent past, I believe that prevalent behavior patterns of extant human foragers can be assumed to have been prevalent also in the Upper Paleolithic. I am thinking in terms of behavioral central tendencies as opposed to taking specific societies as models, and I must quickly specify which extant foragers should be used as a collective model, and which cannot.

Of the more than three hundred ethnographically described foraging cultures identified by Binford (2001), I have elsewhere (see Boehm 2002) conservatively selected 154 societies that at the time of study: (1) were mobile in the sense of not living in permanent settlements; (2) were politically and economically independent; (3) still spoke their own language; and (4) domesticated no animals or plants aside from dogs and possibly tobacco. If one factors out the very considerable cultural diversity found among such people (see Kelly 1995) and looks, rather, for central tendencies, then these 154 societies can be given a profile as follows.

With respect to demography, people in these societies live in smallish, fission-fusion multifamily bands. Politically, they are predictably egalitarian, in that band-level moral communities suppress individuals' tendencies to behave in an alpha fashion. They favor large game and accord special prestige to those who acquire it, and they divide this special type of food more or less equally among the families in the band, some of whom are unrelated. Their moral code includes not only prohibitions against bullying behavior and against cheating on the sharing system, but also sanctions that range from mild criticism and social pressure to ridicule, ostracism, exile, and execution (see Boehm 1993, 2000).

We now have two distinct methodologies. The first is cladistic, and it provides behavioral starting points for moral evolution as of either approximately 7 or 5 million years ago, depending on whether one is dealing with the CA or the *Pan*-Human Ancestor. The second operates by ethnographic analogy, and it provides us with a terminal point for moral evolution in the Upper Paleolithic. By that time humans had become both Anatomically and Culturally Modern, and we must assume that they were living in modern-type small moral communities.

A THESIS

With these tools we will examine the advent of large-game meat consumption (Stanford 1998; see also Stanford and Bunn 2001) in the human evolutionary career, and assess its probable effects on the development of group life and social control in particular. I should emphasize at the start that we are not necessarily dealing with a *"hunting* hypothesis." The emphasis placed on actively taking large game has waxed and waned, not only in the light of archaeological evidence and its varying scientific interpretations, but in the name of political correctness with respect to the foraging efforts of females vs. males (see Stanford 2001). How people acquired their large game, be this individually or collectively, through ambush hunting or pursuit hunting or by active or passive scavenging, is not terribly relevant to the arguments to be made here. What is relevant is how often large acquisitions of animal fat and protein were put to use by multifamily human groups, and, in particular, how the meat was shared.

Stated baldly, a major subthesis will be that an innate attraction to flesh (and to fatty flesh in particular) was important to the evolutionary development of human sociality, which in foraging bands involves significant cooperation above the family level. The facts are simple. Humans are innately disposed to savor fatty foods (see Cordain et al. 2001), and this seems to apply also to the chimpanzees (Stanford 1998). Large game contains far more fat than most other foods available in prehistoric environments, and more fat than small game (see Cordain et al. 2001). Furthermore, in many situations, because of their size, large-game acquisitions may offer the option of consuming desirable fatty portions of carcasses, such as marrow, brain, liver, and subcutaneous fat (ibid.) and leaving less desirable items uneaten. Thus, whenever it made sense for reasons of subsistence to concentrate on acquiring "meat" in large packets, an innate liking for fat tended to "tweak" the practical motivations that facilitated such behavior. As a result, large game was a favored food source in spite of its being relatively difficult, and sometimes dangerous, to acquire.

This innate propensity helps to explain the fact that today bands all over the world place a special emphasis on acquiring large-game meat, be this subsistence practice obligate as in the Arctic, or optional as in certain environments that are very rich in plant foods and small game. Even in areas where large-game resources are scarce or difficult to obtain, because of this preference hunter-gatherer diets seldom dip below 20 percent large-game intake (Kelly 1995). In short, all hunter-gatherers place a special cultural emphasis on large game, and they tend to accord a special status to the men (rarely women) who obtain it, as opposed to the work of collecting plant food or small game, which is engaged in by both women and men.

Making substantial use of large game appears to have been around for at least 500,000 years (Kuhn and Stiner 2001; see also Stiner 2002), and as behavioral ecologists like Winterhalter (1986) and Smith (1988) have pointed out, wherever large game is relied upon and is acquired only sporadically, as is the case most of the time in most environments, it is very useful to practice "variance reduction." The statistical assumptions involved are understood by hunter-gatherers themselves, for they are highly competent at certain types of actuarial thinking (e.g., Wiessner 1977; see also Boehm 1997a). Nutritionally, it makes more sense for everyone in a band to share a given hunter's meat than for the family acquiring the meat to selfishly overeat for a short time and then endure protracted a fat-and-protein "drought," or for that family to invest energy in processing the meat for limited storage when bands are nomadic (see also Smith and Boyd 1990).

Is variance reduction, accomplished by equitable sharing within bands, an ancient practice? Winterhalter (2001) believes this to be the case. I have suggested that as of 35,000 years ago, Culturally Modern large-game hunters of the Upper Paleolithic would easily have had the brain power to see the advantages of "averaging" (see Boehm 1997a), and also the cultural capacity to institutionalize equitable meat-sharing (Boehm 1997b). However, there is more involved in such hypothesizing than mere brain power and cultural capacity. There is also the question of "politics."

IMPORTANCE OF EGALITARIANISM

Whallon (1989) has made a convincing case that equitable meat-sharing, suitable for really effective variance reduction, would have been prehistorically impossible or extremely improbable if a chimpanzee-like alpha-male system remained in place. Chimpanzees do share meat (see Stanford 1998), but this is far from being equitable. At Gombe, high-ranking males confiscate the carcasses taken by adolescent males or females, and take the lion's share for themselves even as they share with preferred partners. They do the same with carcasses they have taken on their own. At Tai Forest in West Africa, hunting seems to be more cooperative, possibly because of differing ecological circumstances, or possibly because of differing cultural traditions. In any event, sharing is somewhat more equalized; but rank still determines an individual's slice of the protein-and-fat cake and individuals who did not contribute to the hunt may receive little or no meat (see Boesch 1994). Bonobo large-packet food-sharing is also limited and generally uneven (Wrangham and Peterson 1996; see also Kuroda 1980).

For the moral-origins hypotheses I am developing here, a key assumption is that as reliance on large-game meat came to play an increasingly important part in human subsistence over the past five hundred millennia, concomitantly our ancestors were developing sophisticated egalitarian cultural institutions of the type that prevail today among hunting bands wherever they remain nomadic. In effect, the would-be alpha males found their powers and "perks" being routinely neutralized by a moral community that was determined to outlaw any serious male domination among the families in a band (Boehm 1993, 1999b). This made it easier for band members to share large game on an essentially equalized basis, whenever such game was scarce enough to make variance reduction useful.

Judging by today's foragers—the 154 societies that qualify as models for the Upper Paleolithic—the Upper Paleolithic multifamily band was not only highly moralistic, but adamant that meat-sharing be done by the rules. Thus, for the best hunter to assume special privileges over the meat he acquired was highly immoral, and so was stealing the meat of others, or cheating by failing to make public a kill (see Boehm 2000, 2002). These people understood human selfishness all too well, for often they developed institutions that actually took the meat out of the hands of those who acquired it and assigned it to others to distribute (e.g., Kelly 1995; Wiessner 1996; see also Erdal and Whiten 1996).

This was done because it is human to want your full share when someone else kills the animal, but also to want a lion's share if you killed it yourself. If a system of pooling large-game meat for variance reduction is to work, people can't have it both ways. Your moral community will see to it that you don't turn into an alpha-male meat monopolizer who also confiscates the meat of others, and this will also reduce conflict among the band's families. The management of conflict is found in all human groups, and also in the other three African apes (Boehm 1999b).

This equalized-sharing plan pertains only to meat that comes in large packets, and sporadically. If there is a temporary glut of large game, the sharing pattern temporarily disappears (e.g., Binford 1978). Likewise, in times of protracted famine, cooperative institutions are likely to break down (Laughlin and Brady 1978). But most circumstances dictate that meat-sharing will take place. By contrast, plant foods are gathered and distributed only within families most of the time, unless they are rare or particularly desirable (see Gould 1982). Similarly, small game taken by family members is shared only within the family.

Most of the time, then, large game is apportioned out in accordance with local rules for meat distribution, and the techniques and effects of distribution are amazingly similar wherever hunter-gatherers continue to be not only politically and economically independent, but nomadic enough so that long-term storage is not feasible. These egalitarian techniques involve morally based cultural rules that are reinforced through social control, and it is my thesis that evolution in the field of political behavior made this development possible by offering a definitive means of heading off alpha-male hegemony.

THE ROLE OF NATURAL SELECTION

Is this sharing explainable in terms of natural selection? Bands today tend to contain both related and unrelated families (see Kelly 1995), so nepotism based on kin selection cannot explain all of the sharing that takes place. Nor does Trivers's (1971) well-known cooperating-pair "reciprocal altruism" model work for explaining hunter-gatherers' meat-sharing. For one thing, over lifetimes individual contributions to band subsistence tend to be seriously out of balance. For another, over time individuals may live in several different bands.

Alexander (1987) has suggested that hunter-gatherer cooperation is based not on tit-for-tat "reciprocal altruism," but on generalized reciprocity in which people see the returns as being far more long-term and probabilistic, with moral reputation as an auxiliary source of reinforcement. Some such explanation is needed, because most hunting bands are sustained mainly by the efforts of just a few excellent hunters, while other hunters in the group are less vital, less experienced, less astute, less skilled, or less lucky, and in any event less productive (see Kelly 1995).

It could be argued that variance reduction leads to benefits even for the best hunter, since he too gains a steady (if moderate) supply of high-quality protein and fat, which is nutritionally more useful than the occasional feast. Furthermore, the constant physical "training" provided by hunting may provide a major health benefit for a good hunter compared with a lazy one. On the other hand, in a nonegalitarian situation the most robust hunters would be gaining major reproductive benefits by controlling meat. Thus, it is rather difficult to conceive of any *precise* cost-benefit analysis that could accurately measure the apparent individual disparities of risk and effort that must be accounted for when a group routinely takes over the large game and distributes it evenly.

To bring in a new dimension, it is possible that cooperation of this type is receiving some support from "human nature." Of course, the social sharing of large game is so well reinforced by moral rules and by social control that there is no *necessary* reason to assume that altruistic genes are involved in such human cooperation; indeed, a substantial degree of cooperation can be *compelled* by social control (see Boyd and Richerson 1992). However, were such genes present, this would lessen the burden of social control in channeling behavior in "unselfish" directions, as when one great hunter feeds his entire band without eating a lion's share or turning his control over meat into power to control other people.

In fact, I believe there could have been some moderate but socially significant genetic group selection taking place in Upper Paleolithic populations. In that epoch humans were already Anatomically and Culturally Modern, and therefore may be assumed to have developed moral communities like those found today—communities that enforced rules about meat-sharing. In spite of Wilson's (1975) and Alexander's (1989) arguments to the contrary, I make this suggestion about gene selection at the between-group level because living in moral communities definitely changed selection mechanics in ways that might have favored some group selection taking place, and also because something like warfare was likely at times (Boehm 1999a).

The assumption is that earlier species in the human line had "despotic," alpha-male-dominated societies like those of gorillas, chimpanzees, bonobos, and many modern humans, and that based on rank there were major discrepancies of reproductive success among group members. In contrast, Culturally Modern Humans of the Upper Paleolithic surely were egalitarian. My suggestion will be that the levels at which natural selection operated were affected profoundly once the political transition to definitive, morally based egalitarianism had taken place (see Boehm 1997b).

Why would this be the case? Let us begin with the fact that it is selection *between* groups that can support genes called "altruistic"—genes that dispose individuals to give assistance to nonkin at reproductive cost to themselves, and to do so without full reciprocation (see Wilson 1975; Alexander 1989). Next, consider the fact that with respect to altruistic genes, selection taking place *within* groups rewards individual genetic selfishness on the basis of inclusive fitness, and works directly against such altruistic traits. Now consider that at either level of selection, the force of natural selection is determined by phenotypic variation and extinction rates. With respect to phenotypic variation, at least, moral communities have had a profound impact on what Sober and Wilson (1998) call "levels of selection."

It is because hunting bands are so persuasive as moral communities that they actually tend to shape the degrees of variation in phenotypic behavior, a basic engine that drives natural selection both within and between groups (see Boehm 1997b). If one thinks about phenotypic variation among individuals *within* a group, egalitarian bands that go out of their way to make alpha-type behaviors untenable will be reducing such variation drastically because the reproductively significant "perks" that used to go with high rank are largely being eliminated. This substantially debilitates the power of within-group selection as it acts as a damper on the retention of altruistic traits in gene pools.

At the same time, a typically egalitarian, morally based consensus process tends to make groups stick together more as they come to group decisions and act as units (see also Boehm 1996). Because humans acting as entire groups can be highly inventive, particularly in ecological emergencies when natural selection processes are intensified, this corporate unpredictability augments phenotypic variation at the between-group level and gives that level of selection more force. Thus, it is better able to support behavioral traits that are genetically altruistic—or otherwise good for the group.

If these morally induced adjustments in phenotypic variation favor the between-group level of selection, there is another effect of living morally that has a similar effect. I have in mind "free-riders" (see Williams 1966), and when it comes to modeling levels of natural selection the free-rider is the mortal enemy of group selection's possibilities for supporting altruistic genes. This is because *in theory* the genetic altruists should always lose to the genetic free-riders. However, at the level of phenotype, humans have invented a special antidote that all but erases this discrepancy. Hunting bands crack down moralistically on most types of free-riders, and do so very effectively whenever the free-rides taken are substantial—and detectable (Boehm 2000). In a small band, virtually every pattern of seriously opportunistic selfish behavior eventually becomes known, and predictably the behavior is sanctioned—all the way up to death by execution. Thus, at the level of phenotype, genetically disposed free-riders are able to take very few major free rides.

With this important mechanical obstacle substantially removed, and with within-group selection seriously debilitated and between-group selection moderately augmented because of socially induced changes at the level of phenotypic variation, the possibility that altruistic genes were being modestly supported in the Upper Paleolithic must at least be entertained (see Bowles et al. 2003). If so, like the love of fat, such genetic tendencies would have made cooperative meat-sharing arrangements easier to arrive at, and easier to maintain.

THE ROLE OF LARGE-GAME ACQUISITION

Elsewhere (Boehm 2000, 2002), I have suggested that the very first instance of morally based social control would have been not the incest taboo (e.g.,

Freud 1938), but the rank-and-file's suppression of alpha behaviors. The hypothesis is that humans had not one but, eventually, two reasons to suppress alpha-male-type behavior. One was innately prompted, and direct. Subordinate resentment of being dominated is present today and was present ancestrally, as a predictable attitudinal side effect of innate tendencies to status rivalry, which mostly tend to make domination attractive and submission unattractive (Boehm 1999b). Ancestral coalitions made it possible for subordinates to act on these feelings, and at the level of protomorality this could easily have taken place with *Homo erectus* or even earlier, especially if symbolic communication capacity was growing (Bruce Knauft, personal communication).

In a sense, the second reason for alpha suppression was less direct, for it had to do with love of flesh, and with variance reduction as discussed above. Much of the fattier meat available to nomadic human foragers does come in large packets, and it is clear that people are motivated to obtain it in spite of its cost in effort, hardship, and danger (Kelly 1995; Stanford and Bunn 2001). They may do so partly because of reputation and prestige (Alexander 1987; see also Hawkes 1991), but more importantly the entire band understands that frequent meat-eating is good for nutrition, and aesthetically they appreciate eating fatty flesh regularly.

With respect to variance reduction there was already the love of flesh, and of fatty flesh in particular, that made large game especially attractive. There was also a preexisting and ancient large-game acquisition pattern that made it easy for meat to become a greater focus as subsistence priorities changed. There were also some innate tendencies to share meat, present in the *Pan*-Human Ancestor. All of these behavioral potentials provided natural selection with some very likely raw materials, and these were useful to the blind genetic selection processes that brought us to the point where we were ready to invent moral communities as we know them today.

The evolutionary end point is the morally based hunting syndrome I have described above, which relates to 154 modern hunting societies and also to all (or most) Upper Paleolithic hunters. While this active hunting syndrome surely applies to the Upper Paleolithic forward, the earlier stages of moral evolution probably involved a combination of passive and active scavenging, with pursuit and ambush hunting becoming more prominent as the Upper Paleolithic approached.

In my opinion, large-game variance reduction based on group social control is likely to have been associated with *any* pattern of regular but sporadic intake of large-packet meat that was highly important to nutrition or critical to survival because cultural group selection (e.g., Soltis, et al. 1995) would have supported this practice. By this I mean that members of egalitarian bands or band-clusters that leveled out their intake of nutritious fat and protein would have survived better in hard times than those who could not do this because they remained despotic, and that this cultural practice could have spread rather quickly on that basis, independently of natural selection.

WHICH CAME FIRST?

We have two possible hypotheses with respect to sequencing. Did acquiring large game become more prevalent in the archaeological record because some version of *Homo* was already sufficiently in control of the alpha-domination problem that bands could easily meet variance reduction requirements, whenever these became perceptually obvious? Or did early humans first become *obliged* to turn to large game because of environmental conditions, and then, because of this, quickly became motivated to start controlling their alphas so as to effectively even out meat distribution? Such causal splitting of hairs obviously will be speculative, and the answer may well be something more like *both*, but the question is of interest in its possible ramifications for the timing and nature of moral origins.

What we do know is that if people live in quite cold environments, then they may be facing tundra conditions in which gathering sufficient edible plant foods and small animals poses difficulties, whereas large game is relatively abundant. We also know that in truly *Arctic* conditions, large game may offer the only viable subsistence. This makes it possible to ask some relatively specific questions about how morally based variance reduction might have appeared, and when.

According to Kuhn and Stiner (2001), large-packet meat acquisition was not unknown even before *Homo erectus,* but about 500,000 years ago we begin to have evidence pointing to a more significant focus on large game. How important this early pattern was to overall basic subsistence is difficult to say, but eventually we begin to have evidence of Neanderthals, and also Anatomically Modern Humans, living in quite cold climates, where very substantial reliance on large game was likely to have been mandatory.

There is still another factor. Archaeologists are really just beginning to take into account the all but unbelievably frequent cyclical climatic perturbations that were particularly prevalent from about 150,000 to 12,000 years ago (see Richerson and Boyd 2001). These cycles would have presented human populations with recurrent and crucial subsistence dilemmas as their environments either changed enough to force new subsistence patterns, or simply made subsistence impossible and migration necessary as an alternative to extinction. Recurrently, large-game hunting would have become useful or necessary at colder phases of these cycles. It was in these dynamic regional environments that humans became Anatomically and then Culturally Modern (see Potts 1996). And it was within this time period that we know people were successfully adapting to colder and colder environments, in which heavy reliance on large game could have been mandatory as opposed to optional.

Now let us consider full-blown moral communities, which went beyond mere alpha suppression to devise complicated rules of conduct that were up to the task of efficiently and smoothly distributing large game to the entire group. Keep in mind that genetically based nepotism could not support such distributions because bands were composed in part of unrelated families, whose basic tendencies were to compete with one another. This type of enlightened cooperation required not only an actuarial brain, but a formidable degree of symbolic communication, for in such complicated matters finding an effective moral consensus requires effective communication (Boehm 2000). As we have seen, an element of genetic altruism would have helped as well; in addition, cultural group selection would have favored group traditions that were more cooperative.

This era, during which humans began sometimes to adapt cyclically to cold climates as opposed to migrating or else becoming extinct in place, provided conditions that would have been appropriate to such a social development. And even though the meat-eating habit would have been mandatory only at certain junctures, the innate disposition to favor fatty flesh would have helped to keep this culturally reinforced habit in place over time. Thus, as human populations experienced these recurrent cycles of environmental change, they would have retained favorable cultural memories of substantial meat-eating even during cyclical phases when meat-eating was not important (or necessary) for survival.

My tentative conclusion is that the first acts of social control were political, in the sense that subordinates banded together so as to deny alpha types the power to control others and remove their various "perks," and that this set the stage for morally based variance reduction to develop through a combination of cultural and genetic selection at individual and group levels. Early on, such practices could have been "optional." By this I mean that fatty flesh was attractive in its own right and ancestral tendencies to hunt were already in place, but because of environmental factors large game had not yet become critical to subsistence. However, with increasingly abrupt and extreme climatic cycles, the practice of seeking large game could have become "compulsory" in that periodically situations would have arisen in which reliance on large mammals became obligatory for adequate nutrition or even for survival.

DISCUSSION

I have argued that an innately based tendency to resent being politically dominated as an adult was an important factor in the evolutionary development of morality, as was love of fatty food. It is perfectly possible that some crude type of social control developed early on, on the basis of raw political resentments that would have produced a purely *political* type of egalitarianism. However, at some point in the line that became human, and possibly rather late in its evolution, this new and deliberate approach to shaping group political patterns was refined, possibly transformed, by the arrival of actuarial intelligence sufficient to motivate and implement even-handed redistributions of meat. At this point, presumably with some help from language, morally based egalitarianism became economic as well as political.

If we assume that variance reduction practices were applicable to meat scavenged by early *Homo erectus* or still-earlier hominids, this moral community scenario could have arrived very gradually, over millions of years. So it is possible that some version of meat-sharing and egalitarianism arrived together fairly early, if a special motive for originally getting rid of alpha types was that they hogged the meat or used it as an instrument of personal power. On the other hand, humans beginning to subsist in cold environments may be a better archaeological marker for the origin of an advanced version of social control, which reached beyond generalized "alpha suppression" to include systematic sharing of large game. It is even conceivable that the definitive social control of alphas actually began at that point.

Another likely marker would be the emergence of symbolic communication, which is likely to have coevolved with band-level moral communities because, as societies that were subject to fission-fusion patterns, they had to not only define their rules but keep tabs on deviants by gossiping (Boehm 2000). Unfortunately, temporal sequencing for language origins remains highly inferential, but the radical increases in cranial capacity that show up with *Homo erectus* could be a general marker for both symbolic language and early types of moral communities.

It is unlikely that we can sort out these various scenarios in any way that is really definitive. However, I believe that these two "original" components of human nature, the strong ambivalence of subordinates over being dominated and the innately based attraction to large game as a fatty food source, are key factors that must be considered in any theory of moral origins. We may add ancestral tendencies to hunt and share meat, at least at the rather modest bonobo level, for such dispositions surely contributed to the use of large-packet meat—whether this was mandatory for subsistence, or optional.

I am proposing that there is likely to have been an intimate relationship between the rise of a modern, fully moralized version of hunter-gatherer egalitarianism and the rise of dependency upon large game. And because the requisite political egalitarianism suppresses competition, this made possible other major developments with respect to band-level equality or cooperation. For instance, consider human mating patterns. Advanced cultural capacity and decisive, institutionalized political egalitarianism could have paved the way for monogamous pair-bonding as a concomitant to equitable meat-sharing. Indeed, the principles involved are quite similar insofar as with alpha males eliminated, treasured and limited resources become available to all adults on a much more equalized basis.

With respect to presumptively innate human propensities to conform to community rules (see Simon 1990, Waddington 1960) and thereby cooperate (see Campbell 1975), the morally sanctioned collaboration that existed in meat acquisition and distribution made it reproductively advantageous for individuals to fit into the overall system, lest they be identified as free-riders. This I hypothesize independently of the group selection possibilities that were discussed above, which could have produced a separate and additional genetic impetus for cooperation.

Another potential on our original Common Ancestral list (Boehm 1999b; see also Boehm 1982) would be propensities to manage conflicts. As concerned moral communities developed, the rough edges that are inevitable in any system of equalized distribution of critical resources like meat or mates were smoothed over by individuals who tried to directly ameliorate the conflicts of others. This was accomplished by means of pre-emptive conflict management, by direct intervention, and by postconflict peacemaking (see also de Waal 1989). For humans, one way of managing conflicts *preemptively* was to create decisive and unambiguous rules for the sharing of large game.

Likewise, marriage institutions, which indirectly reduced male competition over females, had the effect of making it easier for a less powerful or less attractive male to hold onto a female because cultural rules of marriage and divorce regulated pair-bonding. This "solution" was far from being perfect, for male competition over females still presents major problems in forager societies (see Knauft 1991), despite the fact that in other spheres dominant aggression is basically suppressed. In spite of these problems, however, a basic pattern of pair-bonding does prevail in bands, with monogamy prevalent and a few polygamous exceptions.

The fact that political egalitarianism, equalized meat-sharing, familial pair-bonding, active conflict management, and moralistic social sanctioning are universal among nomadic foragers today, and all but surely were present among hunter-gatherers 35,000 years ago, makes it likely at present that human nature has become closely linked to such behaviors. There have been at least a thousand generations for natural selection to do this job. Consider also that important precursors for hunting, sharing, conflict intervention, and even for group social control in the form of antihierarchical coalitionary behavior (see Boehm 2000) were present in the *Pan-*

Human Ancestor, and this adds to the likelihood that human nature strongly supports social control as this exists in today's hunter-gatherer moral communities.

Here, I have endeavored to be as specific as possible in suggesting how large-game acquisition and the sharing of large game could have entered into a moral-origins scenario, and the conclusion is that people's needs to share large-game, along with preexisting social, political, and subsistence tendencies, could have helped natural and cultural selection to arrive at moral communities as we know them today. Hopefully, this set of hypotheses will contribute to the advancement of further or competing hypotheses about the evolutionary development of human sociality.

REFERENCES

Alexander, Richard D. 1987. *The Biology of Moral Systems.* Hawthorne, NY: Aldine de Gruyter.

Alexander, Richard D. 1989. "Evolution of the Human Psyche." Pp. 455–513 in *The Human Revolution,* edited by Paul Mellars and Chris Stringer. Princeton, NJ: Princeton University Press.

Binford, Lewis. 1978. *Nunamiut Ethnoarchaeology.* New York: Academic Press.

Binford, Lewis. 2001. *Constructing Frames of Reference: An Analytical Method for Archeogical Theory Building Using Hunter-Gatherer and Environmental Data Sets.* Berkeley: University of California Press.

Boehm, Christopher. 1982. "The Evolutionary Development of Morality as an Effect of Dominance Behavior and Conflict Interference." *Journal of Social and Biological Structures* 5:413–22.

Boehm, Christopher. 1993. "Egalitarian Society and Reverse Dominance Hierarchy." *Current Anthropology* 34:227–54.

Boehm, Christopher. 1996. "Emergency Decisions, Cultural Selection Mechanics, and Group Selection." *Current Anthropology* 37:763–93.

Boehm, Christopher. 1997a. "Egalitarian Behavior and the Evolution of Political Intelligence." Pp. 341–64 in *Machiavellian Intelligence II,* edited by R. W. Byrne and A. Whiten. Cambridge: Cambridge University Press.

Boehm, Christopher. 1997b. "Impact of the Human Egalitarian Syndrome on Darwinian Selection Mechanics." *American Naturalist* 150:100–21.

Boehm, Christopher. 1999a. "Forager Hierarchies, Innate Dispositions, and the Behavioral Reconstruction of Prehistory." Pp. 31–58 in *Hierarchies in Action: Cui Bono?* edited by Michael W. Diehl. Center for Archaeological Investigations, Occasional Paper No. 27. Carbondale, IL: SIU Press.

Boehm, Christopher. 1999b. *Hierarchy in the Forest: The Evolution of Egalitarian Behavior.* Cambridge, MA: Harvard University Press.

Boehm, Christopher. 2000. "Conflict and the Evolution of Social Control." *Journal of Consciousness Studies* 7:79–183. Special issue on Evolutionary Origins of Morality; Leonard Katz, guest editor.

Boehm, Christopher. 2002. "Variance Reduction and the Evolution of Social Control." Conference on Co-evolution of Behaviors and Institutions, Santa Fe Institute. (Available at SFI website.)

Boesch, Christoph 1994. "Cooperative Hunting in Wild Chimpanzees." *Animal Behaviour* 47:1135–48.

Bowles, Samuel, J.-K. Choi, and Astrid Hopfensitz. 2003. "The Coevolution of Individual Behaviors and Group Level Institutions." *Journal of Theoretical Biology* 223:135–47.

Boyd, Robert and Peter J. Richerson. 1992. "Punishment Allows the Evolution of Cooperation (or Anything Else) in Sizable Groups." *Ethology and Sociobiology* 13:171–95.

Campbell, Donald T. 1975. "On the Conflicts between Biological and Social Evolution and between Psychology and Moral Tradition." *American Psychologist* 30:1103–26.

Cordain, Loren, Bruce A. Watkins, and Neil J. Mann. 2001. "Fatty Acid Composition and Energy Density of Foods Available to African Hominids: Evolutionary Implications for Human Brain Development." Pp. 144–61 in *Nutrition and Fitness: Metabolic Studies in Health and Disease,* Vol. 90, edited by A. P. Simonpoulus and K. N. Pavlou. Basel: Karger.

de Waal, Frans. 1989. *Peacemaking among Primates.* Cambridge, MA: Harvard University Press.

Erdal, David and Andrew Whiten. 1996. "Egalitarianism and Machiavellian Intelligence in Human Evolution." Pp. 139–50 in *Modeling the Early Human Mind,* edited by P. Mellars and K. Gibson. Cambridge: MacDonald Institute for Archeological Research.

Foley, Robert 1988. "Hominids, Humans, and Hunter-Gatherers: An Evolutionary Perspective." Pp. 207–21 in *Hunters and Gatherers,* Vol. I: *History, Evolution, and Social Change,* edited by T. Ingold, D. Riches, and J. Woodburn. Oxford: Berg.

Freud, Sigmund. 1938. "Totem and Taboo: Resemblances between the Psychic Life of Savages and Neurotics". Pp. 807–930 in The *Basic Writings of Sigmund Freud,* A. A. Brill, translator and editor. New York: Modern Library.

Gould, Richard A. 1982. "To Have and Have Not: The Ecology of Sharing among Hunter-Gatherers." Pp. 69–91 in *Resource Managers: North American and Australian Hunter-Gatherers,* edited by N. M. Williams and E. S. Hunn. Boulder, CO: Westview.

Hawkes, Kristen. 1991. "Showing Off: Tests of an Hypothesis about Men's Foraging Goals." *Ethology and Sociobiology* 12:29–54.

Kelly, Robert L. 1995. *The Foraging Spectrum: Diversity in Hunter-Gatherer Lifeways.* Washington, DC: Smithsonian Institution Press.

Knauft, Bruce M. 1991. "Violence and Sociality in Human Evolution." *Current Anthropology* 32:391–428.

Kuhn, Steven L. and Mary C. Stiner. 2001. "The Antiquity of Hunter-Gatherers." Pp. 99–142 in *Hunter-Gatherers: Interdisciplinary Perspectives,* edited by C. Panter-Brick, R. H. Layton, and P. A. Rowley-Conwy. Cambridge: Cambridge University Press.

Kuroda, Suehisa. 1980. "Social Behavior of the Pygmy Chimpanzees." *Primates* 21:181–97.

Laughlin, Charles D. and Ivan A. Brady (Eds.). 1978. *Extinction and Survival in Human Populations.* New York: Columbia University Press.

Potts, Richard. 1996. *Humanity's Descent: The Consequences of Ecological Instability.* Hawthorne, NY: Aldine de Gruyter.

Richerson, Peter J. and Robert Boyd. 2001. "Built for Speed, Not for Comfort: Darwinian Theory and Human Culture." *History and Philosophy of the Life Sciences* 23:423–463.

Simon, Herbert. 1990. "A Mechanism for Social Selection and Successful Altruism." *Science* 250:1665–68.

Smith, Eric Alden. 1988. "Risk and Uncertainty in the 'Original Affluent Society': Evolutionary Ecology of Resource Sharing and Land Tenure." Pp. 221–51 in *Hunters and Gatherers,* Vol. I: *History, Evolution, and Social Change,* edited by T. Ingold, D. Riches, and J. Woodburn. Oxford: Berg.

Smith, Eric Alden and Robert Boyd. 1990. "Risk and Reciprocity: Hunter-Gatherer Socioecology and the Problem of Collective Action." Pp. 167–92 in *Risk and Uncertainty in Tribal and Peasant Economies,* edited by Elizabeth A. Cashdan. Boulder, CO: Westview.

Sober, Eliott and David S. Wilson. 1998. *Unto Others: The Evolution and Psychology of Unselfish Behavior.* Cambridge, MA: Harvard University Press.

Soltis, Joseph, Robert Boyd, and Peter J. Richerson. 1995. "Can Group-Functional Behaviors Evolve by Cultural Group Selection? An Empirical Test." *Current Anthropology* 36:473–94.

Stanford, Craig B. 1998. "The Social Behavior of Chimpanzees and Bonobos: Empirical Evidence and Shifting Assumptions." *Current Anthropology* 14:399–420.

Stanford, Craig B. 2001. "The Ape's Gift: Meat-eating, Meat-sharing, and Human Evolution." Pp. 96–117 in *The Tree of Origin: What Primate Behavior Can Tell Us about Human Social Evolution,* edited by Frans B. M. de Waal. Cambridge, MA: Harvard University Press.

Stanford, Craig B. and J. S. Allen. 1991. "On Strategic Storytelling: Current Models of Human Behavioral Evolution." *Current Anthropology* 32:58–61.

Stanford, Craig B. and Henry T. Bunn (Eds.). 2001. *Meat-Eating and Human Evolution.* New York: Oxford University Press.

Stiner, Mary C. 2002. "Carnivory, Coevolution, and the Geographic Spread of the Genus *Homo*." *Journal of Archeological Research* 10:1–63.

Tattersall, Ian. 1998. *Becoming Human: Evolution and Human Uniqueness.* New York: Harcourt Brace.

Trivers, Robert L. 1971. "The Evolution of Reciprocal Altruism." *Quarterly Review of Biology* 46:35–57.

Waddington, C. H. 1960. *The Ethical Animal.* Chicago: University of Chicago Press.

Whallon, Robert. 1989. "Elements of Cultural Change in the Later Paleolithic." Pp. 433–54 in *The Human Revolution: Behavioural and Biological Perspectives on the Origins of Modern Humans,* edited by P. Mellars and C. Stringer. Edinburgh: Edinburgh University Press.

Wiessner, Polly. 1977. *Hxaro: A Regional System of Reciprocity for Reducing Risk Among the !Kung San.* Ann Arbor, MI: University Microfilms.

Wiessner, Polly. 1996. "Leveling the Hunter." Pp. 171–91 in *Food and the Status Quest: An Interdisciplinary Perspective,* edited by P. Wiessner and W. Schiefenhövel. Oxford: Berghahn.

Williams, George C. 1966. *Adaptation and Natural Selection: A Critique of Some Current Evolutionary Thought*. Princeton, NJ: Princeton University Press.

Wilson, David S. 1975. "A General Theory of Group Selection." *Proceedings of the National Academy of Sciences* 72:143–46.

Winterhalter, Bruce. 1986. "Diet Choice, Risk, and Food Sharing in a Stochastic Environment." *Journal of Anthropological Archaeology* 5:369–92.

Winterhalter, Bruce. 2001. "Intragroup Resource Transfers: Comparative Evidence, Models, and Implications for Human Evolution." Pp. 279–301 in *Meat-Eating and Human Evolution*, edited by Craig B. Stanford and Henry T. Bunn. Oxford: Oxford University Press.

Wrangham, Richard. 1987. "African Apes: The Significance of African Apes for Reconstructing Social Evolution." Pp. 51–71 in *The Evolution of Human Behavior: Primate Models*, W. G. Kinzey. Albany: SUNY Press.

Wrangham, Richard and Dale Peterson. 1996. *Demonic Males: Apes and the Origins of Human Violence*. Boston: Houghton-Mifflin.

14

Cooperation, Commitment, and Communication in the Evolution of Human Sociality

John M. Watanabe and Barbara B. Smuts

Any inquiry into the evolution of human sociality must specify what exactly has evolved, how far, and with what consequences. The human capacity for and dependence on language, and by implication culture, represent an obvious choice for such inquiry. Discussion, however, often turns on how broadly or narrowly to define language and culture. This in turn usually reflects what investigators already think on empirical or theoretical grounds about the uniqueness (or not) of human sociality. Empirically, the impossibility of directly observing early hominid behavior means that evidence for the evolution of human sociality must depend largely on comparison, most often with living nonhuman primates as the closest analogues we have for the capacities and conditions from which human sociality derived. For those who define language and culture broadly, the contrasts here remain a matter of degree; for those who define them narrowly, a matter of kind, especially in the capacity to recognize, use, and create symbols. To see here only continuities or discontinuities, however, belies the evolutionary process itself (cf. King 1999) because Darwin's notion of "descent with modification" clearly demands that we attend to both.

In this chapter, we address the relationship between continuities and transformations in the evolution of human sociality through discussion of the social cooperation and commitment we see intrinsic to both human and nonhuman primate communication. Rather than choose between broad or narrow definitions of language and culture that either obscure or accentuate evolutionary differences across living primates, we begin by asking, not What is language, but rather, What difference does having language and culture make in human sociality? We focus in particular on the understanding and use of symbols and how this both presupposes and

288

intensifies capacities for social cooperation already present in nonhuman primates. We argue that use of symbols evolved in relation to problems of fostering long-term social commitments between otherwise willful individuals, and it perhaps only incidentally and incrementally resulted in the eventual emergence of language as a transformed, as well as transformative, mode of communication (Deacon 1997).

This approach to symbolic communication as intensified social cooperation in turn draws attention to distinctive features of linguistically informed human sociality and how, despite its innovations, language still leaves us grappling with age-old problems of cooperation and social commitment between self and other already found in nonhuman primates and other social mammals. We argue for the deep phylogenetic origins of this nexus between cooperation, commitment, and communication by reviewing and extending evidence we have presented elsewhere on ritual greetings among olive (savanna) baboons (Smuts and Watanabe 1990; Watanabe and Smuts 1999), then reflecting on how the perils and payoffs of long-term social commitments may have helped select for symbolic forms of communication that enable partners to transcend the here and now of short-term failings in their relationship.

LANGUAGE AS SOCIAL COOPERATION

Human sociality differs from the social behavior of other animals in the degree to which human beings imbue their own and others' actions with subjective meanings. As sociologist Max Weber (1947) noted long ago, human beings engage in "social action," not mere reflexive behavior. We presume that to do something often means to say something—it expresses an intent—and we act and interact with others according to what we think they are up to, and we assume they are doing the same. Even though we may not always fully or appropriately apprehend what they intend, we approach others with a "theory of mind" (Premack and Woodruff 1978). While other animals may also do this, we impute a "said" to our own and others' actions more often and extensively than they do (Heyes and Huber 2000). In this sense, language not only facilitates—and complicates—human sociality. It also fundamentally transforms it into languagelike (that is, semiotic) exchanges in which objects, actions, and utterances become signs that social actors routinely endow with socially derived but ultimately subjective meanings that go beyond the immediate instrumentality of the signals themselves.

Here we find useful Charles Sanders Peirce's theory of signs (1932:134–38, 156–73). Peirce uses the term "sign" to refer most generally to anything that comes to stand for something to someone who chooses to interpret it as such. Signs come to stand for something else in three ways (cf. Burks 1949). An "icon" is a sign that represents what it signifies by similarity or analogy, for instance, a diagram. An "index" is a sign that holds a direct existential connection to what it signifies, as a dog's snarl that conveys a threat while itself being part of the aggression threatened. Finally, a "symbol" is a sign in which the relationship to what it signifies remains purely conventional or habitual, agreed upon by "symbol-using mind[s]" (Peirce 1932:168–69), the most ready example being words. With this typology, Peirce provides a continuum of how signs signify, from perceived likeness or actual connection, both of which remain largely rooted in the here and now, to association by agreed-upon conventions that free signs from the literal presence of what they symbolize—as long as interlocutors are there to interpret them as such.

As Deacon (1997) has argued, the ability to understand and use symbols, not necessarily language, distinguishes human communication from that of other animals. Before one can engage in language, one must first understand that something can indeed stand for something else in a purely conventional fashion. This in turn entails the capacity to reconceptualize direct relations between signs and their objects as part of more abstract patterns of relationships between signs. The semioticized sociality precipitated by such symbolic understanding thus presupposes two kinds of conventional constraints on individual action. The first is pragmatic, related to problems of signification (the relationship of signs to what they represent). The second is semantic, regarding the relation of signs to each other (cf. Tambiah 1968:188–89).

Pragmatically, since no necessary relationship binds a symbol in Peirce's sense to what it symbolizes, symbolic meanings depend on mutual agreements between interlocutors about what is going to stand for what and when. This means that in order to communicate symbolically, interlocutors must conform however fleetingly to standards of conduct beyond their own volition. They cannot simply make up new signs or usages and expect others automatically or appropriately to follow them. Instead, like the proverbial tree that falls in the forest, signals only become meaningful signs when others apprehend them as such. Even so-called ostensive definitions based on pointing at and naming something require the "intelligent cooperation [of interlocutors] for catching the meaning of the demonstration" (Polanyi 1966:5–6). In coming to stand for something else to someone else, symbols thus pragmatically entail a complex set of contingent agreements—not only between sign and significance, but also between sender and receiver—over which no interlocutor ever has exclusive, or final, say. While interlocutors need never have equal say in negotiating such meanings, the pragmatic constraints in symbolic communication demand some minimal degree of mutuality.

These mutually agreed-upon conventions in turn produce a second, semantic constraint on individual action. Semantic constraints occur because pragmatic agreements about signs and their associations never work piecemeal but always within the constraining presence of other signs (or more precisely, in relation to each interlocutor's awareness and understanding of other signs and the oppositions and associations between them). Signs thus come to serve as unspoken (but never static or mechanically determinant) normative groundings for other signs (cf. Hanks 1990) that together constitute the semantic constraints on individual symbolic action.

Far from incidental, these semantic constraints address two threats that the power of signification itself ironically poses for symbolic communication (cf. Rappaport 1999). First, as many have noted, the conventional rather than necessary relationship between sign and significance vastly facilitates the transmission of arbitrary or false meanings, most concretely, lies. Whatever their immediate consequences, lies undermine the reliability of the sign associations that convey them (the lesson learned by the boy who cried "Wolf"—and even more ruefully by those who ignored him). In the face of such potential arbitrariness, signs require some means of reaffirming their truth value despite the inevitable vagaries in their usage. As Roy Rappaport (1999) has argued, this ultimately involves embedding signs in the sanctifying constancy of ritual form and performance, as we discuss below regarding baboon greetings.

Conversely, and perhaps less often noted, once interlocutors grasp that one thing can in fact stand for something else, anything has the potential to become a sign to someone and stand for anything else. Signification thus also threatens to infest the world with a surfeit of contagious meaningfulness. To rein in such runaway meanings, interlocutors must work out tacitly (through usage) or explicitly (through rules) ways to sort signs semantically into categories, whether by meaning or function, similarity or contrast, opposition or equivalence. These categories can then order (without ever absolutely determining) the arrays of signs interlocutors might expect to use for what purposes within which contexts, how they might combine different kinds of signs into utterances, and how they might interpret new signs in relation to already agreed-upon signs given the innovator's apparent intent.

Thus, in addition to the pragmatic constraints of social usage that require interlocutors to agree on what particular signs mean, signs in human sociality also entail semantic constraints in relation to other signs. Like pragmatic constraints, semantic constraints emerge through conventional usage and the limits of mutually negotiated understandings, but once established they redound back on usage by structuring the choices interlocutors make about which signs they will use for what purposes.

Both semantic and pragmatic constraints in human sociality presuppose another capacity enabled by symbolic communication. That is, once interlocutors grasp that signs can apply to things in the world, it also becomes possible for them to apply signs to other signs. This metacommunicative possibility of using signs to talk about signs gives human symbolic communication its notable generative productivity. Pragmatically, it allows interlocutors to negotiate more effectively and extensively conventional agreements about what is going to mean what under which circumstances—or at least to ask for amplification when they do not understand. Semantically, a metacommunicative capacity also enables interlocutors to categorize signs relative to other signs through simultaneous, multidimensional, cross-cutting relations of opposition and association.

As Goodenough (1990:597–99) notes, perception itself may well presuppose categorization because without some minimal way—whether innate or acquired—of associating and discriminating sensory inputs, experience of the world would remain little more than an endless succession of undifferentiated occurrences. While other animals can and do categorize perception and experience in complex ways, the metacommunicative property of human symbolic communication greatly enhances such capacities by enabling interlocutors to define, refine, and defend more rigorously the "cherished classifications" by which they order the world (cf. Douglas 1966). Such classifications often consist of configurations (gestalts) that link associated signs into ever larger "arrangements of arrangements" based increasingly on analogical similarities in the way their components are related rather than necessarily on concrete resemblances between the component elements themselves. Such abstraction "is greatly facilitated by the kinds of mental operations that language (or a comparable system of representational signs) makes possible and must remain largely undeveloped without it" (Goodenough 1990:599; cf. Deacon 1997).

It is important to note, however, that despite its social constraints on behavior and immense communicative power, language works in neither absolutely determinant nor totally transparent fashion. Because meaning emerges socially through collaborative—and therefore contingent—agreements and mutual orientations, the sense a sign conveys never resides entirely in the sign itself or in its semantic relations with other signs. Rather, what a sign may mean changes in different contexts depending pragmatically on interlocutors' familiarity with each other, their respective knowledge of the matter—and relevant signs—at hand, their memories of previous interactions—all those things that inform them of the nature and scope of their current interaction and suggest its potential outcomes and implications (Hanks 1990). On the semantic level, familiarity with the other signs interlocutors might have reasonably used in an exchange instead of the ones they actually did use makes their choices all the more clearly intentional to those who know the language. This in turn further delimits the meaning of the signs chosen. In both their pragmatic and semantic dimensions, signs alone never absolutely determine what gets communicated independently of the interlocutors doing the communicating.

Similarly, for all its richness, symbolic communication never makes interlocutors totally transparent to each other. While language enables interlocutors to engage each other more extensively and intensively, it never eliminates the existential distances between them. Even semioticized human sociality remains subjectively experienced and interindividually mediated. As anthropologist Loren Eiseley once observed,

> Since the first cell created a film about itself and elected to carry on the carefully insulated processes known as life, the creative spark has not been generalized. Whatever its principle may be it hides magically within individual skins. To the day of our deaths we exist in an inner solitude that is linked to the nature of life itself. Even as we project love and affection upon others we endure a loneliness which is the price of all individual consciousness—the price of living. (1970:48)

When viewed in light of language's pragmatic and semantic constraints, metacommunicative capacities, and the attendant potentials and pitfalls these hold, human sociality appears all the more improbable and paradoxical. We say improbable because such semioticized sociality depends so heavily on contingent agreements between subjective interlocutors about ultimately arbitrary associations between signs and what they symbolize— meanings that interlocutors must in any event improvise through inventive usage (Wagner 1981). Furthermore, the stability and reliability of these signs derive largely from their articulation with other signs in semantic categories that interlocutors themselves must accept and internalize even as they negotiate them. At the same time, no one speaker ever masters all of a language, and no two speakers ever internalize exactly the same semantic understandings in the same way. Given these multiple indeterminacies, it is a wonder we ever make sense to anyone at all.

Similarly, human sociality remains paradoxical because it presupposes pragmatic and semantic constraints, yet far from dictating any strict conformity between interlocutors, much less existential communion, these constraints ultimately widen the latitude of individual social action. Interlocutors' awareness that signs and their meanings have a conventional rather than necessary relationship creates opportunities within the shifting bounds of mutually imagined intelligibility (and accountability) to improvise, invent, or purposefully misrepresent signs. Having language (and culture) paradoxically makes human beings at once more conventionally social yet more self-consciously individual, with each propensity dependent upon, yet clearly serving to delimit, the other. We remain forever caught halfway between our social and self-conscious selves, with each side made possible and mediated by our linguistically informed, compulsively semioticized sociality.

At first glance, such improbabilities and paradoxes would appear to afford us little purchase for explaining the emergence of a form of sociality that, to borrow an illusion from Clifford Geertz about religion (1968:97), looks not unlike trying to hang a picture from a nail driven into its frame. The very improbability of linguistic communication, however, indicates an evolutionary continuity between human and nonhuman sociality because, no matter how pragmatically constrained and semantically elaborated, human sociality still demands pragmatic agreements between self-possessed individuals, with all the social indeterminacies this implies. While language may well reflect uniquely human neurological capacities for "infinite recursion" (the ability to produce endless utterances from the combination and recombination of a finite set of sign elements) (Hauser, Chomsky, and Fitch 2002), the fact that we must acquire any given language from others, yet can and do learn multiple—and diverse—languages in variable ways (Snowdon 1999), means that human sociality continues to depend at least as much on establishing collaborative relations between individuals, as on species-wide capacities innate within human individuals separately (Deacon 1997).

Such pragmatic continuities suggest that the evolution of human sociality might well be characterized as a cumulative intensification of patterns of social cooperation already found in nonhuman primates (de Waal 1996; King 1999; Goodenough and Deacon, 2003), but indeed a process that ultimately resulted in biological and behavioral transformations. Precisely how these transformations occurred must address the collaborative dimension of symbolic communication in at least two ways. First, in the absence of language, what nonlinguistic social mechanisms might have served to foster the peculiar juxtaposition of conventional conformity yet individual creativity so characteristic of language users? Second, what selective pressures in the social ecology of behaviorally complex yet non-linguistic early hominids might have favored individuals with enhanced tendencies for symbolic communication? To suggest how we might begin to answer these two questions, we turn to our findings about greeting rituals among savanna baboons.

RITUAL AND THE FOSTERING OF TRUST AND TRUTH

Ritual is one social mechanism that may have helped foster the intense sociality requisite to language (Rappaport 1999). In two previous papers (Smuts and Watanabe 1990; Watanabe and Smuts 1999), we reported and discussed our findings about adult male greeting rituals in a troop of olive baboons (*Papio cynocephalus anubis*) near Gilgil, Kenya, previously studied by Smuts (1999). Our research sought to determine how, if at all, male greeting behavior reflected other important aspects of male social relationships, including dominance rank, age, and length of residence in the troop, and especially coalition formation between adult males.

At the time of our study, the troop consisted of 150 members, including twelve fully adult males who served as the subjects of our study. Because olive baboons have a female-bonded social organization in which females, their daughters, and granddaughters form the permanent core of the troop, males transfer out of their natal troops and into others as adolescents or young adults. Since closely related males seldom if ever appear to transfer to the same troop, male olive baboons spend much of their adult lives with unrelated and initially unfamiliar others. Mutual antagonism characterizes most interactions between males, including threats, chases, and fights. When one male approaches another, the other usually avoids the approach or threatens the approacher. Fights routinely occur, as do minor wounds, and serious wounds are not uncommon.

The twelve fully adult males in our troop fell into two classes. Four "old males" included nonnatal males past their physical prime who had lived in the troop for at least two years, most for much longer. The remaining eight "young males" in their physical prime had either recently transferred into the troop or were natal males who had not yet left for other troops. Based on the outcomes of dyadic agonistic encounters, all young males individually outranked all old males. Normally for baboons, the higher-ranking younger males would have mated more often with estrous females than the lower-ranking males, but in this troop, and in the larger population of which it was a part, no correlation existed between male dominance rank and consort activity (cf. Berkovitch 1987; Strum 1987). Thus, in our study group the median sexual consort score for old males was in fact slightly higher than the median score for young males.

Lower-ranking old males achieved this unexpectedly high mating success in part by forming coalitions with each other jointly to harass males in consort with fertile females until a consort turnover occurred. All of these challenges targeted young males; we never saw old males challenge each other's consortships. After the turnover, the female almost always ended up with one of the old males involved in the coalition. Over the long term, old males gained females in rough proportion to the frequency with which they participated in coalitions. Thus, although young males individually possessed superior fighting ability that enabled them to take females away from old males, old males compensated for this disadvantage by cooperating with each other to take females away from young males—but never to defend each other's consortships from a young male.

Forming coalitions, however, poses a dilemma for any middle-aged male baboon with worn canines and slowing reflexes. If he wishes to continue mating with fertile females, he must confront ever younger and stronger rivals at growing risk of injury or even death. Alternatively, he can choose to settle into a tranquil but mostly celibate maturity—or find himself some allies and together challenge the individually dominant younger males. The second alternative would appear more desirable, but a male's most likely allies are other older males, often the very rivals he has long harassed, intimidated, bluffed, and occasionally wounded. The question becomes how, given their history of mutual antagonism and fierce competition, can these males convey to each other their readiness to establish mutually beneficial, cooperative relationships?

Here, we hypothesized that male greetings might play a decisive role. We found that greetings commonly occurred in a neutral social context as males peacefully foraged, traveled, or rested. They occurred twice as frequently as the next most common male interaction of supplants or avoids. In marked contrast to other male interactions, they usually lacked overt aggression, if not necessarily tension.

A greeting typically began when one male approached another with a distinctive, exaggerated gait. The approaching male would look directly at the other and often lip-smack while making the "come hither" face (ears back and eyes narrowed)—both friendly gestures. The second male sometimes avoided the approach or turned away, in which case the first male ceased his approach; other times, the approaching male himself veered away. More commonly, the second male accepted the approach by establishing eye contact and often lip-smacking and making the come hither face in return.

Upon approaching, the males would exchange a series of gestures that usually involved one presenting his hindquarters while the other either grasped them with one or both hands, mounted him, touched his scrotum or pulled his penis. Less often, one greeter would nuzzle the other, or, rarely, they embraced. The gestures used during a single greeting most often remained asymmetrical, with one male taking the more active role. Occasionally, a mutual exchange occurred when each touched the other's scrotum simultaneously or in rapid succession. Immediately after the exchange, one (or occasionally both) of the males would move rapidly away, often using the same exaggerated gait characteristic of the approach. The entire sequence took no more than a few seconds.

Initiation of a greeting never guaranteed its completion. Either male could break off at any time simply by moving away, and nearly half the time, one male would pull away before completing the exchange, resulting in what we called an "incomplete greeting." In only 7 percent of the greetings we recorded did attempts to greet end in threats, chases, or fights. Remarkably, of the 637 greetings documented in our study (as well as roughly 400 additional male-male greetings in another olive baboon population recorded on videotape by Smuts in 1993), not one resulted in a discernible injury. This contrasts dramatically with the agonism of virtually all other male baboon interactions.

Significantly, patterns in greeting tended to reflect coalitional behavior. Greetings between young males almost always displayed considerable tension, if they occurred at all. Young males had the lowest percentage of completed greetings (one-third) because they often circled each other, jockeying over who was going to do what to whom, often without success; incomplete greetings resulted. Similarly, young males never formed coalitions with one another and, except for their attempted greetings, they mostly avoided each other and even refrained from associating with the same females (Smuts 1999).

In contrast, old males who formed coalitions with each other tended to have relatively relaxed greetings, and they completed most (two-thirds) of them. Old and young males also greeted, with the younger, higher-ranking male almost always initiating the approach and taking the more active role, usually by mounting or grasping the other male's hindquarters, and occasionally by touching the other male's genitals. While this asymmetry in roles may have simply reflected the already clear-cut dominance relationships between young and old males, it seemed to us that, at least some of the time, young males sought to greet older males who, by virtue of their affiliative relationships with females, infants, and other old males in the troop, could prove potentially valuable allies.

Two examples illustrate these patterns. The two highest-ranking young males, at the time engaged in a tense standoff for dominance, repeatedly attempted to greet, but we never saw them complete a greeting because neither appeared willing to take on the receiving role. Conversely, one pair of old males greeted much more often than any other pair in the troop. Unrelated but familiar to each other after seven years of living together in this troop, they had the longest-standing, most consistent alliance of any pair of males and routinely helped each other take fertile females away from younger rivals. Unlike all other male pairs, neither tried to dominate the other, and they remained the only pair of males observed defending each other in fights with other males. Unlike the asymmetry of greeting roles characteristic of most male-male dyads, their greetings reflected near-perfect symmetry. Indeed, whenever we saw them greet twice in rapid succession, the male who took the active role in the first greeting initiated the subsequent greeting by inviting his partner to adopt the active role. Their greetings thus paralleled their "turn-taking" in coalitions.

To explain how these greetings might help foster cooperation, we looked to Roy Rappaport's formalist account of ritual (1979a, 1979b, 1999).

Rappaport argues that ritual has two obvious aspects from which flow "certain logically necessary entailments" (1979a:173). First, ritual consists of more or less invariant sequences of acts and utterances that the participants themselves do not create but to which they must conform. Second, in order to have a ritual at all, participants must actually perform it rather than simply invoke or describe it. For Rappaport, conformity to ritual's formalism constitutes "*the* basic social act" (ibid.:197; emphasis in original) because ritual at once stipulates conventions while inducing mutual compliance to them. More importantly, the presumed invariance of ritual can itself become a conventionalization of truthfulness: that which never changes conveys certainty; performing such certainty fosters acceptance; acceptance implies unquestionableness—and thus for all practical purposes a conventionally established truth. However tautological, performing a ritual amounts in purely formal terms to making a promise of truthfulness vouchsafed by the kept promise of the ritual completed. While this guarantees neither the sincerity nor the actual keeping of such promises, mutual participation in a ritual makes promises possible, and the public nature of such promising makes ritual participants liable for any bad faith.

Rappaport remains most concerned with how these logical entailments of ritual form and performance serve to counter problems of lying and alternative inherent in symbolic communication where the very nature of conventionality makes truth—and therefore the trustworthiness of communications—problematic. Ritual addresses these ills by making interlocutors' mutual conformity to outward conventions behaviorally explicit and socially undeniable: participants either do the ritual or not. It is precisely in its literal and behavioral, not symbolic or virtual, aspects that ritual defines a social context in which interlocutors can negotiate relations of trust by enacting their commitment to its shared conventions, whatever these might be. As Rappaport concludes, "The invariance of ritual, which antedates the development of language, is the foundation of convention, for through it conventions are not only enunciated, accepted, invested with morality, and naturalized, but also sanctified [that is, made unquestionable]. Indeed, the concept of the sacred itself emerges out of liturgical [ritual] invariance" (ibid.:211).

Although baboon greetings clearly lack the semiotic complexities of human rituals, they nonetheless conform to Rappaport's formalist account of ritual and, we would argue, appear to serve much the same purpose he proposes. The greetings consist of a limited set of possible gestures clearly framed by the distinctive combination of exaggerated approach, vocalization, and facial expression. In order even to attempt a greeting, both males have to accept the constraints of the greeting's outward form so that each can recognize the other's actions as a greeting in the first place. Accepting the greeting's form, however, by no means obliges either male to accept a particular role in any given greeting. Indeed, the variability of when to greet and with whom, the indeterminacy of who should do what (if anything) to whom, and the neutral social context of the greetings all clearly frame an actual greeting as the willing choice of both greeters to greet in the face of alternative possibilities.

On the one hand, the possibility of breaking off a greeting at any time provides an incremental mechanism for testing the other male's willingness to cooperate while minimizing one's own investment in the relationship. On the other hand, in the absence of language, males who want to cooperate can use the greetings as a way of expressing their good intentions in a world of otherwise suspicious, highly competitive individuals. In either case, a greeting's clarity of form yet contingencies of performance enable greeters to perceive each other's actions all the more clearly as choices, and therefore as rudimentary social action in Weberian terms— that is, as acts that convey a meaningful intent while taking into account the presumed intentions of others. Completing a greeting becomes a powerful nonverbal way of forging mutual trust—or demonstrating its lack— between the greeters.

Regarding the actual form of the greetings, the component gestures appear to derive from two primary social relationships in baboons—lip-smacking, embracing, and nuzzling from the mother-infant bond; and presenting hindquarters, grasping hips, mounting, and genital contact from mating interactions. Rather than making the greetings literally about nurturance or sex, these gestures may have been incorporated into greetings because of their broader social meanings. First, mothering and sex represent universally experienced, and thus readily recognizable, relations of mutual interest and affiliation in baboon society. Second, these relationships involve a significant degree of cooperation and trust. Third, they also embody power distinctions—mothers dominate their infants, and in sexual relations, male baboons dominate their female partners.

To the extent that baboon greetings draw on gestures from other experiential contexts to convey related, but by no means identical, social meanings, they might be said to represent symbolic communication. At the very least, they suggest how social relations may have strongly motivated what devices first came to serve as symbols—and what they symbolized. As King and Shanker have noted in their discussion of "co-regulation" in ape communication, "the most appropriate unit of analysis is not the gesture or vocalization itself, but instead the ongoing, ever-changing socio-affective relationship" between the interlocutors involved (2003:10).

In addressing the larger question of the evolution of human sociality, these greeting rituals provide a plausible mechanism by which socially complex yet nonlinguistic early hominids might have conveyed intentions and perhaps even promises to one another in the absence of language. As such, they suggest how human symbolic communication could have emerged as an intensification of already-existing patterns of primate social cooperation. Deacon (1997) has argued that symbolization evolved specifically out of ritualized promises of fidelity that early hominids developed to stabilize pairbonding based on male provisioning of females in the face of competing reproductive interests (and opportunities) in multimale, multifemale groups. While we strongly support his general position that symbolic communication emerged from inherently social (and perhaps ultimately reproductive) concerns, we find his exclusive focus on hominid hunting and pairbonding unnecessarily narrow. We focus instead on the more general social substrate of multilateral cooperation and competition found in nonhuman primate societies today (and presumably in hominid societies in the past) out of which could have evolved any number of adaptive behaviors and social capacities.

To continue this discussion, we now turn to the question of what in this social—as opposed to purely physical or utilitarian—environment might have selected for the other phylogenetic and behavioral changes that ultimately led to the evolution of symbols, language, and human culture (Dunbar 1996; Deacon 1997).

COOPERATION AND COMMITMENT

If, as we suggest, semioticized human sociality represents an intensified form of nonhuman primate social cooperation, our baboon example highlights a further dynamic of cooperation that may have helped select for such intensification. With coalitional cooperation arises the question of how willful individuals like male baboons can sustain cooperative relationships over the long term when potential partners always remain free to act alternatively as friend or foe, depending on circumstances that can vary from day to day or moment to moment. Even for the most cooperative pair of male baboons, occasions occur when either or both will find that treating their partner as a competitor reaps immediate benefits, and their partnership will at least temporarily break down. In the next moment, however, circumstances might change so that, once again, each male benefits from cooperation, provided the other also cooperates. Because circumstances rarely shift in such unambiguous, clear-cut fashion, individuals may often have great difficulty determining when it serves their best interests, or those of their partner, to cooperate or compete. Short-term switching back and forth between being friend and foe thus always runs the risk that cooperation will break down permanently.

One solution to this difficulty would be for partners to develop ways to communicate to each other that the situation has shifted. This may in fact explain why male baboons at times exchange greeting gestures, if not full-blown greetings, when in the midst of aggressive coalitions against other males (Smuts and Watanabe 1990: 165). Another solution involves ignoring those instances in which immediate self-interest leads to defection in favor of a long-term commitment to a cooperative relationship even when this entails short-term costs (cf. Stephens, McLinn, and Stevens 2002). Nesse (2001) calls this "the problem of commitment" and argues that trust remains essential to such relationships. Specifically, partners must have confidence that their benevolent behavior today, however costly, will reap greater gains over the long term because each trusts that their partner shares a similar commitment to mutual benefit and a clear, if not always equal, degree of fairness. By definition, the benefits of such commitment accrue over time, which means that individuals can neither communicate their own commitment, nor measure that of others, through short-term tit-for-tat-like exchanges rooted in the here and now. Some other way of conveying such commitments must emerge.

One way of sustaining long-term commitments involves third-party enforcement, what Nesse calls "objective commitments," but these remain perhaps unique to humans. Nesse also notes, however, what he terms "subjective commitments" expressed through publicly recognizable promises, reputation, and "irrational displays"—what we would prefer to call "noninstrumental displays"—that have no intrinsic material consequences. Essentially unenforceable by others, these subjective commitments "involve a continuing option for reneging" (ibid.:18). Although Nesse's discussion of commitment derives from observations of human behavior, his description echoes our analysis of male baboon greetings. It also converges with Rappaport's argument that questions of trust entail the problem of truth, and ritual as both promise and noninstrumental display elegantly and powerfully addresses both.

All three perspectives suggest how strategically valuable it might be for individuals to be able to communicate long-term commitments to cooperation even when short-term expediencies make it impossible to reaffirm, much less prove, such commitment in the here and now. Solving this problem requires social sophistication well beyond that of nepotism or tit-for-tat reciprocity (ibid.). Not surprisingly, it is precisely these kinds of committed relationships (and their demise) that define human sociality, and indeed the human ability to name relationships and social groups vastly extends the possibilities of dyadic or group identification beyond immediate association in the here and now (cf. Davidson 1999:249–51). We therefore argue that ever intensifying selection for the capacity to convey commitments beyond the here and now served as a driving force in the evolution of human sociality, especially language. We further contend that rudimentary solutions to the problem of commitment evolved in at least some nonhuman primates, including our prehominid ape ancestors, and probably in some other socially complex animals as well. Such solutions in turn constituted the evolutionary substrate on which early hominids built increasingly complex social cooperation and commitments, including those necessary for language.

To explore this issue further, we turn briefly to three kinds of communicative exchanges that exemplify Nesse's subjective commitments in nonhuman animal relationships. While by no means exhaustive, these examples illustrate how other animals may address the problem of conveying commitment in the absence of language. These examples also suggest pressures in the social environment of early hominids that may have selected for increasingly languagelike modes of communication that could transcend the immediacies of the here and now.

CONVEYING COMMITMENT

One type of communicative exchange about long-term commitments consists of greetings between male and female baboons. Typical of the baboons we studied, as well as of many other baboon populations, adult males form long-term affiliative relationships or "friendships" with a subset of adult females in their troop, along with the infants of those females (Seyfarth 1978; Altmann 1980; Strum 1987; Smuts 1999). In some populations, such friendships involve protection of infants the male most likely fathered (Palombit et al. 1997). In other populations, however, males invest considerable time and energy protecting infants they are unlikely to have sired from other baboons and from predators (Smuts 1999). Evidence from one population (ibid.) indicated that when a female friend resumes estrus a year or two later, she often prefers to mate with her male friends, but nothing guarantees the male's investment will pay off in this way.

This raises the question: How does a female indicate to a male that she will follow through and mate with him? Similarly, how does a male convey to a female that if she mates with him, he will be a good friend to her and her offspring in the future? Lacking the ability to communicate beyond the here and now, baboon friends engage in several behaviors that could be interpreted as evidence of long-term commitment, including frequent grooming and frequent greetings. Male-female greetings, like those between males, are ritualized but vary from one relationship to another. Males are significantly more likely to acknowledge a hindquarters present by a female friend than by a nonfriend (ibid.), and greetings among friends often involve greater physical intimacy, such as hugging and prolonged eye contact, than greetings among nonfriends (Smuts, in preparation). Greater physical intimacy, especially for the female, entails a potential risk of injury and so may serve as a reliable indicator of her commitment to the relationship, similar to the way genital touching may function among males.

Reconciliation and other mechanisms for conflict resolution represent a second example of fostering long-term commitments. Frans de Waal first described reconciliation among chimpanzees in 1979 (de Waal and van Roosmalen 1979), and since then researchers have documented similar behaviors in numerous nonhuman primates and some nonprimate social mammals (Aureli and de Waal 2000). Reconciliation occurs when, shortly after a confrontation, two individuals engage in affiliative contact at a rate higher than expected based on their overall rates of affiliation. Often, this affiliative contact takes a ritualized form involving signals derived from other contexts. Bonobos, for example, use sexual behaviors in their reconciliations. Multiple studies indicate that reconciliation tendencies vary across dyads and are strongest when the prior conflict involves a partner of value, such as an ally (Cords and Thurnheer 1993; Cords and Aureli 2000; de Waal 2000; van Schaik and Aureli 2000). A number of studies also indicate that reconciliation helps maintain relationships by reducing anxiety and decreasing the likelihood of a subsequent altercation (Aureli and Smucny 2000; Cords and Aureli 2000). Short-term partnerships specific to particular situations presumably do not require reconciliation, but the ability to make up after conflict may be essential to maintaining long-term cooperative relationships (de Waal 1989; Cords and Aureli 2000).

A third example comes from role reversal and self-handicapping during social play. Among most mammals, social play remains restricted to juveniles, but among some primates such as chimpanzees, and other highly social mammals like bottlenose dolphins and wolves, play persists into adulthood. Most social play entails play-fighting, in which, as in real fighting, individuals apparently try to "win." Researchers have noted, however, that during play-fighting, in contrast to real fighting, individuals sometimes make themselves extremely vulnerable to an opponent ("self-handicapping"), and that dominant animals sometimes willingly adopt a subordinate position ("role reversal"). Some researchers argue that self-handicapping and role reversal make play more attractive to the subordinate or weaker partner and thus help to sustain the play interaction (Bekoff, this volume; Biben 1986).

Furthermore, Bekoff (2001, chapter 3 in this volume) has proposed that social play may provide a critical opportunity during development for individuals to learn how to behave "fairly" toward social partners. Among both juveniles and adults, play may sometimes establish a special context in which individuals can negotiate their relationships with little risk of injury, a function similar to what we have proposed for greetings among male baboons. It is striking that play, like greetings, begins with distinctive behaviors such as the "play bow" in canids that clearly differentiate play from aggression (Bekoff 1995). Also, again like greetings, play involves behaviors derived from other contexts, such as chasing and biting, that are understood to have different meanings in play than those conveyed during real agonism (Bekoff, this volume). Some have argued that to sustain play, behaviors like self-handicapping and role reversal must render the partners equal (Aldis 1975). In a detailed study of social play in domestic dogs, however, Bauer and Smuts (2002) found that rates of self-handicapping and role reversal vary dramatically across dyads. It appears that equalizing behavior remains the most marked in the closest relationships, for example, between dogs that live together.

Despite the radically different contexts, these examples, along with our example of greetings among male baboons, have three elements in common. First, they all involve a degree of ritualization, such that the behaviors shown mean something different to the interactants than they would in the primary context in which these behaviors normally occur. Second, they typically involve behaviors that entail some risk to the actor and thus can function as honest signals of commitment to the relationship (Zahavi 1977, 1993; Smuts and Watanabe 1990). Third, although in each instance the interaction follows a prescribed pattern that makes it recognizable as a greeting, reconciliation, or play, a great degree of variability and flexibility exists in how the interaction unfolds or even whether it will occur at all.

In addition to these common elements, relationships based on subjective commitment may well presuppose at least some minimal capacity for metacommunication (Bekoff 1995)—for example, in the case of self-handicapping or role reversal, the ability to convey as well as recognize vulnerability as a commentary on the relationship rather than as the outcome of the interaction itself. This would also presume sophisticated assessments about when and where to exercise (or impute) such vulnerability appropriately. In any case, the strategic advantage to both willing cooperators and opportunistic defectors of long-term social commitments would have selected strongly for any ability to communicate about—and beyond—the here and now as these emerged in the evolution of our otherwise highly improbable, intensely semioticized human sociality.

EVOLUTIONARY CONTINUITIES AND TRANSFORMATIONS

To summarize, language represents more than simply an incremental addition to human sociality. Rather, it constitutes a pragmatically constrained and semantically elaborated process of symbolization that has transformed human sociality into highly semioticized encounters of intention and meaning that nonetheless depend on contingently negotiated conventions of trust and truth. As this kind of intensified social cooperation, language evolved out of preexisting forms of primate sociality, and we have argued that ritual provides a plausible, evolutionarily available mechanism by which it did so. Finally, the ever problematic yet potentially advantageous nature of long-term, committed social relationships may have served importantly to select for such intensified social cooperation, especially the linguistic ability to transcend the here and now, as such capacities emerged evolutionarily.

Thinking about the evolution of human sociality as a progressive intensification of social cooperation does three things. First, it shifts evolutionary questions away from a narrow accounting of features unique to human sociality, such as those internal to language like syntax or grammar, and toward identifying more general patterns of social cooperation and commitment—not just communication—in other social animals that might serve as analogues for both the adaptive context that selected for ever greater cooperation among our hominid ancestors, and the social mechanisms by which they came to intensify that cooperation to the level required so improbably by language. Centrally, this concerns the fostering of trust between otherwise willful individuals, and truth in the otherwise arbitrary conventions they as interlocutors must negotiate with each other. This in turn prompts us to ask what motivated such cooperation and what advantages it conferred on individuals so engaged to make it an evolutionarily stable strategy. Here we find particularly suggestive the possible advantages in being able to sustain long-term social commitments in the face of inevitable short-term impetuosities.

Second, such inquiry into the social pragmatics of communication as cooperative behavior keeps us from reifying human sociality—in particular language, but also culture—into some self-contained abstraction independent of the individuals who practice it. Instead, it suggests we think about the evolution of human sociality as a multilevel selection process (Sober and Wilson 1998) involving a range of disparate components and nested contexts. Minimally, such components would include individuals of varied and changing cognitive capacities and social dispositions, relations between them both within and between social groups, and the larger ecology and energetics of the morphological changes, especially larger brains, that made intensified social cooperation like language possible. At the individual level, what, other than the not-yet-evolved benefits of language, reduced selection pressure against energetically expensive, larger brains that made possible the expanded cognitive capacities and neural motor control essential to memory, abstraction, finer vocal discrimination and articulation, and hence language (cf. Dunbar 1996; Deacon 1997; Davidson

1999)? On the social level, since interlocutors, no matter how primed for language, need at least one other individual similarly predisposed in order to start a conversation, what conditions of group size, composition, and distribution might have brought (or kept) together such individuals, and what kind of intergroup relations enabled both morphological and behavioral changes to spread?

Third, thinking of the evolution of human sociality as a progressive accumulation of pragmatic and semantic features contingent on changes in individual cognition and social cooperation suggests that these features emerged incrementally, rather than necessarily (but all the more improbably) all at once. If indeed language constitutes a set of interacting systems, not a unitary whole (Goodenough 1990:603), it makes little sense to predicate theories of its evolution on some overarching communicative logic that in fact could have only applied once all its parts had emerged. In this sense, it is perhaps not so cut and dried that something "either is or is not a symbol" that stands for something else (Davidson 1999:234, 266). As Peirce noted long ago, no symbol in itself ever stands for anything until a "symbol-using mind" perceives a relation there and interprets it. Given the thousands of generations of coevolution between brain growth and social intensification that produced symbol-using minds, crossing the threshold of language may appear unequivocal and irrevocable only in retrospect (Deacon 1997). Instead, the need to have individuals with the right aptitude, the right interlocutors—and the right thing to say—all coincide at the right moment for language to begin suggests that at least some, if not all, of these conditions may well have been met repeatedly in the evolution of human sociality, but interlocutors failed to connect or failed to reproduce their achievement socially. Even after speakers existed, they had to form themselves into a community of speakers capable of reproducing itself and colonizing other groups—a process perhaps as chancy as establishing speech in the first place, undoubtedly impeded by small group size and intergroup transfer of one sex or the other at maturity that left initial speech communities vulnerable to accident and continual loss of speakers to nonspeaking populations. Not surprisingly, what most unequivocally distinguishes human and nonhuman primate societies lies in the ongoing bonds that transferring individuals maintain with their natal group, a characteristic enabled—indeed, demanded?—by language (cf. Rodseth et al. 1991).

In the end, for all their magical power, words retain pragmatic indeterminacies much like those in nonhuman sociality. Questions about the evolution of human sociality thus might begin most profitably, not with language or other things human and work backwards, but with the pragmatic complexities of nonhuman sociality, and then ask what enabled—and drove—intensification of such social cooperation toward the increasingly improbable, semioticized human sociality that we know evolved even if we cannot yet say exactly how.

REFERENCES

Aldis, O. 1975. *Play-Fighting*. New York: Academic Press.

Altmann, J. 1980. *Baboon Mothers and Infants*. Cambridge, MA: Harvard University Press.

Aureli, F. and F. B. M. de Waal (Eds.). 2000. *Natural Conflict Resolution*. Berkeley: University of California Press.

Aureli, F. and D. Smucny. 2000. "The Role of Emotion in Conflict and Conflict Resolution." Pp. 199–224 in *Natural Conflict Resolution*, edited by F. Aureli and F. B. M. de Waal. Berkeley: University of California Press.

Bauer, E. B. and B. B. Smuts. 2002. "Role Reversal and Self-Handicapping during Playfighting in Domestic Dogs, *Canis familiaris*." Paper presented at the meetings of the Animal Behaviour Society, University of Indiana.

Bekoff, M. 1995. "Play Signals as Punctuation: The Structure of Social Play in Canids." *Behaviour* 132:419–29.

Bekoff, M. 2001. "Social Play Behaviour, Cooperation, Fairness, Trust, and the Evolution of Morality." *Journal of Consciousness Studies* 8(2):81–90.

Berkovitch, F. B. 1987. "Reproductive Success in Male Savanna Baboons." *Behavioral Ecology and Sociobiology* 21:1163–72.

Biben, M. 1986. "Individual- and Sex-Related Strategies of Wrestling Play in Captive Squirrel Monkeys." *Ethology* 71:229–41.

Burks, A. W. 1949. "Icon, Index, and Symbol." *Philosophy and Phenomenological Research* 9(4):673–89.

Cords, M. and F. Aureli. 2000. "Reconciliation and Relationship Qualities." Pp. 177–98 in *Natural Conflict Resolution*, edited by F. Aureli and F. B. M. de Waal. Berkeley: University of California Press.

Cords, M. and S. Thurnheer. 1993. "Reconciliation with Valuable Partners by Long-Tailed Macaques." *Ethology* 93:315–25.

Davidson, I. 1999. "The Game of the Name: Continuity and Discontinuity in Language Origins." Pp. 229–68 in *The Origins of Language: What Nonhuman Primates Can Tell Us*, edited by B. J. King. Santa Fe, NM: School of American Research Press.

de Waal, F. B. M. 1989. *Peacemaking among Primates*. Cambridge, MA: Harvard University Press.

de Waal, F. B. M. 1996. *Good Natured: The Origins of Right and Wrong in Humans and Other Animals*. Cambridge, MA: Harvard University Press.

de Waal, F. B. M. 2000. "The First Kiss: Foundations of Conflict Resolution Research." Pp. 15–31 in *Natural Conflict Resolution*, edited by F. Aureli and F. B. M. de Waal. Berkeley: University of California Press.

de Waal, F. B. M. and A. van Roosmalen. 1979. "Reconciliation and Consolation among Chimpanzees." *Behavioral Ecology and Sociobiology* 5:55–66.

Deacon, T. W. 1997. *The Symbolic Species: The Co-Evolution of Language and the Brain*. New York: W. W. Norton.

Douglas, M. 1966. *Purity and Danger: An Analysis of the Concepts of Pollution and Taboo.* London: Routledge and Kegan Paul.

Dunbar, R. I. M. 1996. *Grooming, Gossip, and the Evolution of Language.* Cambridge, MA: Harvard University Press.

Eiseley, L. 1970. *The Invisible Pyramid.* New York: Charles Scribner's Sons.

Geertz, C. 1968. *Islam Observed: Religious Development in Morocco and Indonesia.* Chicago: University of Chicago Press.

Goodenough, U. and T. Deacon. 2003. "From Biology to Consciousness to Morality." *Zygon* 38(4):801–819.

Goodenough, W. H. 1990. "Evolution of the Human Capacity for Beliefs." *American Anthropologist* 92(3):597–612.

Hanks, W. F. 1990. *Referential Practice: Language and Lived Space Among the Maya.* Chicago: University of Chicago Press.

Hauser, M. D., N. Chomsky, and W. T. Fitch. 2002. "The Faculty of Language: What Is It, Who Has It, and How Did It Evolve?" *Science* 298:1569–79.

Heyes, C. and L. Huber (Eds.). 2000. *The Evolution of Cognition.* Cambridge, MA: MIT Press.

King, B. J. 1999. "Introduction: Primatological Perspectives on Language." Pp. 3–19 in *The Origins of Language: What Non-Human Primates Can Tell Us*, edited by B. J. King. Santa Fe, NM: School of American Research Press.

King, B. J. and S. G. Shanker. 2003. "How Can We Know the Dancer from the Dance?: The Dynamic Nature of African Great Ape Social Communication." *Anthropological Theory* 3(1):5–26.

Nesse, R. M. 2001. "The Evolution of Subjective Commitment." Pp. 1–44 in *Evolution and the Capacity for Commitment*, edited by R. M. Nesse. New York: Russell Sage Press.

Palombit, R. A., R. M. Seyfarth, and D. L. Cheney. 1997. "The Adaptive Value of 'Friendships' to Female Baboons: Experimental and Observational Evidence." *Animal Behaviour* 54:599–614.

Peirce, C. S. 1932. *Collected Papers of Charles Sanders Peirce*, vol. 2. Cambridge, MA: Harvard University Press.

Polanyi, M. 1966. *The Tacit Dimension.* Garden City, NY: Doubleday.

Premack, D. and G. Woodruff. 1978. "Does the Chimpanzee Have a Theory of Mind?" *Behavioral and Brain Sciences* 1:515–26.

Rappaport, R. A. 1979a. "The Obvious Aspects of Ritual." Pp. 173–221 in *Ecology, Meaning and Religion*, by R. A. Rappaport. Richmond, CA: North Atlantic.

Rappaport, R. A. 1979b. "Sanctity and Lies in Evolution." Pp. 223–46 in *Ecology, Meaning and Religion*, by R. A. Rappaport. Richmond, CA: North Atlantic.

Rappaport, R. A. 1999. *Ritual and Religion in the Making of Humanity.* New York: Cambridge University Press.

Rodseth, L., R. W. Wrangham, A. Harrigan, and B. B. Smuts. 1991. "The Human Community as a Primate Society." *Current Anthropology* 32(3):221–54.

Seyfarth, R. M. 1978. "Social Relationships among Adult Male and Female Baboons, II: Behaviour Throughout the Female Reproductive Cycle." *Behaviour* 64:227–47.

Smuts, B. B. 1999. *Sex and Friendship Among Baboons*, with a new preface. Cambridge, MA: Harvard University Press.

Smuts, B. B. In preparation. "Coordination and Synchrony in Olive Baboon Greetings."

Smuts, B. B. and J. M. Watanabe. 1990. "Social Relationships and Ritualized Greetings in Adult Male Baboons (*Papio cynocephalus anubis*)." *International Journal of Primatology* 11(2):147–72.

Snowdon, C. T. 1999. "An Empiricist View of Language Evolution and Development." Pp. 79–114 in *The Origins of Language: What Nonhuman Primates Can Tell Us*, edited by B. J. King. Santa Fe, NM: School of American Research Press.

Sober, E. and D. S. Wilson. 1998. *Unto Others: The Evolution and Psychology of Unselfish Behavior.* Cambridge, MA: Harvard University Press.

Stephens, D. W., C. M. McLinn, and J. R. Stevens. 2002. "Discounting and Reciprocity in an Iterated Prisoner's Dilemma." *Science* 298:2216–18.

Strum, S. C. 1987. *Almost Human: A Journey into the World of Baboons.* New York: Random House.

Tambiah, S. J. 1968. "The Magical Power of Words." *Man* (n.s.) 3(2):175–208.

van Schaik, C. P. and F. Aureli. 2000. "The Natural History of Valuable Relationships in Primates." Pp. 307–33 in *Natural Conflict Resolution*, edited by F. Aureli and F. B. M. de Waal. Berkeley: University of California Press.

Wagner, R. 1981. *The Invention of Culture*, revised and expanded ed. Chicago: University of Chicago Press.

Watanabe, J. M. and B. B. Smuts. 1999. "Explaining Religion without Explaining It Away: Trust, Truth, and the Evolution of Cooperation in Roy A. Rappaport's 'The Obvious Aspects of Ritual.'" *American Anthropologist* 101(1):98–112.

Weber, M. 1947. *The Theory of Social and Economic Organization*, edited with an introduction by Talcott Parsons. A. M. Henderson and T. Parsons, translators. New York: Oxford University Press.

Zahavi, A. 1977. "The Cost of Honesty (Further Remarks on the Handicap Principle)." *Journal of Theoretical Biology* 67:603–5.

Zahavi, A. 1993. "The Fallacy of Conventional Signaling." *Philosophical Transactions of the Royal Society London B* 340:227–30.

VI
Philosophical Overview

15

Primate Sociality and Natural Law Theory

A Case Study on the Relevance of Science for Ethics

Stephen J. Pope

In the mainstream paradigm of evolutionary theory, animals are pitted against one another in a competitive battle to gain resources and reproduce. The paradigm is rooted in some typically modern philosophical attitudes summarized generally as "atomistic individualism" (Taylor 1989). It gives rise to a focus on animal competition and aggression and away from cooperation, affiliation, and other expressions of sociality. The mainstream paradigm holds that animals are in an odd predicament. They are forced to live in groups for their own survival and protection, but joining a group increases the number and proximity of competitors who desire access to important resources, including sexual partners. These animals have to cooperate to survive, but cooperation is secondary and is in the service of egoistic goals, goals that are also served by aggression, deception, manipulation, etc. In this paradigm, primate reconciliation is corrective of a deeper pattern of aggression that allows for some degree of social order within the group. Conflict resolution is reactive to conflict.

The authors of this volume offer good reasons for rethinking the presumed primacy of competition and for giving greater focus to the cooperative and affiliative capacities of the social primates. Since our view of humanity is now, one way or another, related to our understanding of other animals, the differences between these positions bear important implications for the field of ethics. The point of this chapter is to show some ways in which our increased understanding of primate sociality can function as a corrective to the dominant paradigm. It will also suggest, if not explicate fully, ways in which recent primatological insights into sociality confirm and extend some of the traditional convictions of the natural law tradition of moral reflection.

PRIMATOLOGY AND ETHICS

Primatology and ethics seem to be worlds apart. Scientific examination of animal behavior, even of animals close to us on the evolutionary tree, appears to be at great remove from the attempt to identify right and wrong kinds of behavior. Both fields of study are separated by the kinds of questions they entertain, the methodologies they employ in attempting to pursue answers to their central questions, and the evidence that they find relevant.

Yet the history of ethics has shown many attempts to understand human action in light of the behavior of other animals. This comparison has often been drawn either to show human superiority to animals or to legitimate human exploitation of animals. Historical references to animals have often been used to identify the distinctively human nature of action, in contrast to the behavior of "brutes." Medieval bestiaries often depict the damned either as trapped in their places of torment by demonic-looking beasts or as themselves having been deformed into some kind of bizarre mixed-breed wild animals themselves.

Recent animal studies, including primatology, have promoted a much more complex view of animals and have given a much more helpful account of the diversity among species in this regard. The course of modern history reinforces the acknowledgment that we are the most deadly of all species on the planet. Indeed, ecologists have increasingly revealed the many ways in which we pose lethal danger to the survival of whole species (including of course some of the higher primates). Whereas philosophers and theologians in the past contrasted humans and animals to underscore our distinctiveness and superiority, evolutionists today tend to draw comparisons to make more apparent our commonalities.

The reductionistic thrust of sociobiology and evolutionary psychology sparked a strong reaction in both philosophy and theology that reasserts the autonomy of ethics as a discipline and the independence of morality as a cultural institution (see Rose and Rose 2000). The field of ethics concerns the free acts of rational agents, the argument runs, and any scientific information regarding genes is at best preethical, dealing with the necessary conditions of moral action but unable to explain or justify ethical choices as such (e.g., Gewirth 1993). The natural history of the human race and its predecessors, similarly, helps us to understand the context of human action but does not "explain" it in any strong sense.

The popular imagination has of course been captivated by news regarding the genetic roots (sometimes even described as "causes") of human behavior from gender differences and sex to violence and war. Popularized evolutionary accounts of human behavior often resonate with, and reinforce, the atomistic individualism that has rapidly gained strength in our culture over the course of the last two hundred years (see Bellah et al. 1985; Taylor 1989).

THE DOMINANT PARADIGM

Various forms of neo-Darwinism, from sociobiology to evolutionary psychology, have so dominated public conversations about evolution, animal behavior, and human nature that we have come to expect that almost all human conduct is selfish and driven by an intense drive to compete with others (see, *inter alia*, Wilson 1978; Wright 1994; Ridley 1996). Readers today have become used to hearing about "selfish genes," intraspecific competition, strategies of deception and counterdeception, violence and aggression, "arms races" and "retaliatory strikes," etc. These competitive traits are associated with primates. We can easily be led to the impression that, within social primate species, prosocial behavior is, if not simply an aberration, derivative from self-centered, competitive, aggressive, and antisocial inclinations.

HOBBES TO DARWIN

The dominant paradigm has a history (one that cannot be adequately traced here). The founder of modern English-speaking ethics was Thomas Hobbes, a philosopher of such profundity that he continues to exert a powerful influence on the field today. Hobbes employed the language of "natural law" while changing it radically (see Rommen 1946; Tuck 1979; Taylor 1989; but Braybrooke 2001). He regarded human beings as inherently selfish, competitive, and aggressive—"man is a wolf to man," in the famous phrase. Hobbes knew of course that some animals, especially bees and ants, are social and do not need to be compelled to cooperate with one another. But he was quick to point out that these creatures are radically different from human beings, who compete for honors, prefer their private good to the public good, blame others for their own mistakes, allow themselves to be deceived by false information, and are easily offended and prone to vainglory. Human beings, unlike bees and ants, are not social by nature but made social by contract. We do not naturally cooperate according to instinct, so must be made to conform by a powerful sovereign (see Hobbes, *De Cive*, V; *Leviathan* 17). Individuals will act in their self-interest unless prevented from doing so by a stronger force. It is not a far leap to move from this position to the "selfish gene." Hobbes explicitly rejected the Aristotelian focus on whole organisms and the groups within which they act; Hobbes was committed to explaining the parts of organisms in isolation from their larger wholes.

Government for Hobbes and his heirs is not, as it was for his ancient and scholastic predecessors, a "natural institution" in the sense of meeting some fundamental human needs and therefore essential for human flourishing. Hobbes lowered the goals and standards of government by seeing it as an artificial construct created to protect life and property. It is established by social contract to bring some degree of security and safety in the midst of pervasive human evil and unremitting threats to survival presented by the "state of nature."

In this view, observations of the behavior of animals seem to replicate what is thought about human behavior. Individuals in the "state of nature" are like beasts. More moderate views were held by Locke and Rousseau, but both continued to regard sociality as derivative from and useful for instrumental purposes. All three modern philosophers viewed the self as egoistic, calculating, and essentially competitive with others. They rejected natural teleology—the desire for truth, goodness, and friendship—all of which they regarded as expressive of a more fundamental self-centeredness. Antisociality implies that these values cannot be realized and ethics built in pursuit of larger goods is hopelessly idealistic. We may be embedded in the natural world, the argument runs, but striving for virtue and the common good is an exercise in futility. Better to seek to constrain aggression and manage the costs of conflict within acceptable limits. Adherents of the Hobbesian position such as Bernard Mandeville ([1714] 1970) invoked animal behavior to make the case that selfishness ("private vices") can generate an enormous amount of conformity and cooperation ("public benefits").

DARWINIAN SOCIALITY

All this changed in the nineteenth century with the Darwinian revolution. Darwin's own rather nuanced position was partially influenced by a stream of Hobbesian thinking about human behavior mediated, and in some ways moderated, by the natural "moral sense" theories of Hume and Smith, and especially by their notion of "sympathy" (see Arnhart 1998: Chapter 3). The exact influence of Hobbes on Darwin is a matter of scholarly dispute, but Ridley is too simplistic when he characterizes the dependency as linear and uncomplicated: "Thomas Hobbes was Charles Darwin's direct intellectual ancestor. Hobbes (1651) begat David Hume (1739), who begat Adam Smith (1776), who begat Thomas Robert Malthus (1798), who begat Charles Darwin (1859)" (1996:252). Ridley also argues, less controversially, that "the Hobbesian diagnosis—though not the prescription—still lies at the heart of both economics and modern evolutionary biology" (ibid.).

Darwin confessed in his *Autobiography* that one of the most important insights into what became the theory of natural selection came from reading Thomas Malthus's *Essay on the Principle of Population* (1798). Yet along with the Malthusian principle of competition, Darwin also famously recognized in a profound way that human beings, like many other mammals, have a social nature. *The Descent of Man* ([1871] 1936) argued that morality and the "social instincts," including the capacity for "sympathy" and need for social "approbation," are the products of the evolutionary process and rooted in human nature. He also of course recognized that the rudiments of these capacities and needs are displayed in other social species as well. He offered an explanation for the evolutionary benefits conferred on human beings by social cooperation and group living as outgrowths of more primitive bonds established first in alliances of mating and kinship.

Darwin's most adamant public supporter, T. H. Huxley, emphasized in a most strident manner the Hobbesian stream of Darwin's view of human nature. The "cosmic process" that permeates the natural world generates an "intense and unceasing competition of the struggle for existence" ([1893] 1993:36). Life is a relentless "struggle for existence"—indeed, he describes it in explicitly Hobbesian terms as a constant *bellum omnium contra omnes*—that requires a "gladiatorial" spirit (ibid.:67). In order to prevail against the natural world, human societies employ law to suppress the self-assertion natural to each individual. Yet human beings are marked by "the insatiable hunger for enjoyment" (ibid.:42), and passions are essentially selfish, irredeemably partial, proud, quick to take offense, and relentless in seeking revenge. These are the elements of the animal in human nature that persist despite the best efforts of educators and public officials to socialize us.

Social Darwinians like Spencer tended to downplay the social thrust of human nature or to reduce it to a function of sublimated selfishness (see Hofstadter [1944] 1955). Popularizing Darwinians a century later were not always properly criticized for being social Darwinians. But authors like Robert Ardrey, Desmond Morris, and then, later in the century, E. O. Wilson and Richard Dawkins continued the same stream of Hobbesian Darwinism seen in T. H. Huxley (see Sussman 1999).

Yet historians have pointed out the dangers of reading Tennyson's vision of "Nature, red in tooth and claw" in a simplistic way. Darwin used the phrase "struggle for existence," but he did not mean by this phrase that individual animals are in a constant state of violent combat with one another. Darwin explained, "I should premise that I use the term Struggle for Existence in a large and metaphorical sense, including dependence of one being on another, and including (which is more important) not only the life of the individual, but success in leaving progeny" ([1859] 1968:116).

Contemporary neo-Darwinians tend to be more literal on this issue. Matt Ridley argues that sociobiology captures in contemporary language the kind of struggle seen in Hobbes and Darwin. He explains, "Just as Hobbes argued that the state of nature was not one of harmony, so Hamilton and Robert Trivers, two pioneers of selfish-gene logic, argued that the relationships between parents and offspring, or between mates, or between social partners was [sic] not one of mutual satisfaction, but one of mutual struggle to exploit the relationship" (Ridley 1996:22). This struggle extends, Ridley argues, to the "greedy foetus" in the womb trying to extract sugars from its "thrifty mother" who must preserve food stores for her own use. All pregnancies, he believes, are "tugs of bitter war between enemies" (ibid.:23).

Robert Wright (1994), in *The Moral Animal*, for example, argues that human desires are "ruthlessly" subordinated to the individual's self-interests and consistently run counter to the good of others. Each person's morality is ultimately self-serving rather than truly directed to the good of all people. Society is essentially a sphere in which the self-interest of one individual competes and clashes with the self-interest of all other individuals; morality is a tool used to negotiate this constant competition for one's own benefit.

Wright, however, fails to explain what exactly he means by "self-interest" or how it is to be distinguished, if at all, from what purportedly benefits copies of genetic material. Like other sociobiologists, he massively downplays the ways in which members of communities thrive through the ongoing patterns of cooperation and interdependency that are mutually beneficial rather than competitive in nature. In natural law terms, he consistently ignores the "common good" that promotes the well-being of all members of a community.

One finds a much more complex view of behavior in the writings of Frans de Waal. Arguing in *Good Natured* that "we are moral beings to the core" (1996:2), he insists that we need to acknowledge the prosocial as well as antisocial capacities of human nature. We are capable of aggression, callousness, and opportunism, but also of caregiving, generosity, empathy, reconciliation, and tolerance. Our biologically based emotional reward system takes pleasure in prosocial behavior. Like some other social primates, we enjoy successful social interaction and healthy emotional bonds with friends and do not value them only as instrumentally valuable. Our natural sense of equity, argues de Waal, is probably rooted in some important way in the "economy of sharing and exchange" employed by our hominid forebears and seen today in "chimpanzee politics." This sense of equity may be the natural root of what natural law refers to as our rudimentary natural knowledge of right and wrong. Rather than an artifact of calculating self-interest, morality is rooted in our social nature: "The fact that the human moral sense goes so far back in evolutionary history that other species show signs of it plants morality firmly near the center of our much-maligned nature. . . . It is neither a recent innovation nor a thin layer that covers a beastly and selfish makeup" (ibid.:218). De Waal, like Darwin, provides a more helpful account of the relation between sociality and morality than does Wright, but ultimately his view reflects the dominant paradigm. Thus he understands "community concern" as "the stake each individual has in promoting those characteristics of the community or group that increase the benefits derived from living in it by that individual and its kin" (ibid.:207).

SOCIALITY AND COMPASSION

This shift in perspective has some important implications for our understanding of sociality. Here I would like to consider its implications for how we understand the term "compassion." In antiquity, a person was said to have "pity" (*eleos*) when witnessing the undeserved suffering of another person. It was made possible by the natural imaginative and affective capacities conjoined to our social nature. Christians later developed a strong ethic of *agape*, which includes compassion. The good Samaritan had compassion for the man who had been robbed, beaten, and left by the road to die. The Samaritan's pity was entirely natural, flowing from a recognition of what the "neighbor" had gone through and from which he continued to suffer (Gospel of Luke 10:27).

The modern period introduced a new egoistic way of thinking about compassion. Hobbes defined pity as "grief, for the calamity of another . . . [which] ariseth from the imagination that the like calamity *may befall himself*" (Hobbes 1977:53; emphasis added). Pity or compassion is a very uncomfortable experience because it creates in us a feeling of our own vulnerability, but it is always directed to those who suffer from pains that the agent could also suffer. When a person acts out of compassion, Hobbes thought, he or she does so to be removed from the experience of the pain associated with pity. The good Samaritan, according to Hobbes, acted in order to relieve his own anxiety or emotional discomfort, not directly to help the victim for his own sake.

Hobbes's position continues to be the most influential view of human sociality, altruism, and morality. His egoistic views were also held by Jean-Jacques Rousseau, a philosopher usually depicted as Hobbes's intellectual opponent. Rousseau, like Hobbes, held that we care about others because we care about ourselves. We care about others because doing so serves our own interests, and in fact we care about others to the extent that doing so contributes to our own self-interest. The truly happy person, Rousseau held, is the solitary person, completely alone: "It is man's weakness which makes him sociable" (Rousseau [1762] 1979:221). Pity is thus "sweet" because we feel the pleasure of not suffering what the other has gone through. Moreover, Rousseau argued, "when the strength of an expansive soul makes me identify myself with my fellow and I feel I am, so to speak, in him, it is in order not to suffer that I do not want him to suffer; I am interested in him for love of myself, and the reason for the precept [of the Golden Rule] is in nature itself, which inspires in me the desire of my well-being in whatever place I feel my existence. . . . Love of men derived from love of self is the principle of human nature" (ibid.:235).

The same general view of compassion has been advocated by neo-Darwinians. Behavioral ecologist Richard Alexander argues in *The Biology of Moral Systems* that "the most general principle of human behavior" is that "people in general follow what they perceive to be their own interests" (1987:34). What, then, of society? Darwinism tells us, Alexander argues, that societies are simply "collections of individuals seeking their own self-interests" (ibid.:3). People do not constantly choose to act on the basis of extensive analyses of self-interest—we are not computers, after all—but nature provides emotional predispositions that do the rough calculations for us. Morality evolved as a social institution because it tended to promote fitness. The costs of group living are extensive but they are worthwhile in the face of between-group competition. For Alexander, the antisocial relations between groups fuel the tendency of human beings to cooperate within groups and to compete within groups in a reasonably controlled manner. Antisociality between groups generated human evolution: "In no other species do social groups have as their main jeopardy other social groups of the same species—therefore the unending selective race toward greater social complexity, intelligence, and cleverness in dealing with one another" (ibid.:80). Aggression and fear thus drove both biological evolution and the expansion of social complexity, primarily in the form of systems of "indirect reciprocity" (ibid.:81, 85ff.). Within systems of indirect reciprocity, individuals often engage in "social hustling" to gain maximum benefits while paying minimally acceptable costs. Good Samaritanism works, Alexander argues, when it provides a net benefit to the agent. Individuals want to appear to be willing to engage in indiscriminate social beneficence so that they can extract such beneficence from others (ibid.:102). Benevolence is socially rewarded, so "it can pay to give the impression of being an altruist even if you are not" (ibid.:114). Alexander thus explicitly rejects Darwin's universalism whereby natural sympathies are extended from loved ones "to the men of all nations and races" (cited from *The Descent of Man*, ibid.:173). On the contrary, we have evolved to be preferential toward family members and ethical toward reciprocators within the group, but to be immoral, or at best indifferent, toward outgroups and their members. Because of our animal nature, we are naturally inclined to ingroup morality and outgroup hostility.

The depth to which this egoistic presupposition has penetrated our culture is revealed to the extent to which we take these citations from Rousseau and Alexander to be self-evidentially true and beyond remark. The claims of "kin altruism" and "reciprocal altruism" are neo-Darwinized versions of what were already proposed by Hobbes and Rousseau: cooperation and affiliation justified in egoistic terms. Fear plays a large role in compassion because we only have compassion when we fear that what happened to the victim could happen to us. This explains, they would say, why people sometimes can experience simultaneously both pity for the sufferer and relief that it is not they who suffer.

SOCIALITY AND PRIMATOLOGY

The primatologists featured in this book offer an alternative to the mainstream view of sociality—why animals gather and live in groups, form and maintain affiliations, protect and care for one another. They offer good reasons for rethinking the presumed primacy of competition and for giving greater attention to prosocial traits of primates. These findings also bear important implications for humanity, especially for the field of ethics. Richard Potts (this volume) helpfully reminds us (and especially those of us who are not scientists) that behavior varies from species to species, and within species from context to context, and within contexts from individual to individual. He offers a salutary and much needed warning about our common tendency to make naive overgeneralizations about animals, including primate species, that are in fact distinctive. He also provides a helpful warning to resist the attempt to draw quick inferences from generalizations about other primates to human behavior. All this is taken up by those who regard human nature as essentially egoistic and in need of modification, curbing, and inhibition by the counteregoistic, group-centered forces generated by culture.

Animals can be aggressive under some conditions, but they can also cooperate under other circumstances. The primatologists contributing to this book do not suggest that the primates they study are purely altruistic but only that sociality needs to be taken seriously in its own right as a fundamental and pervasive feature of the lives of many primates. Primates in the wild show a remarkable diversity in their patterns of sociality. They display sociality that is not the product of culture (by which is meant here "conventional understandings manifested in art and artifact") and so call into question the common assumption that human sociality is produced by the "counterhedonic" pressures of culture (see Campbell 1975). If closely related primates display rich patterns of sociality without extensive verbal instruction, why should human beings not also be inclined by nature to sociality?

In this volume, Robert Sussman and Paul Garber argue that in all the free-ranging primates they studied, affiliative behavior is much more frequent than aggressive behavior. Indeed, they point out that generally less than 1 percent of the day is spent in any type of aggressive interaction. Agustin Fuentes concurs, arguing that overt conflict is much less common than affiliation for most primates. Zihlman and Bolter explore the social significance of mother-infant bonds. Strier's chapter, in a complementary way, explores ways in which primates maintain bonds within and beyond their natal groups. She speaks of kin relationships, mate bonds, and even "friendships" beyond kin. These chapters thus offer a much more prosocial view of primate behavior than is usually found in public presentations of primate behavior by advocates of the dominant paradigm (e.g., Wrangham and Peterson 1996).

THE NATURAL LAW TRADITION

Mainstream primatology is consistent with the atomistic individualism found in mainstream philosophy, especially among the heirs of Hobbes. Yet one particular ancient stream of ethics, the classical natural law tradition, understands the human person as powerfully embedded within, and an integral part of, the natural world. This ethical tradition finds important affinities and gains helpful insights with the primatological studies of natural sociality provided here.

Natural law regards the human person in two primary ways that will be examined in order, though in fact they are interdependent and mutually related descriptions: we are *"rational* animals" and *"social* animals." Thomas Aquinas, the classic expositor of natural law ethics, followed Aristotle in regarding human beings as irreducibly natural beings embedded in the natural world. We learn through the senses. We share with plants and other animals the vegetative faculties of nutrition, growth, and reproduction. Like other animals, we function in the world by exercising the "sensitive faculties" of our exterior senses (sight, smell, hearing, taste, and touch) and our interior senses (including imagination and memory). Our "sensitive appetites" naturally incline us to what we perceive to be attractive or good (e.g., to food and drink when we are hungry or thirsty) and away from what is sensed to be harmful (e.g., fire or snakes); it also incites responses to objects that appear to be threatening (e.g., the fight or flight response). The sensitive appetites generate the range of basic emotions that we share with other animals: love and hatred, desire and delight, sorrow, hope and despair, fear and daring, and anger. Animality is not of marginal significance for our humanity.

Natural law, then, recognizes the significance of the fact that we are equipped with the full repertoire of needs and desires found in other animals. There is within each person sets of biological needs—"an inclination to certain more special things according to the nature he has in common with other animals" (Aquinas [c. 1270] 1946:Vol. 2, 1009; I-II, 94, 2). In virtue of this inclination, those things "which nature has taught to all animals, such as sexual intercourse, education of offspring, and so forth," pertain to the natural law (ibid.). These inclinations are the "raw material" of the moral life that must be shaped by training in childhood, nurtured by later education, and supported throughout life by deliberate moral self-discipline.

Aristotle, Aquinas's authority on this matter, believed that human sociality began with sex and childrearing and was then extended to the local group. The family unit is important but by no means self-sufficient. People thrive within communities, and therefore the proper and most characteristic expression of sociality is the political community. We are "political animals" and are incomplete without one another. Sociality is primary and not simply derivative from instrumental purposes (though it is indeed instrumentally valuable); it is essential to human well-being, rooted in biology as well as intelligence, and not a dispensable addition of culture.

Zihlman and Bolter (chapter 2 in this volume) argue that kinship is a "cross-cultural universal and may be a fundamental component of human society that specifies roles and obligations among individuals" (p. 44). If so, morality has important roots in kinship and the emotional connections that bind us to various kin (if not genetic relations, which do not overlap perfectly with various socially construed kinship systems). Knowledge of evolved species-wide emotional proclivities and natural emotional predispositions, including kin preference, can help us reflect on both the moral goals for which we strive and on the means we use to attain them (see Kossel 2002). Justice, for example, acknowledges the general primacy of responsibilities to close kin but does not give moral legitimacy to unfair nepotism, blindly loyal tribalism, or xenophobia.

Sociality has biological roots, but so do a variety of other inclinations. In human beings, animality can play a role in actions that are either beneficent or malfeasant. Behavioral plasticity and flexibility, primatologists tell us, is one of the most important features of human nature (Zihlman and Bolter, this volume). Moral training brings a settled order to how the agent acts upon natural desires, and specifically empowers the mature adult to transcend the appeal of immediate gratification in pursuit of the higher

good. It allows one, for example, to resist the emotional appeal of immediate retaliation as a response to aggression or betrayal. Reason guides the
expression of emotions rather than vice versa, but it does not suppress or
destroy them. Thus Aquinas, citing Aristotle, wrote that law can bring out
the best in the animal dimension of human nature:

> for, as the Philosopher says (*Polit*. I.2), "as man is the most noble of animals,
> if he be perfect in virtue, so is he the lowest of all, if he be severed from law
> and righteousness"; because man can use his reason to devise means of sat-
> isfying his lusts and evil passions, which other animals are unable to do.
> (Aquinas [c. 1270] 1946:Vol. 2, 1014; I-II, 95, 1)

Though detailed knowledge of primates was beyond his historical
reach by many centuries, Aquinas not only acknowledged our animal
inclinations but even incorporated an account of them into the heart of his
natural law ethics. Natural law is a normative enterprise in that it strives
to provide an intelligible account of the basic norms governing human
conduct in light of what constitutes the true good for human beings. Since
human conduct ought to strive for the ordered pursuit of fundamental
human goods, no balanced and comprehensive ethical position can ignore
the kinds of goods pursued by animals. These include, Aquinas argued,
goods naturally pursued by all animals: existence, nourishment, sex and
reproduction, appropriate social relations, etc. Primatology can provide a
significant resource for ethics in that it sheds light on these kinds of goods
and the tendencies through which animals pursue them. In principle, the
more we know about animal orientations to these goods, the more we will
know about human orientations to these goods—because this knowledge
allows us better to understand both what is *distinctive* to us and what we
share in common with other animals.

Natural law ethics, then, finds it essential to connect reflection on
human nature to available information regarding the behavior of other
animals. A brief example can illustrate this important point. Aquinas distinguished animal species in which males depart from their mates after
coitus from those that remain with them (Aquinas [1265] 1956:SCG III, 2,
Ch. 122, 142 ff.). Females from the former species are equipped by nature
to raise their young on their own. In contrast, both males and females from
the other species (especially mammals) are involved in raising their
young. In the case of some animals, the male remains long enough to give
some assistance, even if only in provisioning, in the process of raising the
young.

We are a species that, other things being equal, benefits when both male
and female are involved in the rearing of children. The developmental
process requires formation of the heart and mind in addition to providing

for physical necessities. This is reinforced by the fact that, unlike all other animals, we are not bequeathed with a strong set of instincts but rather need to be taught how to live. Herein lies one of the most profound features of human nature: we need one another, and a wider communal context, not only to exist but also to grow and develop into mature and constructive members of our communities.

Primatologists confirm this claim when they explain that there is, across species, a "correlation between the length of life stages, especially the closeness between mother and young, and social complexity" (Zihlman and Bolter, this volume, p. 27). Prolonged infant immaturity and a lengthy period of intimate physical association of mother and child provide the kind of extended time frame needed for the formation and development of significant emotional connections and for instruction in social skills, especially those involving communication, which are necessary for successful interaction within the group (ibid.). These skills, seen in many primate groups, promote social bonds and encourage affiliation (ibid.). Skilled communication also allows for social learning and the formation of "traditions" in the broad sense of learned practices being passed on through some form of (nonverbal) instruction, e.g., tool use among chimpanzees.

What obtains for social primates applies to human beings as well. Primatology assists ethics by providing information on the roots and conditions of sociality, but it cannot give direct normative guidance. It cannot tell us how human sociality *ought* to be informed by moral norms. One cannot, then, move directly from generalizations about primate sociality, or even human sociality, to moral standards. This means that natural law ethics is not to be confused with "laws of nature." Classical "laws of nature" account for the movement of objects by forces over which they have no control. The "laws" examined by Isaac Newton and Robert Boyle account for uniform natural processes and do not supply a set of obligations explaining how nature "ought" to act. Karl Popper (1968) spoke similarly of "laws" as invariant patterns in nature. This use of "law" stands at a considerable distance from how it is used in ethics. Human laws are drafted precisely because of the variability of human conduct, and they strive to supply known standards in the face of disruptive behavior. Natural law, on the other hand, attempts to identify moral standards that can be used to direct human agents faced with choices between different courses of action, and these standards may or may be adhered to by human agents. These standards are said to be "natural" in several ways: they attempt to promote genuine human well-being or flourishing, they identify kinds of attitudes and acts of which human beings are capable, and they guide the expression of natural human inclinations.

It is perhaps worth pointing out, though, that natural law ethics applies to human beings *alone* because it is derived from a normative account of human nature rather than from nature in general. This is why Aquinas defined the "natural law" as the "rational creature's participation in the eternal law" ([c.1270] 1946:Vol. 2, 996; I-II, 91, 2). Natural law is the way in which human beings precisely as intelligent and free agents order their actions in the world.

This position is premised on the assumption that human beings, among all animals, are in unique possession of intelligence and special capacities for self-determination. This general position accords with the name *sapiens*. It was made possible by the significant enlargement of the brain—nearly tripling in size—over the course of the last two million years. It is confirmed by our relatively large brain-to-body ratio and the uniquely complex organization (more than the simple size) of the modern human brain.

Aquinas did not have access to the insights we now have about the different forms of primate social intelligence—their capacities for nonverbal communication, information processing and problem-solving, monitoring the behavior of other animals, deliberate deception, food gathering, the use of tools, and various skills of problem-solving. He knew that some kinds of animals showed signs of intelligence, but he did not regard this as identical to possession of "reason." Our capacity to use symbolic communication, and especially language, gives us a unique intellectual capacity that is not even remotely approximated in the lives of other primates (see Ahn et al., Potts, Strier, Tattersall, Watanabe, and Smuts, all in this volume).

Practical rationality is made possible by language. A hungry dog knows to eat meat and not wood. The dog does not, however, have the ability to advert to the fact that meat is food and wood is not, and he cannot verbally identify and reflect on the food as means to sustaining life as the end of eating. It is thus, Aquinas would hold, impossible for the dog to make a deliberate choice—to choose after having gone through a process of deliberation over the relative value of various potential means to a given end and to make an informed judgment about the most appropriate means to that end. Human beings have the capacity to distinguish means from ends and to choose the proper means, and secondary ends, to attain an ultimate end (Aquinas [c.1270] 1946). As Larry Arnhart explains, "Human beings use their unique capacity for rational deliberation to formulate ethical standards as plans of life for the harmonious satisfaction of their natural desires over a complete life" (2001:2). Nature itself provides the principles of behavior in the dog's nature and so makes deliberation and choice unnecessary.

Some scholars have accused Aquinas of promoting a radical dualism between humans and animals (e.g., Rachels 1990:86–91). Yet Aquinas intended no such dualism. He believed that the natural world constitutes a whole within which all creatures are located. In fact, the dualism that sep-

arates humanity from animality emerged most forcefully in early modernity with Bacon, Hobbes, and Descartes (see Midgley 1978). Descartes
famously rejected Aquinas's view that animals have souls, and even
desires, and instead held that animals are completely lacking in consciousness and function purely as machines. Marc Bekoff, and other authors in
this volume, show the untenability of this blindly mechanistic view of animals. Primates and other animals clearly have feelings, emotions, and consciousness—traits proper to all creatures that possess what Aquinas and
Aristotle called "sensitive souls."

Primatological investigations of sociality are especially valuable for the
natural law tradition because of their consistently strong focus on the
human person as a "social animal." According to Aquinas, we exist by
nature as parts of larger social wholes on which we depend for our existence and functioning, and these provide instrumental reasons for political community. Yet it bears repeating that, for natural law ethics, political
community has a more noble purpose than the simple meeting of physical
needs—it is the only way in which we can satisfy our natural desire for
mutual love and friendship. We naturally belong in groups but we are not
simply parts of the group. We cannot be—or at least ought not be—completely subordinated to the group, as the worker bee is to the hive
(Aquinas [c. 1270] 1946:Vol. 1, 688, I-II, 21, 4, ad 3). This is for two reasons:
first, we possess human dignity. Second, we are all members of a much
larger and more important body, the universal community of all creation
(see ibid.:1125; ST I-II, 109, 3).

Natural sociality has played an important role in religious ethics. A
famous teaching of the Second Vatican Council summarized the contemporary expression of this belief in the claim that "by our innermost nature
the person is a social being" (O'Brien and Shannon 1992:*Gaudium et spes*,
par. 12, p. 173). The natural law tradition holds that human beings exist for
the purpose of love, communion, and friendship—without these goods any
human life is deeply impoverished. For this reason it holds that money, pos-
sessions, and status, while goods, ought not be overvalued at the expense
of interpersonal goods. It also suggests that life within community entails
social responsibilities proportionate to the agent's power—life is not, as
Huxley sees it, primarily a matter of "combat" against nature and society.
Natural sociality also supports an array of human rights that protect every
person and promote the good of all within any given community. The dignity of the person is not realized in isolation but in community with others.
Indeed, "Human dignity can be realized and protected only in community"
(O'Brien and Shannon 1992: *Economic Justice for All*, par. 28, p. 584). Natural
sociality thus provides anthropological support for viewing political institutions and economic practices as means for the flourishing of persons,
including the development of human freedom, growth in solidarity, and

strengthening the bonds of community. For those who are less inclined to be persuaded by religious ethics, the same ethical conclusions are drawn from natural human sociality by neo-Aristotelian philosophers Larry Arnhart (1998, 2001) and Mary Midgley (1978); Martha Nussbaum (1988, 1990) develops a similar ethical perspective in light of natural sociality, but by relying primarily on Aristotelian philosophy and classical literature rather than contemporary scientific resources.

Behavioral ecologists have shown in great detail how local contexts can either encourage or discourage healthy sociality, affiliation, and cooperation in various primate species. Local ecology also affects mating strategies and parenting strategies. For all their differences, what is true of other primates also applies to human beings. In the case of human beings, or at least modern human beings, the challenges considered by behavioral ecologists are magnified enormously by the vastly expanded size of human groups, especially when compared to the small scale of groups within which our premodern human precursors lived (see Boehm, this volume). This fact points up why ethical reflection must always be *social* ethics—moral reflection on life in societies and not only on interpersonal dilemmas or personal moral integrity. Modern social interactions take place within large anonymous groups, not only in intimate, face-to-face contexts, and if ethics is to account for the entire range of human conduct it must acknowledge and address the extraordinary complexity of human social life in modern society. Individuals in our society form and maintain an enormous variety of alliances, have to face competition from many different quarters, and must confront radically different kinds of challenges—many of which are overwhelming in the absence of intragroup mutual support.

Natural law ethics seeks to promote social norms and policies that are most likely to contribute to a kind of social ecology that encourages prosocial behavior and discourages aggression. The modern development of the natural law tradition thus offers a socially responsible defense of private property, just wages and fair unions, healthy occupational conditions for workers, sound marriages and intact families, respect for religion, and a deep sense of civic responsibility and social justice. Immigrants, for example, cannot be regarded as simply "others" who are on their own. The most needy and vulnerable deserve special priority because the well-being of each individual within a community is related to the well-being of everyone else in a community.

CONCLUSION

The primate studies traced in this book offer a helpful alternative to the dominant paradigm. If other primates are prone to social behavior more often than antisocial behavior, perhaps pity, empathy, and other prosocial feelings do not have to be laid on top of a substrate that is essentially antisocial. If this is also the case for human beings, then perhaps our first-person experience of caring for others is neither illusory nor derivative from self-concern. In this case social institutions have more to build on than atomistic individualism typically recognizes. Social life is not reducible to a cease-fire called between antagonistic rivals. Perhaps we do not need to think of society as a collection of competing individuals who enter political community in order to reduce the chances that they will be killed by other, more powerful individuals. An alternative position, and one that retrieves Aristotle's notion of the human being as a "political animal," can draw some help from this prosocial view of primatology in viewing society as a network of communities that make a positive contribution to human well-being. In this vision, social ethics takes its bearings from an account of what contributes to the common good, and politics works for its promotion through laws and public policy.

None of this is to deny that certain kinds of social conditions can increase the likelihood of aggression and make necessary the engagement of reconciliatory behaviors studied so carefully by de Waal. Yet as Fuentes (this volume) points out, some of these may simply be "friendly and affiliative" behaviors rather than "reconciliatory." One interesting question this poses is whether competition and agonistic behavior is generated from the breakdown of social order rather than its underlying basis. If we assume, as do Sussman and Garber, "that interactions between individuals are neutral . . . in relation to evolutionary phenomena at any particular point in time, then competition over food and mates may not be directly responsible for driving . . . sociality" (chapter 8 in this volume, p. 166). The alternative theory being developed here suggests that proximate explanations account for how group members are able to "live in relatively stable social groups and solve the problems of everyday life in a generally cooperative fashion" (ibid:178), and perhaps, at the very least, we should not be too quick to accept the "dominant view" as the best way of accounting for the behavior of all primates, including that of human beings. We cannot, of course, reject simple analogies from primate behavior to human behavior that are suggested by the dominant paradigm and then propose other equally simple analogies from the prosocial approach to primate behavior to human behavior. But the latter can help us to be clearer about the importance of specificity—the need to be very careful about making any broad analogies from one primate species to human behavior—as well as to be more critical of any attempts to move from generalizations about the aggressive basis of primate behavior to proposals about social ethics. If we human beings have a more prosocial nature than the dominant paradigm acknowledges, then at the very least we must consider ways in which to create social environments that will increase the likelihood that these social capacities will be developed rather than, as is currently too often the case, suppressed and subordinated to more aggressive capacities that also lie within our behavioral repertoire.

REFERENCES

Alexander, Richard D. 1987. *The Biology of Moral Systems*. Hawthorne, NY: Aldine de Gruyter.

Aquinas, Thomas. [1265] 1956. *Summa contra Gentiles*. Translated by Vernon J. Bourke. Vol. 3: *Providence*. Garden City, NY: Doubleday.

Aquinas, Thomas. [c. 1270] 1946. *Summa Theologiae*. Translated by Fathers of the English Dominican Province. New York: Benziger Brothers.

Arnhart, Larry. 1998. *Darwinian Natural Right: The Biological Ethics of Human Nature*. Albany: SUNY Press.

Arnhart, Larry. 2001. "Thomistic Natural Law as Darwinian Natural Right." *Social Philosophy and Policy* 18(Winter):1–33.

Bellah, Robert N., Richard Madsen, William M. Sullivan, Ann Swidler, and Steven M. Tipton. 1985. *Habits of the Heart: Individualism and Commitment in American Life*. Berkeley: University of California Press.

Braybrooke, David. 2001. *Natural Law Modernized*. Toronto: University of Toronto Press.

Campbell, Donald T. 1975, "On the Conflicts between Biological and Social Evolution and between Psychology and Moral Tradition." *American Psychologist* 30:1103–26.

Darwin, Charles. [1859] 1968. *The Origin of Species by Means of Natural Selection*. New York: Penguin.

Darwin, C. [1871] 1936. *The Descent of Man and Selection in Relation to Sex*. New York: Random House.

de Waal, Frans. 1996. *Good Natured: The Origins of Right and Wrong in Humans and Other Animals*. Cambridge, MA: Harvard University Press.

Gewirth, Alan. 1993. "How Ethical is Evolutionary Ethics?" Pp. 241–58 in *Evolutionary Ethics*, edited by M. H. Nitecki and D. V. Nitecki. Albany: State University of New York Press.

Hobbes, Thomas. 1977. *Leviathan: Or the Matter, Forms and Power of a Commonwealth Ecclesiasticall and Civil*. Edited by Michael Oakeshott. New York: Collier.

Hofstadter, Richard. [1944] 1955. *Social Darwinism in American Thought*. Philadelphia: University of Pennsylvania Press.

Huxley, T. H. [1893] 1993. "Evolution and Ethics." Pp. 29–80 in *Evolutionary Ethics*, edited by M. H. Nitecki and D. V. Nitecki. Albany: State University of New York Press.

Kossel, Clifford G., S.J. 2002. "Natural Law and Human Law (IaIIae, qq. 90–97)." Pp. 169–93 in *The Ethics of Aquinas*, edited by Stephen J. Pope. Washington, DC: Georgetown University Press.

Mandeville, Bernard. [1714] 1970 . *The Fable of the Bees*. New York: Penguin.

Midgley, Mary. 1978. *Beast and Man: The Roots of Human Nature*. Ithaca, NY: Cornell University Press.

Nussbaum, Martha C. 1988. "Nature, Function, and Capability." *Oxford Studies in Ancient Philosophy* 1(Suppl.):145–84.

Nussbaum, Martha C. 1990. "Aristotelian Social Democracy." Pp. 203 in *Liberalism and the Good,* edited by R. B. Douglass, G. Mara, and H. Richardson. New York: Routledge.

O'Brien, David J. and Thomas A. Shannon (Eds.). 1992. *Catholic Social Thought: The Documentary Heritage.* Maryknoll, NY: Orbis.

Popper, Karl. 1968. *The Logic of Scientific Discovery.* New York: Harper and Row.

Rachels, James. 1990. *Created from Animals: The Moral Implications of Darwinism.* New York: Oxford University Press.

Ridley, Matt. 1996. *The Origins of Virtue.* London: Viking.

Rommen, Heinrich A. 1946. *The Natural Law: A Study in Legal and Social Philosophy and History and Philosophy.* Translated by Thomas R. Hanley, O.S.B. St. Louis, MO: B. Herder.

Rose, Steven and Hilary Rose (Ed.). 2000. *Alas, Poor Darwin: Arguments against Evolutionary Psychology.* New York: Harmony.

Rousseau, Jean-Jacques. [1762] 1979. *Emile or On Education.* Translated by Alan Bloom. New York: Basic Books.

Sussman, Robert W. 1999. "The Myth of Man the Hunter, Man the Killer and the Evolution of Human Morality." *Zygon: Journal of Religion and Science* 34(3, September):453–71.

Taylor, Charles. 1989. *Sources of the Self: The Making of the Modern Identity.* Cambridge, MA: Harvard University Press.

Tuck, Richard. 1979. *Natural Rights Theories: Their Origin and Development.* Cambridge: Cambridge University Press.

Wilson, Edward O. 1978. *On Human Nature.* Cambridge, MA: Harvard University Press.

Wrangham, Richard and Dale Peterson. 1996. *Demonic Males: Apes and the Origins of Human Violence.* Boston/New York: Houghton Mifflin.

Wright, Robert. 1994. *The Moral Animal: Evolutionary Psychology and Everyday Life.* New York: Vintage.

Index